Essential Engineering Dynamics

J. C. MALTBAEK
Department of Engineering Science
University of Exeter

Crosby Lockwood Staples London

To Kathleen

Granada Publishing Limited
First published in Great Britain 1975 by Crosby Lockwood Staples
Frogmore St Albans Herts and 3 Upper James Street London W1R 4BP

ISBN 0 258 96965 2 hardback

Made and printed in Great Britain by
William Clowes & Sons, Limited
London, Beccles and Colchester

Contents

Preface

The basic subject of dynamics is taught in many different types of course of varying length. A textbook for one particular course would, therefore, not be very useful for other courses. The present book has been written to provide the general background material for a two-year course in classical dynamics for engineers. It is hoped that the book will be found useful for students on different courses at universities, polytechnics and – in parts – at technical colleges, both for the actual courses and for home study.

The writer has included what he considers to be essential material that should form the basic knowledge of the subject for any graduate in mechanical, civil or electrical engineering or engineering science.

The first eight chapters deal with the main parts of the subject, including sufficient material in kinematics for a first course in that subject. Vector methods have been used wherever their application offers an advantage. In general, the author has used what he considers to be the most useful mathematics in each particular situation. A knowledge of elementary statics is assumed, including force systems and resultant forces and moments, equilibrium conditions, etc. The analytical background is the usual differential and integral calculus up to and including the solution of second order differential equations. Vector algebra, differentiation and integration and the vector dot and cross products are used throughout the first eight chapters.

Chapter 9 deals with Lagrange's equations. The time has come to include this great method in any university course on dynamics.

A knowledge of functions of several variables and partial differentiation is necessary for this chapter.

Because of the importance of a special class of dynamics–mechanical vibrations–a special chapter has been included in this field. For the convenience of students who need to know a little about some topics in the theory of machines, an introductory chapter in this field has also been included.

The writer would like to express his appreciation and thanks to the publishers, especially to Mr D. Fox, for help and encouragement during the writing of the book.

Exeter
January, 1975

J. C. Maltbaek

CHAPTER ONE

Kinematics of a Point

The most elementary type of motion is rest. The science of bodies at rest, their equilibrium and forces acting between them is called statics. Statics is the oldest of the engineering sciences; important contributions were made by Archimedes (287–12 BC), who discovered the principle of the lever and scale balance; the modern science of statics started, perhaps, with the discovery of the law of the parallelogram of forces by Stevin in 1586.

Dynamics is the science of motion of bodies and the forces acting during the motion. This science started much later than statics, with the work by Galileo (1564–1642) on falling bodies. The late start of dynamics was due to the fact that accurate measurements of time are involved in dynamics experiments. In Archimedes' time, the necessary measurements were sometimes attempted by measuring a person's heartbeat, which was, of course, highly unreliable. A step forward was taken by Leonardo da Vinci (1452–1519), who used a leaking water tap as a time-measuring instrument, the time interval between the falling drops being considered constant. Real precision time-measuring devices did not appear until the first pendulum clock invented by Huyghens in 1657, and the balance-wheel watch by Hooke somewhat later.

The scientific basis of dynamics is provided by Newton's laws, published in 1687. The first uses of dynamics were in astronomy, and the new science was hardly used by engineers until the last decades of the last century, the machinery before then being so slow moving that it could be designed by using the principles of statics. With the invention of the steam turbine, the internal combustion engine and the electric motor, a great increase in speed took place, which created design problems that could only be solved by the application of the principles of dynamics.

In some problems only the motion itself, or the geometry of the motion without consideration of the forces acting, is of importance; this branch of dynamics is called *kinematics*. The branch of dynamics dealing with the forces acting during the motion is called *kinetics*. It is necessary to study kinematics in some detail before any useful work can be done in dynamics, because kinematical concepts appear in the basic laws of dynamics.

The only units involved in kinematics are the two basic units of length (metre) and time (seconds); these units are defined in the Appendix.

1.1 Rectilinear motion

One of the most useful and important types of motion of a point is motion in a straight line, and since this is also the most elementary type of motion, we will consider the kinematics of rectilinear motion first.

1.1.1 Displacement, velocity and acceleration

To describe the rectilinear motion of a point P, we take the line of motion, the *path*, as the x-axis (Fig. 1.1). If the distance x of P from

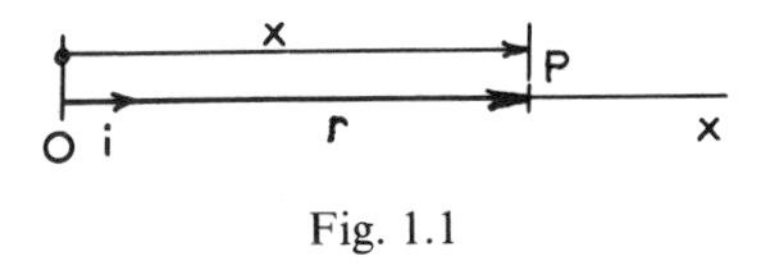

Fig. 1.1

a fixed point O on the x-axis is given as a function of time t by $x=f(t)$, the position of P may be determined at any instant of time.

The distance x is measured in metres (m), and the time t in seconds (s).

The distance x is called the *displacement* of P at the time t, and $x=f(t)$ is called the displacement–time equation. Actually the displacement of P is a *vector quantity*, since the displacement has both magnitude and direction. The position of P is determined by the vector $\mathbf{OP}=\mathbf{r}$; if a unit vector $\mathbf{i}$ is introduced as shown, we have $\mathbf{r}=x\mathbf{i}$, where x is the scalar magnitude of the position vector $\mathbf{r}$. In this simple case it is only necessary to give the displacement x, with proper sign, to determine the position of P, so that vector notation is not necessary. During a time element Δt from the time t to $t+\Delta t$, the point moves through a distance Δx; we define now the *average speed* during the time Δt by the ratio $\Delta x/\Delta t$. Letting $\Delta t \to 0$, we obtain the *instantaneous speed* of P as

$$V = \lim_{\Delta t \to 0} \frac{\Delta x}{\Delta t} = \frac{dx}{dt} = \dot{x}$$

The speed is considered positive if x is increasing with time and negative if x diminishes with time. The unit for speed is the ratio of the units for displacement and time, that is m/s.

Since the displacement is a vector quantity $\mathbf{r}=x\mathbf{i}$, the vector of magnitude V is the *velocity* of P, and

$$\mathbf{V} = \lim_{\Delta t \to 0} \frac{\Delta \mathbf{r}}{\Delta t} = \frac{d\mathbf{r}}{dt} = \dot{\mathbf{r}} = \dot{x}\mathbf{i}$$

If the point P moves along the x-axis with a variable speed $V=F(t)$, the speed is changing with time, and we say that P has an *acceleration*. In the time element Δt the speed has changed by an amount ΔV; the *average* acceleration in the time interval is defined by the ratio $\Delta V/\Delta t$; the *instantaneous* acceleration at time t is defined by

$$a = \lim_{\Delta t \to 0} \frac{\Delta V}{\Delta t} = \frac{dV}{dt} = \dot{V} = \frac{d^2x}{dt^2} = \ddot{x}$$

The acceleration is taken as positive if ΔV is positive in the interval Δt, negative if ΔV is negative in this interval. A negative acceleration is sometimes called a *deceleration*.

The units for acceleration are metres per second per second, or m/s^2. The acceleration has both magnitude and direction and is a vector $\ddot{\mathbf{r}}=\dot{\mathbf{V}}=\ddot{x}\mathbf{i}$.

It will be seen later, from the laws of dynamics, that the second derivative with respect to time of the position vector $\mathbf{r}$, or the acceleration $\ddot{\mathbf{r}}$ is all that is needed; higher derivatives of $\mathbf{r}$ are of no particular importance in dynamics and need not be considered.

Example 1.1. A point P is moving in rectilinear motion with *constant acceleration* a; if the displacement is x_0 and the velocity V_0 when $t=0$, determine the velocity–time and displacement–time functions. Determine also the displacement as a function of the velocity and time, and the velocity as a function of the displacement.

Solution. Taking the x-axis in the direction of motion, we have

$$\ddot{x} = a \text{ (constant)} \tag{a}$$

integrating gives

$$\dot{x} = V = \int a\, dt + C_1 = at + V_0 \tag{b}$$

using the condition that $V = V_0$ when $t=0$.

Integrating eq. (b) gives

$$x = \tfrac{1}{2}at^2 + V_0 t + x_0 \tag{c}$$

since $x = x_0$ when $t=0$.

Equations (b) and (c) give

$$x_2 - x_1 = \tfrac{1}{2}(V_2 + V_1)(t_2 - t_1)$$

so that the distance moved in a certain time is the *average* velocity during that time multiplied by the length of time.

We find also that

$$V_2^2 - V_1^2 = (V_2 + V_1)(V_2 - V_1) = 2\frac{(V_2 + V_1)}{2}a(t_2 - t_1)$$
$$= 2a\,(x_2 - x_1)$$

1.2 Rotation of a radial line in a plane

If a radial line (Fig. 1.2) is rotating about a point O in the plane, the position of the line at any time t may be given by the angle $\theta = f(t)$ between the rotating line and a line with a fixed direction in the plane. The position angle θ is measured in radians (rad).

In a time element Δt the position changes by $\Delta\theta$, the average *angular velocity* of the line in the time Δt is defined as the ratio $\Delta\theta/\Delta t$; the *instantaneous angular velocity* is defined by

$$\omega = \lim_{\Delta t \to 0} \frac{\Delta\theta}{\Delta t} = \frac{d\theta}{dt} = \dot{\theta}$$

The angular velocity is measured in radians per second (rad/s). If, in the time element Δt, the angular velocity changes by an amount $\Delta\omega$, we say that the line has an average *angular acceleration* $\Delta\omega/\Delta t$; the *instantaneous angular acceleration* is defined by

$$\lim_{\Delta t \to 0} \frac{\Delta\omega}{\Delta t} = \frac{d\omega}{dt} = \dot{\omega} = \ddot{\theta}$$

The units for angular acceleration are radians per second per second (rad/s^2). If the angular velocity ω is *constant*, then $\dot{\omega}=0$.

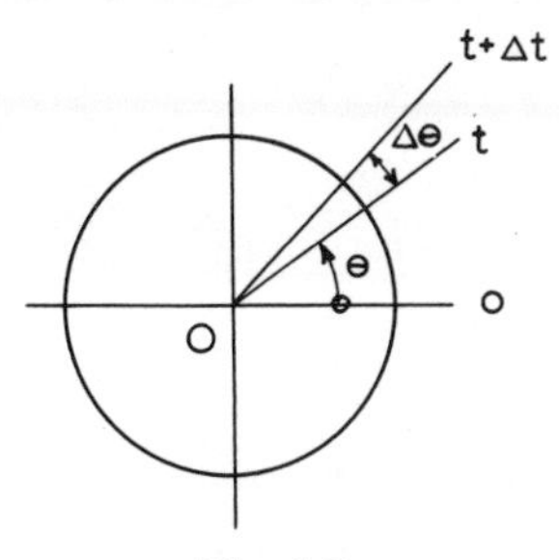

Fig. 1.2

Example1.2. A radial line is rotating in a plane about a point O with *constant angular acceleration* $\ddot{\theta}=\dot{\omega}_c$. Determine the displacement–time and angular velocity–time functions, given $\theta=\theta_0$ and $\omega=\omega_0$ when $t=0$. Determine also θ as a function of ω and t, and ω as a function of θ.

Solution. We have

$$\ddot{\theta} = \dot{\omega}_c = \text{constant} \tag{d}$$

Integrating gives

$$\omega = \dot{\theta} = \dot{\omega}_c t + \omega_0 \tag{e}$$

Equation (e) may be integrated directly, to give

$$\theta = \tfrac{1}{2}\dot{\omega}_c t^2 + \omega_0 t + \theta_0 \tag{f}$$

From eqs. (e) and (f) we find

$$\theta_2 - \theta_1 = \tfrac{1}{2}(\omega_2 + \omega_1)(t_2 - t_1)$$

the rotational displacement in a time $t_2 - t_1$ may then be found by using the *average* angular velocity in the time interval.

We may also state that

$$\omega_2^2-\omega_1^2 = (\omega_2+\omega_1)(\omega_2-\omega_1) = \frac{2(\omega_2+\omega_1)}{2}\dot{\omega}_c(t_2-t_1)$$
$$= 2\dot{\omega}_c(\theta_2-\theta_1)$$

Example 1.3. Figure 1.3 shows a slider–crank mechanism, widely used in internal combustion engines. The radial arm AB, the crank of length r, rotates in the plane about the fixed point A, and the angular position of the crank is given by the angle $\theta=f(t)$ as shown.

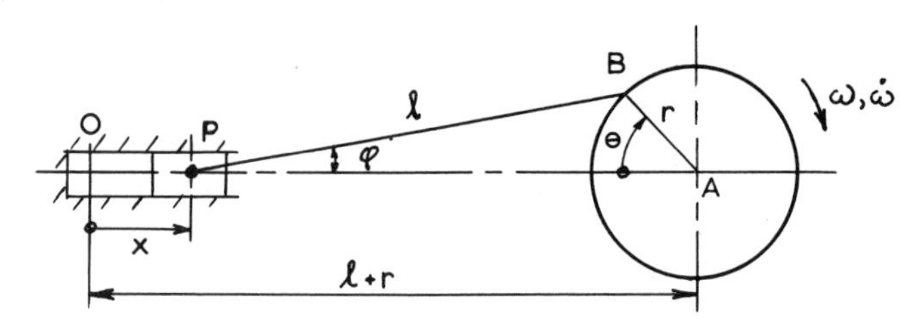

Fig. 1.3

The crank is pin-connected to a rod BP which is called the connecting rod; BP is of length l. The rod BP is pin-connected at P to a piston, which slides in a fixed cylinder; the motion of the piston is then rectilinear motion.

Determine the exact expressions for the piston displacement, velocity and acceleration. Simplify expressions for velocity and acceleration, if the ratio $(l/r)^2$ is assumed large compared to $\sin^2\theta$.

Solution. Taking the x-axis along the line PA and measuring the piston displacement x from the top dead centre O as shown, we find from the figure that

$$x = l+r-r\cos\theta-l\cos\varphi \qquad \text{and} \qquad l\sin\varphi = r\sin\theta$$

introducing the ratio $l/r=n$, we have

$$\cos\varphi = (1-\sin^2\varphi)^{1/2} = \left(1-\frac{\sin^2\theta}{n^2}\right)^{1/2}$$

so that

$$x = r(1-\cos\theta)+l\left[1-\left(1-\frac{\sin^2\theta}{n^2}\right)^{1/2}\right]$$

or

$$x = r[(1+n)-\cos\theta-(n^2-\sin^2\theta)^{1/2}] \tag{g}$$

differentiating eq. (g) gives the results

$$\dot{x} = r\omega\left[\sin\theta+\frac{\sin 2\theta}{2}(n^2-\sin^2\theta)^{-1/2}\right] \tag{h}$$

$$\ddot{x} = r\dot{\omega}\left[\sin\theta+\frac{\sin 2\theta}{2}(n^2-\sin^2\theta)^{-1/2}\right]$$

$$+r\omega^2\left[\cos\theta+\cos 2\theta\,(n^2-\sin^2\theta)^{-1/2} + \frac{\sin^2 2\theta}{4}(n^2-\sin^2\theta)^{-3/2}\right] \tag{i}$$

Since n is generally between 4 and 6, we may neglect $\sin^2\theta$ compared to n^2 in the expression $(n^2-\sin^2\theta)^{-1/2}$, and also neglect the last term in the eq. (i); for all practical purposes, we can then simplify the expressions (h) and (i) to the following:

$$\dot{x} = r\omega(\sin\theta+\frac{\sin 2\theta}{2n}) \tag{j}$$

$$\ddot{x} = r\dot{\omega}(\sin\theta+\frac{\sin 2\theta}{2n})+r\omega^2(\cos\theta+\frac{\cos 2\theta}{n}) \tag{k}$$

In most practical applications, the angular velocity ω is constant, so that $\dot{\omega}=0$; the acceleration of the piston then takes the simple form

$$\ddot{x} = r\omega^2\left(\cos\theta+\frac{\cos 2\theta}{n}\right) \tag{l}$$

1.3 Graphical kinematics. Motion curves

It is sometimes convenient and illustrative to construct motion curves for rectilinear motion, especially in experiments where a series of values of displacements, velocities or accelerations are obtained.

Figure 1.4 shows a displacement–time curve $x=f(t)$. The slope of the tangent is $dx/dt=V$; the speed at any particular position or time may then be obtained from the curve by graphical differentiation.

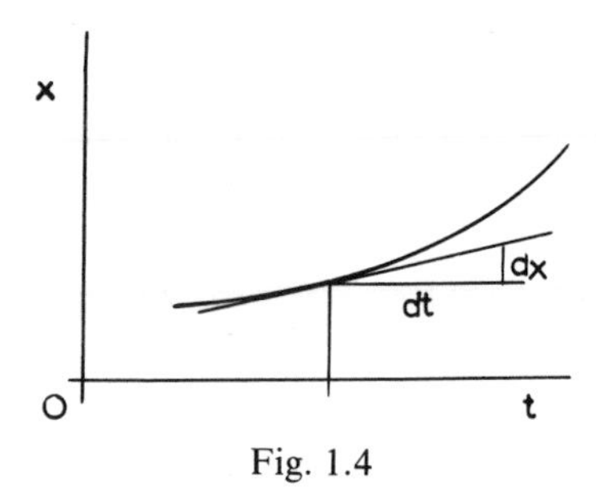

Fig. 1.4

A velocity–time curve is shown in Fig. 1.5. The slope of the tangent is $dv/dt = A$; the acceleration at any time may thus be found by graphical differentiation. The area under the curve between time values t_1 and t_2 is

$$\int_{t_1}^{t_2} V\,dt = \int_{t_1}^{t_2} \frac{dx}{dt}\,dt = x_2 - x_1$$

The area then gives the distance moved in the time interval $t_2 - t_1$; if the function $V = h(t)$ can be established, the integration may be

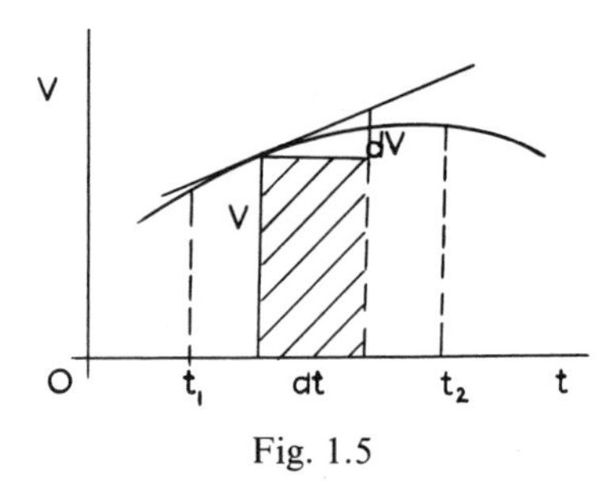

Fig. 1.5

sometimes performed analytically, otherwise the area may be found by using a planimeter or by other means of graphical integration.

Figure 1.6 shows the acceleration–time diagram. The slope is of no particular interest, but the area under the curve from time value t_1 to t_2 is

$$\int_{t_1}^{t_2} a\,dt = \int_{t_1}^{t_2} \frac{dV}{dt}\,dt = V_2 - V_1$$

or the difference in speed between the two times or positions.

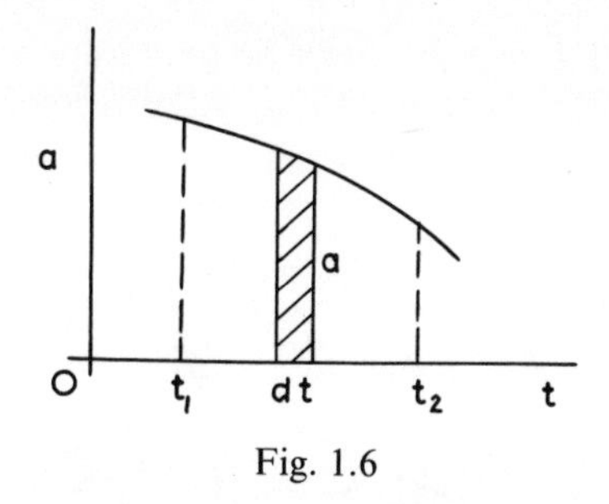

Fig. 1.6

Another curve that is sometimes useful is the acceleration–displacement curve shown in Fig. 1.7. The area under the curve from x_1 to x_2 is

$$\int_{x_1}^{x_2} a\,dx = \int_{x_1}^{x_2} dv\,\frac{dx}{dt} = \int_{v_1}^{v_2} V\,dv = \tfrac{1}{2}(V_2^2 - V_1^2)$$

If the velocity is known at one position, it may be determined at another position by graphical integration of the a–x curve, or by using a planimeter.

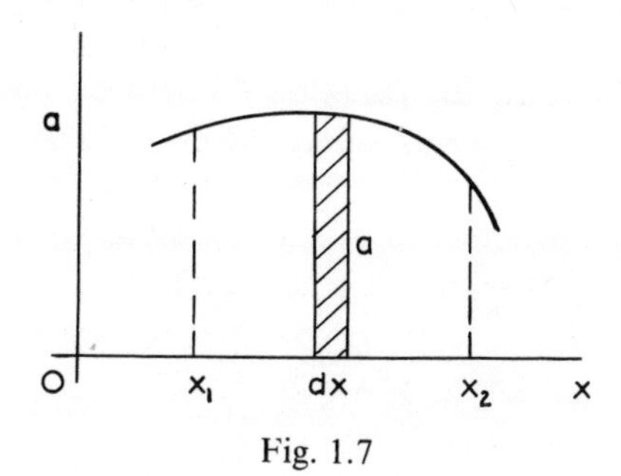

Fig. 1.7

1.4 Curvilinear motion

1.4.1 Displacement, velocity and acceleration in curvilinear motion

The motion of a point P in a coordinate system (x, y, z) (Fig. 1.8), is completely defined if $x=f_1(t)$, $y=f_2(t)$ and $z=f_3(t)$ are given functions of time t. The functions define a curve in space, the *path* of P.

The position of P may be given by the radius vector $\mathbf{r}=\mathbf{r}(t)$, or position vector **OP**. At time t the position vector is $\mathbf{r}$ and at the time $t+\Delta t$ the point P has moved to a new position given by the

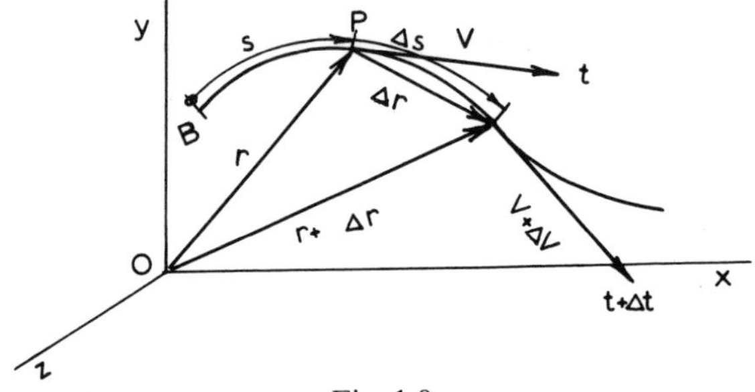

Fig. 1.8

position vector $\mathbf{r}+\Delta\mathbf{r}$. We define now the *instantaneous velocity* of P at time t by the vector

$$\mathbf{V} = \lim_{\Delta t \to 0} \frac{\Delta \mathbf{r}}{\Delta t} = \frac{d\mathbf{r}}{dt} = \dot{\mathbf{r}}$$

The velocity is, therefore, directed *along the tangent* to the path at the position of P, it is *not* perpendicular to the position vector. The scalar magnitude, the *speed* of P may be expressed by considering the function $s=h(t)$, where the position of P on the curve is determined by the distance s measured along the curve from a fixed point B on the curve; the displacement along the curve in the time Δt is then Δs as shown, and the speed is

$$V = \lim_{\Delta t \to 0} \frac{\Delta s}{\Delta t} = \frac{ds}{dt} = \dot{s}$$

If the velocity vectors of Fig. 1.8 are drawn from the same point, the vector diagram looks as shown in Fig. 1.9. The *instantaneous*

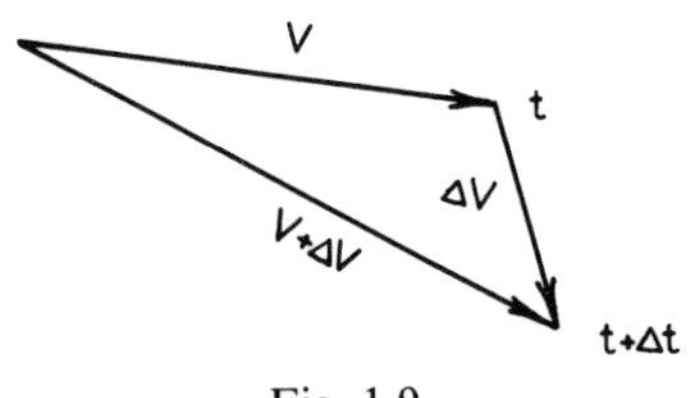

Fig. 1.9

acceleration of P is now defined by the vector

$$\mathbf{A} = \lim_{\Delta t \to 0} \frac{\Delta \mathbf{V}}{\Delta t} = \frac{d\mathbf{V}}{dt} = \dot{\mathbf{V}} = \frac{d^2\mathbf{r}}{dt^2} = \ddot{\mathbf{r}}$$

This vector is *not* tangential to the path. To have $\mathbf{A}=\mathbf{0}$ in a time interval, it is necessary that *both* the *magnitude* and *direction* of $\mathbf{V}$ is unchanged in that time interval.

These definitions of velocity and acceleration may be seen to be in agreement with the definitions previously given for the simple case of rectilinear motion.

It is usually most convenient to work with components of the acceleration vector $\mathbf{A}$. The following three sets of components have been found to be most useful in practical problems: rectangular, normal and tangential, and radial and transverse components.

1.4.2 Rectangular components of velocity and acceleration

Figure 1.10 shows a rectangular coordinate system (x, y, z) with unit vectors $\mathbf{i}$, $\mathbf{j}$ and $\mathbf{k}$ along the axes. A point P is shown with position vector $\mathbf{r}=x\mathbf{i}+y\mathbf{j}+z\mathbf{k}$, where x, y and z are functions of time.

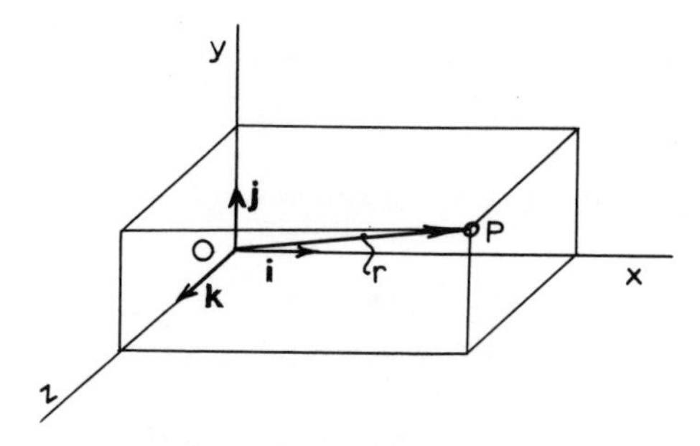

Fig. 1.10

The *velocity* of P, from the previous definition, is

$$\mathbf{V} = \dot{\mathbf{r}} = \dot{x}\mathbf{i}+\dot{y}\mathbf{j}+\dot{z}\mathbf{k} \tag{1.1}$$

The speed of the projections of P on the axes are

$$V_x=\dot{x} \qquad V_y=\dot{y} \qquad V_z=\dot{z}$$

these are the magnitudes of the components of the velocity $\mathbf{V}$.

The acceleration of P is

$$\mathbf{A} = \dot{\mathbf{V}} = \ddot{x}\mathbf{i}+\ddot{y}\mathbf{j}+\ddot{z}\mathbf{k} \tag{1.2}$$

The magnitudes of the acceleration components are

$$A_x = \ddot{x} \qquad A_y = \ddot{y} \qquad A_z = \ddot{z}$$

Example 1.4. The displacement of a point P moving in the xy-plane is given by the functions $x=r\cos\theta$ and $y=r\sin\theta$, where r (m) is a constant length and θ (rad) is a function of time $\theta=f(t)$. Determine the path of the point, its velocity and acceleration components in rectangular coordinates, and the magnitude of the velocity and acceleration.

Solution. the angle θ may be directly eliminated to give the path $x^2+y^2=r^2$, or a *circle* with radius r. The point P may be visualized as the end point of a radial line OP of length r which rotates about O in the xy-plane, as shown in Fig. 1.11. The angle between the x-axis and OP is seen to be θ from the expressions for x and y.

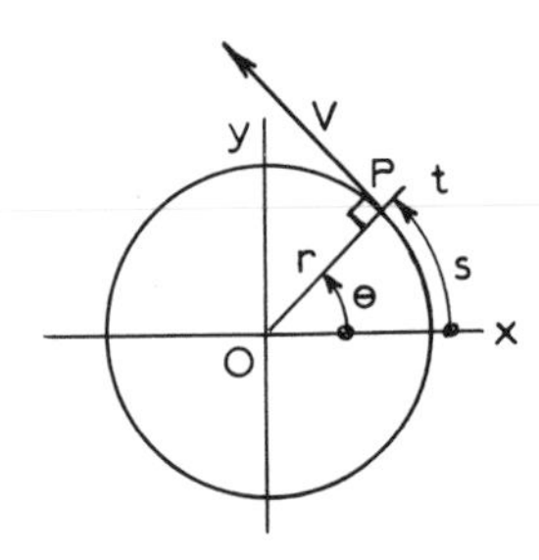

Fig. 1.11

The angular velocity and acceleration of OP is then $\dot{\theta}=\omega$ and $\ddot{\theta}=\dot{\omega}$.

The velocity components are

$$\dot{x} = -r\omega\sin\theta \qquad \dot{y} = r\omega\cos\theta$$

the magnitude of $\mathbf{V}$ is the speed $V=\sqrt{(\dot{x}^2+\dot{y}^2)}=r\omega$ (m/s). The velocity $\mathbf{V}$ is along the tangent to the path, as shown in Fig. 1.11; its components, found by projection on the x- and y-axes, are seen to be in agreement with the expressions for $\dot{x}$ and $\dot{y}$. The speed may also be determined from $s=r\theta$, so that $V=\dot{s}=r\dot{\theta}=r\omega$.

The acceleration components are

$$\ddot{x} = -r\omega^2 \cos\theta - r\dot{\omega}\sin\theta$$
$$\ddot{y} = -r\omega^2 \sin\theta + r\dot{\omega}\cos\theta$$

the magnitude of the acceleration is

$$A = \sqrt{(\ddot{x}^2+\ddot{y}^2)} = r\sqrt{(\omega^4+\dot{\omega}^2)}\ (\mathrm{m/s^2})$$

Example 1.5. Figure 1.12 shows a circular wheel of radius r which rolls, *without slipping*, along the x-axis in the vertical xy-plane. Determine the displacement, velocity and acceleration of a point P on the rim of the wheel.

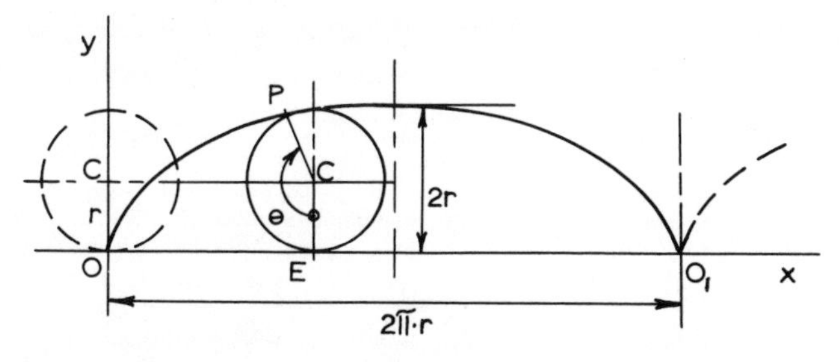

Fig. 1.12

Solution. Introducing the xy-coordinate system as shown, so that the point P starts at O, the position of the radial line CP may be given by the angle $\theta = f(t)$ between CP and the vertical direction. The distance $OE = x_C$ is equal to the arc EP of the wheel, so that $x_C = r\theta$, while $y_C = r$; the velocity components of C are then $\dot{x}_C = r\dot{\theta}$ and $\dot{y}_C = 0$; the centre C of the wheel moves on a horizontal line with speed $V_C = r\dot{\theta}$.

From the geometry of the figure, we find the coordinates of P:

$$x = r(\theta - \sin\theta)$$
$$y = r(1-\cos\theta)$$

Differentiation with respect to time gives the velocity components:

$$\dot{x} = r\dot{\theta}\,(1-\cos\theta)$$
$$\dot{y} = r\dot{\theta}\sin\theta$$

the acceleration components are

$$\ddot{x} = r\dot{\theta}^2 \sin\theta + r\ddot{\theta}\,(1-\cos\theta)$$
$$\ddot{y} = r\dot{\theta}^2 \cos\theta + r\ddot{\theta}\sin\theta$$

The shape of the path is given by the x- and y-expressions, and has the form indicated in Fig. 1.12. It is called a *cycloid*, and clearly repeats itself indefinitely, so that we need consider only the part for $0 \leqslant \theta \leqslant 2\pi$.

When $\theta = 0$, 2π etc., that is when P is the *point of contact* between the wheel and the x-axis, we find $\dot{x} = 0$ and $\dot{y} = 0$, which means that the *instantaneous velocity of the point of contact is zero*; it may be shown that the cycloid has vertical tangents at these points.

When $\theta = \pi$, P is at the top point of the wheel, $\dot{x} = 2r\dot{\theta} = 2V_C$ and $\dot{y} = 0$; the velocity of the top point is horizontal and twice the magnitude of the velocity of the centre of the wheel.

For the contact points, $\theta = 0$, 2π etc., the acceleration components are $\ddot{x} = 0$, $\ddot{y} = r\dot{\theta}^2$; the acceleration of the contact point is of magnitude $r\dot{\theta}^2$, and always vertical, that is always *directed towards the geometrical centre* of the wheel.

1.4.3 Normal and tangential components of acceleration

1.4.3.1 Radius of curvature of a plane curve

Figure 1.13 shows a plane curve $y = F(x)$. The tangents at points P and P_1 on the curve have been drawn, the distance between P and P_1 being ds; the tangent at P is inclined at an angle θ to the horizontal and the angle between the tangents is $d\theta$.

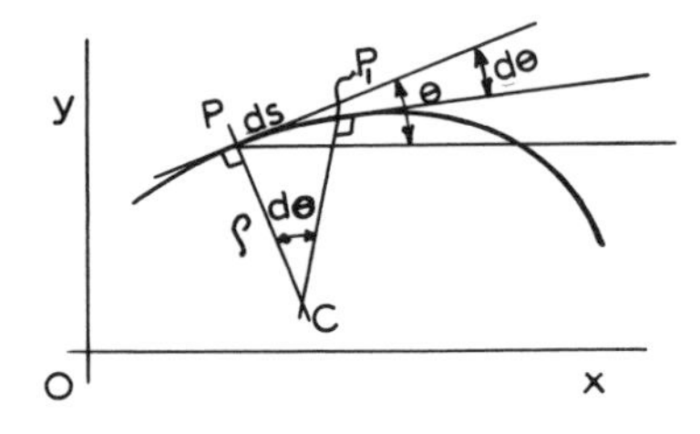

Fig. 1.13

The *curvature* of the curve at P is defined as the ratio $d\theta/ds$. A circle with centre C on the normal at P and with radius ρ has been introduced, and this circle has the same tangent and the same curvature as the curve at P. We have $ds = \rho\, d\theta$, where ρ is the radius of curvature at P, C is the centre of curvature, and $1/\rho = d\theta/ds$. The slope of the tangent is $\tan\theta = dy/dx = y'$, and differentiating with respect to x gives $(1/\cos^2\theta)\, d\theta/dx = y''$, or $d\theta/dx = y'' \cos^2\theta$.

Now $\sin^2\theta/\cos^2\theta=\tan^2\theta=(y')^2$, and substituting in $\sin^2\theta+\cos^2\theta=1$ gives $\cos^2\theta=1/[1+(y')^2]$, so that

$$\frac{d\theta}{dx}=\frac{y''}{1+(y')^2}$$

now

$$\frac{d\theta}{dx}=\frac{d\theta}{ds}\frac{ds}{dx}=\frac{1}{\rho}\frac{ds}{dx}$$

Using $ds^2=dx^2+dy^2$, or $ds/dx=(1+(y')^2)^{1/2}$, gives

$$\frac{1}{\rho}(1+(y')^2)^{1/2}=\frac{y''}{1+(y')^2} \quad \text{or} \quad \frac{1}{\rho}=\frac{y''}{[1+(y')^2]^{3/2}} \tag{1.3}$$

If the curve is given by $x=f_1(t)$ and $y=f_2(t)$, we find

$$y'=\frac{dy}{dx}=\frac{dy}{dt}\frac{dt}{dx}=\frac{\dot{y}}{\dot{x}}$$

$$y''=\frac{d}{dx}\left(\frac{dy}{dx}\right)=\frac{d}{dt}\left(\frac{dy}{dx}\right)\frac{dt}{dx}=\frac{d}{dt}\left(\frac{\dot{y}}{\dot{x}}\right)\frac{1}{\dot{x}}=\frac{\dot{x}\ddot{y}-\dot{y}\ddot{x}}{\dot{x}^3}$$

Substituting in (1.3) gives

$$\frac{1}{\rho}=\frac{\dot{x}\ddot{y}-\dot{y}\ddot{x}}{(\dot{x}^2+\dot{y}^2)^{3/2}} \tag{1.4}$$

For a circle, the radius of curvature is constant and equal to the radius of the circle.

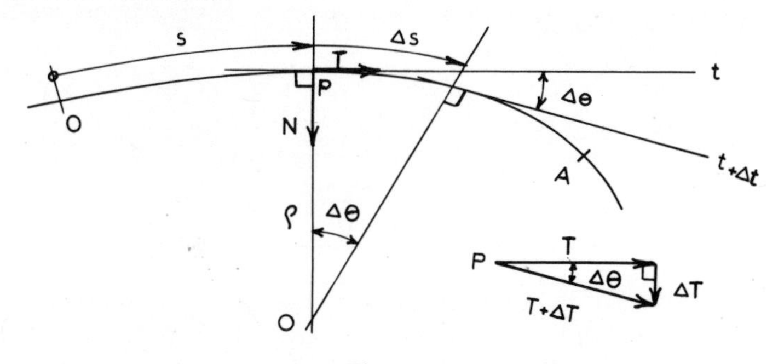

Fig. 1.14

1.4.3.2 Normal and tangential components of acceleration

Figure 1.14 shows a point P of a plane curve. Unit vectors **T** and **N** have been introduced on the tangent and the normal to the curve at

P. The point O is the centre of curvature of the curve at P and the radius of curvature is ρ.

The velocity of P is $\mathbf{V} = V\mathbf{T}$ along the tangent, where V is the speed of P:

$$V = \frac{ds}{dt} = \dot{s} = \rho\frac{d\theta}{dt} = \rho\dot{\theta}$$

The acceleration of P is

$$\mathbf{A} = \dot{\mathbf{V}} = \frac{d}{dt}(V\mathbf{T}) = \dot{V}\mathbf{T} + V\dot{\mathbf{T}}$$

We find from the figure that $\Delta\mathbf{T} = 1 \times \Delta\theta\,\mathbf{N}$, and $\Delta s = \rho\,\Delta\theta$, so that $\Delta\mathbf{T} = 1 \times (\Delta s/\rho)\mathbf{N}$.

We have now

$$\dot{\mathbf{T}} = \lim_{\Delta t \to 0} \frac{\Delta\mathbf{T}}{\Delta t} = \frac{1}{\rho}\mathbf{N} \lim_{\Delta t \to 0} \frac{\Delta s}{\Delta t} = \frac{1}{\rho}\dot{s}\mathbf{N} = \frac{V}{\rho}\mathbf{N}$$

Substituting this in the expression for the acceleration leads to

$$\mathbf{A} = \mathbf{A}_T + \mathbf{A}_N = \dot{V}\mathbf{T} + \frac{V^2}{\rho}\mathbf{N} = \ddot{s}\mathbf{T} + \frac{\dot{s}^2}{\rho}\mathbf{N} = (\dot{\rho}\dot{\theta} + \rho\ddot{\theta})\mathbf{T} + \rho\dot{\theta}^2\mathbf{N} \tag{1.5}$$

The *tangential* component of $\mathbf{A}$ has the magnitude $A_{\mathbf{T}} = \dot{V} = \ddot{s} = \dot{\rho}\dot{\theta} + \rho\ddot{\theta}$, and gives the change in *magnitude* of the velocity only. The *normal* component of $\mathbf{A}$ has the magnitude $A_{\mathbf{N}} = V^2/\rho = \dot{s}^2/\rho = \rho\dot{\theta}^2$, and is *directed towards* the centre of curvature O; it gives the change in *direction* of $\mathbf{V}$ only.

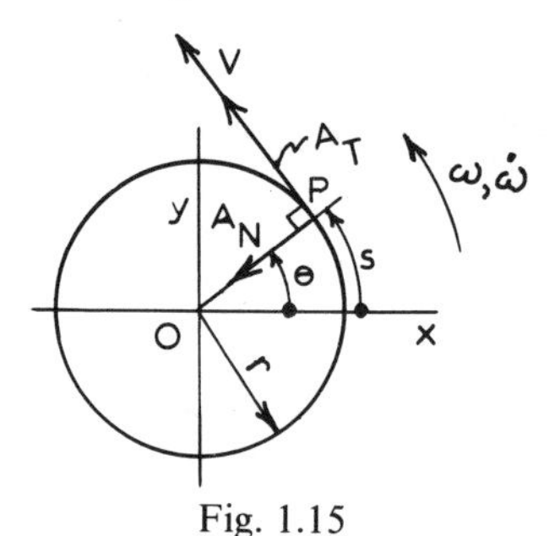

Fig. 1.15

Example 1.6. A point P is moving in a circle as shown in Fig. 1.15. Determine the velocity and the normal and tangential components of its acceleration.

Solution. We have directly $s=r\theta$ so that the speed of P is $V=\dot{s}=r\dot{\theta}=r\omega$, where ω is the angular velocity of the radial line OP. The velocity is $\mathbf{V}$ as shown, this is directed along the tangent at P and is therefore, in this special case, perpendicular to the radius vector OP.

The tangential component of the acceleration is $\mathbf{A_T}=\dot{V}\mathbf{T}$ as shown, the magnitude is $\dot{V}=\ddot{s}=r\ddot{\theta}=r\dot{\omega}$, where $\dot{\omega}$ is the angular acceleration of the radial line OP.

The normal component of the acceleration is $\mathbf{A_N}=(V^2/\rho)\mathbf{N}$ as shown; the magnitude of $\mathbf{A_N}$ is $A_N=V^2/\rho=V^2/r=r\omega^2$.

These two components are of great importance for all work in rigid-body kinematics. Projection on the x- and y-axis gives the rectangular components already found in Example 1.4. The magnitude of the acceleration is $A=\sqrt{(A_T^2+A_N^2)}=r\sqrt{(\dot{\omega}^2+\omega^4)}$ as found before.

Example 1.7. A point P is moving with constant speed V on the ellipse $x^2/a^2+y^2/b^2=1$, as shown in Fig. 1.16. The direction of motion is anti-clockwise. Determine the acceleration components when P passes through the point B.

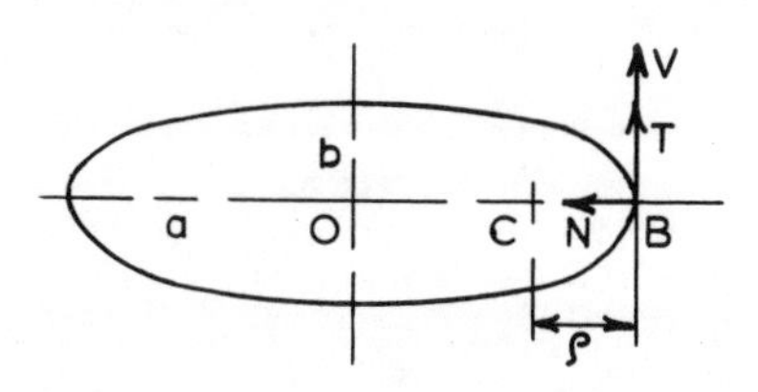

Fig. 1.16

Solution. The $\mathbf{T}$ and $\mathbf{N}$ vectors are shown in Fig. 1.16 at B.

The velocity at B is $\mathbf{V}=V\mathbf{T}$, and the acceleration is $\mathbf{A}=\dot{V}\mathbf{T}+(V^2/\rho)\mathbf{N}$; since V is constant, we have $\dot{V}=0$, so that the tangential component vanishes, that is $\mathbf{A_T}=\mathbf{0}$.

The radius of curvature ρ is given by (1.3):

$$\frac{1}{\rho}=\frac{y''}{[1+(y')^2]^{3/2}}$$

We have

$$y=\frac{b}{a}\sqrt{(a^2-x^2)}$$

$$y' = -\frac{b}{a}\frac{x}{\sqrt{(a^2-x^2)}}$$

$$y'' = -\frac{ab}{(a^2-x^2)^{3/2}}$$

so that

$$\frac{1}{\rho} = \frac{-ab}{[a^2+x^2(b^2/a^2-1)]^{3/2}}$$

At position B, $x=a$, and $1/\rho=-a/b^2$; using the numerical value of $1/\rho$, we have

$$\mathbf{A_N} = \frac{V^2}{\rho}\mathbf{N} = V^2\frac{a}{b^2}\mathbf{N}.$$

1.4.4 Polar coordinates. Radial and transverse components

The position of a point P moving in a plane may be given by the *polar coordinates* (r, φ), as shown in Fig. 1.17. The coordinates r and φ are functions of time t; if the time is eliminated we find the *path* of P in polar coordinates $r=f(\varphi)$.

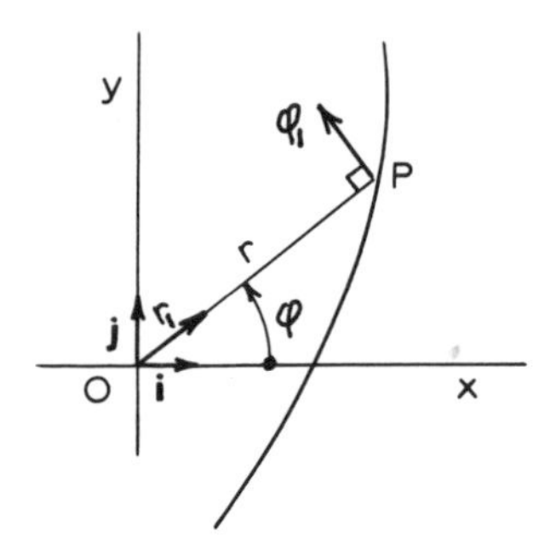

Fig. 1.17

Introducing the unit vectors $\mathbf{r}_1$ and $\boldsymbol{\varphi}_1$ as shown, we may take the components of velocity and acceleration of the point in the directions of these unit vectors.

The position vector of P is $\mathbf{OP}=\mathbf{R}=r\mathbf{r}_1$. From the figure we find that

$$\mathbf{r}_1 = \cos\varphi\,\mathbf{i}+\sin\varphi\,\mathbf{j} \qquad \text{and} \qquad \boldsymbol{\varphi}_1 = -\sin\varphi\,\mathbf{i}+\cos\varphi\,\mathbf{j}$$

differentiating gives

$$\dot{\mathbf{r}}_1 = -\sin\varphi\,\dot{\varphi}\,\mathbf{i} + \cos\varphi\,\dot{\varphi}\,\mathbf{j} = \dot{\varphi}\boldsymbol{\varphi}_1$$

and

$$\dot{\boldsymbol{\varphi}}_1 = -\cos\varphi\,\dot{\varphi}\mathbf{i} - \sin\varphi\,\dot{\varphi}\mathbf{j} = -\dot{\varphi}\mathbf{r}_1$$

Substituting $\dot{\mathbf{r}}_1 = \dot{\varphi}\boldsymbol{\varphi}_1$ in the expression for $\mathbf{V}$ gives

$$\mathbf{V} = \dot{r}\mathbf{r}_1 + r\dot{\varphi}\boldsymbol{\varphi}_1 \tag{1.6}$$

We may write this as $\mathbf{V} = \mathbf{V}_r + \mathbf{V}_\varphi = V_r\mathbf{r}_1 + V_\varphi\boldsymbol{\varphi}_1$, where $V_r = \dot{r}$ and $V_\varphi = r\dot{\varphi}$.

The acceleration of P is $\mathbf{A} = \dot{\mathbf{V}} = \dot{r}\dot{\mathbf{r}}_1 + \ddot{r}\mathbf{r}_1 + r\dot{\varphi}\dot{\boldsymbol{\varphi}}_1 + (r\ddot{\varphi} + \dot{r}\dot{\varphi})\boldsymbol{\varphi}_1$; by substituting $\dot{\mathbf{r}}_1 = \dot{\varphi}\boldsymbol{\varphi}_1$ and $\dot{\boldsymbol{\varphi}}_1 = -\dot{\varphi}\mathbf{r}_1$, we finally obtain the acceleration:

$$\mathbf{A} = (\ddot{r} - r\dot{\varphi}^2)\mathbf{r}_1 + (r\ddot{\varphi} + 2\dot{r}\dot{\varphi})\boldsymbol{\varphi}_1 \tag{1.7}$$

we may write this as $\mathbf{A} = \mathbf{A}_r + \mathbf{A}_\varphi = A_r\mathbf{r}_1 + A_\varphi\boldsymbol{\varphi}_1$, where $\mathbf{A}_r$ is the *radial component* and $\mathbf{A}_\varphi$ is the *transverse component* of acceleration, and

$$A_r = \ddot{r} - r\dot{\varphi}^2 \qquad A_\varphi = r\ddot{\varphi} + 2\dot{r}\dot{\varphi} = \frac{1}{r}\frac{d}{dt}(r^2\dot{\varphi}).$$

If r is of constant length, the motion is circular, and $\dot{r} = \ddot{r} = 0$, $\dot{\varphi} = \omega$, $\ddot{\varphi} = \dot{\omega}$, so that

$$\mathbf{V} = r\omega\boldsymbol{\varphi}_1 \qquad \text{and} \qquad \mathbf{A} = -r\omega^2\mathbf{r}_1 + r\dot{\omega}\boldsymbol{\varphi}_1$$

The direction and magnitude of the velocity and the components of the acceleration are in agreement with the expressions developed in examples 1.4 and 1.6.

Radial and transverse components are particularly useful in planetary motion and for orbiting space vehicles.

Example 1.8. A point P moves in a plane in such a way that its polar coordinates r (m) and φ (rad) are given by the time functions $r = t^2 + 2t$, and $\varphi = t$. Determine the path in polar coordinates and the position, velocity and acceleration of P at the instant when $t = 2$ s.

Solution. We find by eliminating t that the path is determined by $r = \varphi^2 + 2\varphi$. When $t = 2$, we find $r = 8$ m and $\varphi = 2$ rad, which determines the position of P at this instant.

The velocity $\mathbf{V} = \dot{r}\mathbf{r}_1 + r\dot{\varphi}\boldsymbol{\varphi}_1$ from (1.6). We have $\dot{r} = 2t + 2$ and $\dot{\varphi} = 1$ (constant), so when $t = 2$ s, $\dot{r} = 6$ m/s and $\dot{\varphi} = 1$ rad/s; the instantaneous velocity is then $\mathbf{V} = 6\mathbf{r}_1 + 8\boldsymbol{\varphi}_1$ m/s.

The instantaneous speed is $V = \sqrt{(36+64)} = 10$ m/s.

The acceleration is determined from (1.7):

$$\mathbf{A} = (\ddot{r} - r\dot{\varphi}^2)\mathbf{r}_1 + (r\ddot{\varphi} + 2\dot{r}\dot{\varphi})\boldsymbol{\varphi}_1$$

we have $\ddot{r} = 2$ (constant), and $\ddot{\varphi} = 0$, so that

$$\mathbf{A} = -6\mathbf{r}_1 + 12\boldsymbol{\varphi}_1 \text{ m/s}^2$$

The magnitude of the acceleration is

$$A = \sqrt{(36+144)} = 13{\cdot}4 \text{ m/s}^2$$

Two other sets of components are sometimes used in three-dimensional motion. These are cylindrical and spherical components. The definition of those and the formulae for velocity and acceleration are given in Problems 1.8 and 1.10.

Problems

1.1. Water drips from a faucet at the uniform rate of n drops per second. Find the distance x between any two adjacent drops as a function of the time t that the trailing drop has been in motion. Neglect air resistance.

1.2. A car starting from rest increases its speed from 0 to V with a constant acceleration a_1, runs at this speed for a time, and finally comes to rest with constant deceleration a_2. Given that the total distance travelled is s, find the total time t required. If the greatest possible acceleration and deceleration that the automobile may have is a and maximum speed is v, what is the minimum time required to cover a distance s from rest to rest?

1.3. A particle moves in the xy-plane according to the law $\dot{x} = 2t - 6$ and $\dot{y} = 3t^2 - 18t + 27$. If the particle is at $(9, -27)$ m when $t = 0$, determine in rectangular coordinates (a) the acceleration of the particle when $t = 4$ s, and (b) the equation of the path.

1.4. The position of a particle moving on a circular path with a radius of 32 m varies according to $s=3t^2+4t$, where s is the distance in metres from a fixed point to the particle measured along the path. The radial line from the centre to the particle is turning counterclockwise when $t=1$ s, and the particle is at the top of the path when $t=2$ s. Determine the acceleration when $t=2$ s in normal and tangential components.

1.5. Rod AB (Fig. 1.18) rotates in the xy-plane about O with a constant angular acceleration of 1 rad/s^2. A washer slides out along the rod from O. The distance r from O to the washer increases uniformly at the rate of 2 m/s. Determine the velocity and acceleration of the washer in polar coordinates when the angular velocity of the rod is 3 rad/s, $r=2$ m and $\theta=90°$.

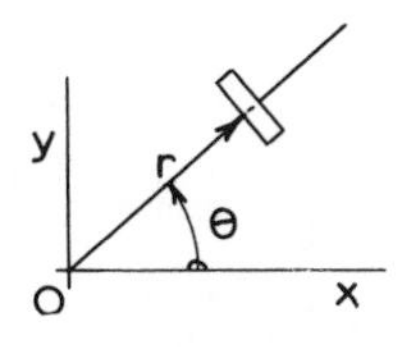

Fig. 1.18

1.6. $AB=BC=6$ m in Fig. 1.19. Block C has a constant velocity of 12 m/s to the right. Determine the angular velocity and acceleration of the rod AB for $\theta=40°$.

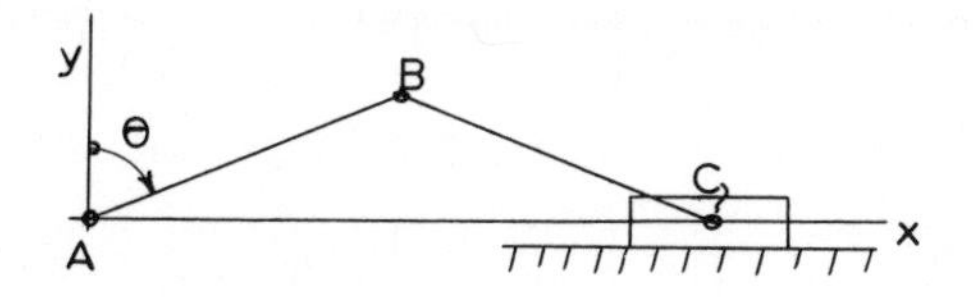

Fig. 1.19

1.7. A point moves in rectilinear motion with velocity–time function $V=3 \sin (\pi t/6)$ m/s. Determine the maximum velocity. Determine the displacement and acceleration at $t=6$ s, given that $x=0$ when $t=0$.

1.8. The position of a point P may be given in *cylindrical coordinates* (r, φ, z) as shown in Fig. 1.20; these coordinates are functions of time. Introducing the unit vectors $\mathbf{r}_1$, $\boldsymbol{\varphi}_1$ and $\mathbf{k}$ as shown, the components of velocity and acceleration may be taken in direction of these unit vectors. Show that the derivatives of the unit vectors are $\dot{\mathbf{r}}_1 = \dot{\varphi}\boldsymbol{\varphi}_1$, $\dot{\boldsymbol{\varphi}}_1 = -\dot{\varphi}\mathbf{r}_1$, and $\dot{\mathbf{k}} = \mathbf{0}$. Show that the velocity and acceleration of P are determined by $\mathbf{V} = \dot{r}\mathbf{r}_1 + r\dot{\varphi}\boldsymbol{\varphi}_1 + \dot{z}\mathbf{k}$; $\mathbf{A} = (\ddot{r} - r\dot{\varphi}^2)\mathbf{r}_1 + (r\ddot{\varphi} + 2\dot{r}\dot{\varphi})\boldsymbol{\varphi}_1 + \ddot{z}\mathbf{k}$.

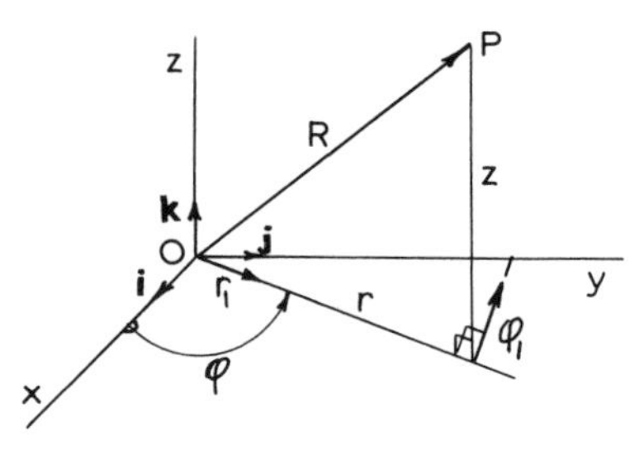

Fig. 1.20

1.9. An aeroplane is cruising at 334 km/h east in straight and level flight. The propeller has a diameter of 3·96 m and rotates at 1035 rev/min clockwise as viewed from the rear. Determine the velocity and acceleration of the tip of the propeller when the tip is in the southernmost position. Take the z-axis to the east along the line of flight, the x-axis vertical and the y-axis to the south. (Cylindrical coordinates.)

1.10. The position of a point P is given by the coordinates θ, φ and r as shown on Fig. 1.21; these are all functions of time and are called *spherical coordinates.*

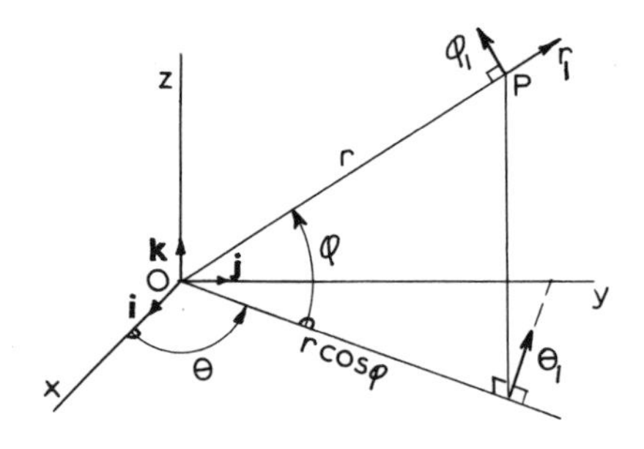

Fig. 1.21

By introducing the unit vectors $\boldsymbol{\theta}_1$, $\boldsymbol{\varphi}_1$, and $\mathbf{r}_1$ as shown, the components of the velocity and acceleration of P may be taken in the direction of these unit vectors.

Show that the derivatives of the unit vectors are $\dot{\mathbf{r}}_1 = \dot{\varphi}\boldsymbol{\varphi}_1 + \dot{\theta}\cos\varphi\,\boldsymbol{\theta}_1$; $\dot{\boldsymbol{\theta}}_1 = -\dot{\theta}\cos\varphi\,\mathbf{r}_1 + \dot{\theta}\sin\varphi\,\boldsymbol{\varphi}_1$, $\dot{\boldsymbol{\varphi}}_1 = -\dot{\varphi}\mathbf{r}_1 - \dot{\theta}\sin\varphi\,\boldsymbol{\theta}_1$. Show that the velocity is determined by $\mathbf{V}_{\mathrm{p}} = \dot{r}\mathbf{r}_1 + r\dot{\varphi}\boldsymbol{\varphi}_1 + r\dot{\theta}\cos\varphi\,\boldsymbol{\theta}_1$, and the acceleration by

$$\mathbf{A}_{\mathrm{p}} = [\ddot{r} - r(\dot{\varphi}^2 + \dot{\theta}^2\cos^2\varphi)]\mathbf{r}_1 + [2\dot{r}\dot{\varphi} + r(\ddot{\varphi} + \dot{\theta}^2\sin\varphi\cos\varphi)]\boldsymbol{\varphi}_1 + [(2\dot{r}\dot{\theta} + r\ddot{\theta})\cos\varphi - 2r\dot{\varphi}\dot{\theta}\sin\varphi]\boldsymbol{\theta}_1$$

1.11. Figure 1.22 shows a *spherical pendulum* of length r. Determine the velocity of the bob in terms of the given coordinates θ, φ and r.

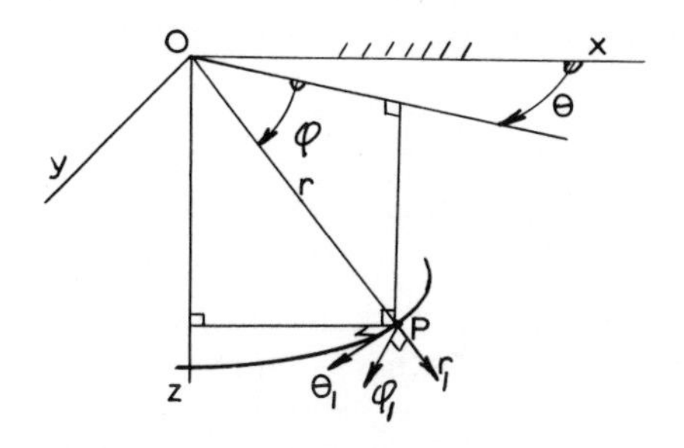

Fig. 1.22

CHAPTER TWO

Dynamics of a Particle

2.1 Newton's laws

2.1.1 Newton's laws of motion

The basic SI units of length (m) and time (s) were used extensively in kinematics in Chapter 1. In dynamics it soon becomes apparent that further concepts and units must be introduced to describe adequately the motion of bodies. To move a body, a certain action must be taken, the body must be pushed or pulled. This action is called to apply a *force* to the body.

A force may, in statics, be defined by the change in configuration it produces in a standard body. In dynamics a force also produces changes in a body, but of far greater importance is the fact that a force changes the motion of a body; this change of motion is used to define a force, and the change is accounted for in Newton's three laws of motion; these laws are the foundations of classical dynamics and were first stated in Newton's famous book *Principia* in 1687; the laws are axioms in dynamics. The first law may be stated as follows:

Newton's first law. A body continues in a state of rest or uniform motion, unless it is acted upon by a resultant force.

This law is really a special case of the second law, but it was included by Newton in his statements and has therefore been retained in works on dynamics. It defines a force as an action which changes the motion of a body.

It is common knowledge that to prevent a body from falling to the ground a force must be exerted on the body, and if the body is moved about it resists any change in its motion; experiments also show that two bodies attract each other with a force, and that the force necessary to accelerate a body is proportional to the acceleration. This property of the matter of which a body consists, to resist any change in motion and attract other bodies, is called the *inertia* or the *mass* of the body.

The mass of a body is defined in Newton's second law as the proportionality factor between the applied resultant force and the acceleration it produces.

The masses of two bodies may be compared directly by attaching them to identical springs on the same location of the earth's surface; if the extension of the springs is the same, the bodies have the same mass. If the experiment is repeated high above the surface of the earth, the extension of the springs is smaller, but it is still the same for both bodies if this is so at the surface; we assume from this that the mass of a body is independent of its location and is constant throughout the universe.

To give a body a certain acceleration, the same resultant force must be applied anywhere in space.

The unit of mass in the SI system is the mass of a prototype body kept in Paris; this mass is one kilogram (kg). If a body of mass m is moving with a velocity $\mathbf{v}$ it is said to have a *linear momentum* $m\mathbf{v}$.

Newton's second law. This is the most important in dynamics. Newton stated that the motion is proportional to the natural force impressed and in the same direction.

This statement needs some clarification. The law was stated for a particle only, and was later extended to include a collection of particles and a rigid body.

In Newtonian mechanics the mass of a particle is assumed constant. For a system of particles the total mass of the system may change, as in the case of a rocket burning fuel or a snowball rolling down a hill.

Newton's statement has been interpreted in two ways: in the first case the statement has been taken to mean that the resultant force is proportional to the time rate of change of the momentum and in the same direction. In the second interpretation the resultant force is taken to be proportional to the acceleration and in the same direction, the scalar mass being the proportionality factor.

In the first formulation the law may be expressed as follows:

$$\mathbf{F} = \frac{d}{dt}(m\mathbf{V}) = m\frac{d\mathbf{V}}{dt} + \mathbf{V}\frac{dm}{dt}$$

For a constant mass m, the result is

$$\mathbf{F} = m\frac{d\mathbf{V}}{dt} = m\ddot{\mathbf{r}} \tag{2.1}$$

and (2.1) expresses the second interpretation of the law. This is a vector equation, for which the three corresponding scalar equations are

$$F_x = m\ddot{x} \qquad F_y = m\ddot{y} \qquad F_z = m\ddot{z} \tag{2.2}$$

The fact that the acceleration is involved in Newton's law explains why it is necessary to discuss kinematics at considerable length before dealing with dynamics.

The acceleration produced by a force is independent of the previous motion of the body and other forces acting; the resultant acceleration is proportional to the resultant force and in the same direction.

Newton's third law. This is the law of action and reaction. It states that the forces acting between two bodies are always equal in magnitude, opposite in direction and directed along the same line. The law holds at any instant for all forces, whether constant or variable and stationary or moving; it means that forces always occur in equal and opposite pairs, one of the forces being called the action, the other the reaction.

Newton's laws are claimed to be valid only in a *primary inertial system*, that is a coordinate system with axis fixed in space without rotation, or moving in a parallel translation with constant speed. It will be shown in Chapter 4 that the acceleration measured in a coordinate system attached to the earth's surface is very nearly the same as the absolute acceleration measured in a primary inertial system.

A coordinate system fixed on the earth may therefore be accepted as a secondary inertial system in which Newton's law may be applied without correction in the great majority of cases. The main exceptions to this are orbital motion, space travel, long-range rocket flight and certain problems in fluid and air flow.

The unit for force in the SI system is now *derived* from the basic units mass, length and time by using Newton's second law; the unit is the newton (N), which is defined as the force which, applied to a mass of 1 kg, gives it an acceleration of 1 m/s^2. A newton is thus equivalent to 1 kg m/s^2.

In classical or Newtonian dynamics the mass of a particle is assumed to be constant, and the particle must be large compared to subatomic particles; it is also assumed that velocities are small compared to the velocity of light (about 299 000 km/s).

In the case of subatomic particles, a newer branch of dynamics—quantum mechanics—has been developed in this century, and for particles moving with velocities approaching the speed of light, the laws of relativistic mechanics must be applied. In this book we consider only those situations in which Newtonian dynamics may be assumed valid.

2.1.2 Newton's universal law of gravitation

This law was also given in 1687, and states that the gravitational force of attraction between two particles of mass m_1 and m_2 (Fig. 2.1) and at a distance r is along the line connecting them and equal to

$$F = \gamma \frac{m_1 m_2}{r^2} \tag{2.3}$$

This is the famous inverse square law of gravitation. The universal gravitational constant γ has been found by experiment to be equal to $6{\cdot}673 \times 10^{-11}$ m^3/kg s^2.

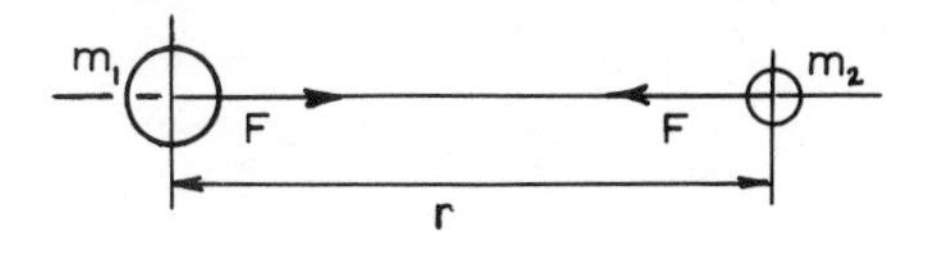

Fig. 2.1

The gravitational attraction on a particle from a homogeneous sphere may be found in the following manner: A thin spherical shell of radius a and thickness da is shown in Fig. 2.2. The shell attracts a particle of mass m at P at a distance r from the centre O of the shell ($r > a$).

The gravitational forces from the ring-shaped element of the shell are directed along the generators of the cone shown; taking components along PO and in the plane perpendicular to PO through P, the components in this plane form a star of concurrent equal forces at P; these components cancel in the summation, and only the components along PO give contributions to the resultant force dS; this force is then directed towards the centre of the shell, since this is the case for all such ring shaped elements, the resultant force from the spherical shell is towards its centre O.

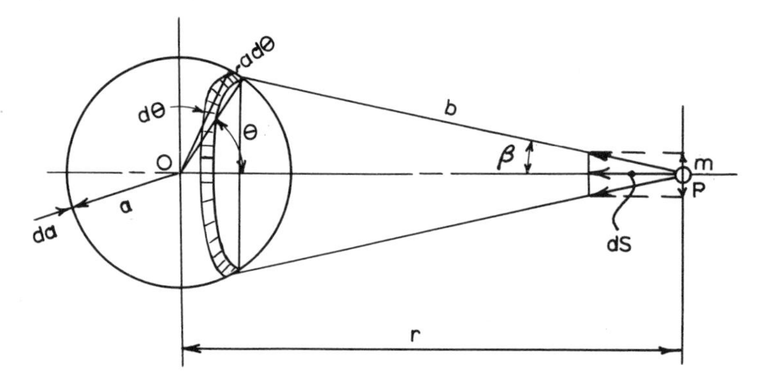

Fig. 2.2

If the density of the shell is ρ, the mass of the ring-shaped element is $(2\pi a \sin\theta)(a\,d\theta)(da)\rho = 2\pi\rho a^2 \sin\theta\,d\theta da$. If the universal gravitational constant is γ, Newton's gravitational law gives

$$dS = \gamma\frac{(2\pi\rho a^2 \sin\theta\,d\theta da)m\cos\beta}{b^2}$$

from the figure

$$\cos\beta = \frac{r - a\cos\theta}{b} \qquad \text{and} \qquad b^2 = a^2 + r^2 - 2ar\cos\theta$$

Therefore the total force from the shell is

$$S = 2\pi\gamma\rho m a^2\,da\int_{\theta=0}^{\pi}\frac{\sin\theta\,(r - a\cos\theta)}{(a^2 + r^2 - 2ar\cos\theta)^{3/2}}\,d\theta$$

towards O. Using $b^2 = a^2 + r^2 - 2ar\cos\theta$, we have

$$\sin\theta\,d\theta = \frac{b}{ar}\,db \qquad \text{and} \qquad r - a\cos\theta = \frac{r^2 - a^2 + b^2}{2r}$$

Substitution now gives

$$S = \frac{\pi\gamma\rho ma\,da}{r^2}\int_{b=r-a}^{r+a}\left(1+\frac{r^2-a^2}{b^2}\right)db$$

$$= \frac{\pi\gamma\rho ma\,da}{r^2}\left[b-\frac{(r^2-a^2)}{b}\right]_{r-a}^{r+a} = \frac{4\pi\gamma\rho ma^2}{r^2}\,da$$

For a solid homogeneous sphere of centre O and radius R, the gravitational attraction may be found by dividing it into thin concentric shells; the attraction from each shell is towards O, so that the total gravitational force on m is towards the centre of the sphere.

The magnitude of the force may be found from the above expression by summing up for all the shells:

$$F = \frac{4\pi\gamma\rho m}{r^2}\int_{a=0}^{R} a^2\,da = (\tfrac{4}{3}\pi R^3)\,\frac{\gamma\rho m}{r^2}$$

The mass of the sphere is

$$M = (\tfrac{4}{3}\pi R^3)\rho$$

Hence $F=\gamma Mm/r^2$. This shows that the total gravitational force from a homogeneous sphere acts *towards the centre of the sphere* and as if the total mass of the sphere were concentrated there. It may be seen that the same result is found if the density ρ is a function of the radius only. This is the actual situation in the case of the earth.

If the earth is considered to be a sphere of radius R_e and mass M_e, the gravitational force on a mass m at a distance $r (r > R_e)$ is towards the centre of the earth and of magnitude

$$F = \frac{\gamma M_e m}{r^2} = \left(\frac{\gamma M_e}{R_e^{\,2}}\right)\frac{m}{r^2}\,R_e^2$$

Introducing a new factor $g=\gamma M_e/R_e^2$, we have $F=gm(R_e/r)^2$; at the earth surface $r=R_e$, so

$$F = mg \qquad (2.4)$$

We call g the acceleration due to gravity; it is not a constant, since the earth is not a sphere and its density varies; g depends on altitude, location and local geography.

The mass of the earth is about $M_e = 5{\cdot}98\times10^{24}$ kg, and the radius is about 6378 km at the equator and about 6357 km at the poles, giving a mean value of about 6368 km; this gives a value of g of about 9·8 m/s^2.

The total variation in g over the earth surface up to an altitude of 5000 m is about $\frac{3}{4}$%; this variation is too small to be of importance in most practical engineering problems, and a *standard value* of g from 45° latitude is used. This value is $g=9{\cdot}80665$ m/s^2; for practical slide-rule work this is rounded off to $g=9{\cdot}81$ m/s^2.

The gravitational force from the earth is called the weight force or the weight of a body. If the mass of a body is m kg, the weight is $W=mg$ N, where $g=9{\cdot}81$ m/s^2.

For a sphere of steel with a radius of 1 m, the weight force is about 325 000 N; the gravitational attraction from a similar sphere on the earth surface at a centre distance of 10 m is about 0·00073 N; for comparison, the attraction from the moon is about 1·1 N and from the sun about 196 N; consequently the gravitational force from the earth is the only gravitational force that need be considered in all practical problems.

Since the gravitational forces acting on the mass particles of a body are all directed towards the earth centre about 6 368 000 m distant, we can consider these forces as *parallel forces*, so that the gravitational force field may be considered to be a uniform parallel force field.

2.2 Centre of parallel forces. Centre of gravity

Figure 2.3 shows a system of parallel forces acting on a rigid body; the resultant is a single force **R**, whose magnitude is found by adding the force magnitudes algebraically.

The point (x_C, y_C, z_C) on the line of action of **R** may be determined as follows: Resolving each force into rectangular components, we have $F_{ix}=F_i \cos\theta_x$, $F_{iy}=F_i \cos\theta_y$, $F_{iz}=F_i \cos\theta_z$, $i=1, 2, 3, \ldots, n$. The cosines are direction cosines for the forces, and in this case they are the same for each of the forces; $R_x=\sum F_{ix}=\sum F_i \cos\theta_x =\cos\theta_x \sum F_i$, $R_y=\cos\theta_y \sum F_i$, $R_z=\cos\theta_z \sum F_i$.

Taking moments of the x-components about the z-axis gives

$$R_x y_C = \sum F_{ix} y_i = \sum (F_i \cos\theta_x) y_i = \cos\theta_x \sum F_i y_i$$

$$y_C = \frac{\cos\theta_x \sum F_i y_i}{\cos\theta_x \sum F_i} = \frac{\sum F_i y_i}{\sum F_i}$$

Similarly

$$x_C = \frac{\sum F_i x_i}{\sum F_i} \quad \text{and} \quad z_C = \frac{\sum F_i z_i}{\sum F_i}$$

Since the direction cosines cancel out, the same result is found if all the forces are rotated through the same angle to remain parallel.

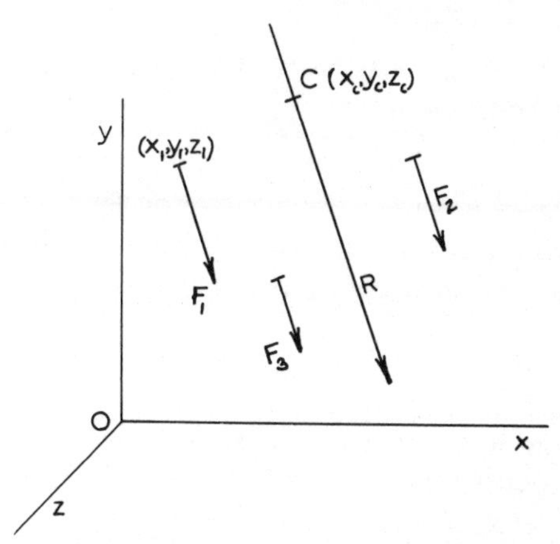

Fig. 2.3

If the acting forces are *gravity forces* $F_i = m_i g$ on a series of rigidly connected particles m_i; the centre C of the gravity forces is then located at:

$$x_C = \frac{\sum m_i g x_i}{\sum m_i g} = \frac{\sum m_i x_i}{\sum m_i} \qquad y_C = \frac{\sum m_i y_i}{\sum m_i} \qquad z_C = \frac{\sum m_i z_i}{\sum m_i}$$

This point C is called the *centre of gravity* of the particles; its location is apparently independent of the orientation of the system of particles in space, since the same point is found if all the forces are rotated through the same angle, or if the body is rotated in any way about C.

For a *rigid body* consisting of infinitely many rigidly connected particles, the centre of gravity is determined by

$$x_C = \frac{\int x \, dm}{M} \qquad y_C = \frac{\int y \, dm}{M} \qquad z_C = \frac{\int z \, dm}{M}$$

where M is the total mass of the body.

2.3 Newton's second law for a rigid body; motion of the centre of mass

For a typical mass particle m_i in a rigid body, the equation of motion, from Newton's second law, is $\mathbf{F}_{ei}+\mathbf{F}_{ii}=m_i\ddot{\mathbf{r}}_i$, where $\mathbf{F}_{ei}$ is the *resultant external force* on m_i, and $\mathbf{F}_{ii}$ is the *resultant internal force* on m_i from the other particles of the body.

Writing the above equation for all particles of the body and adding all the equations gives:

$$\sum \mathbf{F}_{ei}+\sum \mathbf{F}_{ii} = \sum m_i\ddot{\mathbf{r}}_i$$

however, for a rigid body the internal forces between the particles are always in equal and opposite pairs from Newton's third law; this means that $\sum \mathbf{F}_{ii}=\mathbf{0}$; taking the resultant of the external forces $\sum \mathbf{F}_{ei}=\mathbf{F}$, we get $\mathbf{F}=\sum m_i\ddot{\mathbf{r}}_i$; now writing $\sum m_i\mathbf{r}_i=(\sum m_i)\mathbf{r}_C = M\mathbf{r}_C$, we define by this expression a point C by the position vector $\mathbf{r}_C=(\sum m_i\mathbf{r}_i)/M$, where M is the total mass of the body; the point C is called the *centre of mass* of the body.

From the expression for $\mathbf{r}_C$, we find

$$\sum m_i\ddot{\mathbf{r}}_i = M\ddot{\mathbf{r}}_C$$

Therefore

$$\mathbf{F} = M\ddot{\mathbf{r}}_C \tag{2.5}$$

This is Newton's second law for a rigid body of total mass M acted upon by a resultant force $\mathbf{F}$; $\ddot{\mathbf{r}}_C$ is the *absolute* acceleration of the centre of mass, that is the acceleration of C in an inertial system; for practically all problems we can accept a coordinate system fixed at the earth's surface as such a system.

Equation (2.5) is in vector form; analytically it gives the three scalar equations for the motion of the centre of mass:

$$F_x = M\ddot{x}_C \qquad F_y = M\ddot{y}_C \qquad F_z = M\ddot{z}_C$$

The motion of the centre of mass is then determined by accumulating the total mass in the centre, and moving all forces to act at the centre of mass, keeping the directions in which they act unchanged. The centre then moves as a particle of total mass equal to the mass of the body.

Consider now a series of particles m_i, as shown in Fig. 2.4, with position vectors $\mathbf{r}_i$; the *centre of mass* of the particles was *defined* by the position vector:

$$\mathbf{r}_C = \frac{\sum m_i \mathbf{r}_i}{\sum m_i} \tag{2.6}$$

With $\mathbf{r}_C = x_C\mathbf{i} + y_C\mathbf{j} + z_C\mathbf{k}$ and $\mathbf{r}_i = x_i\mathbf{i} + y_i\mathbf{j} + z_i\mathbf{k}$, we find by substitution that $x_C = (\sum m_i x_i)/\sum m_i$, with similar expressions for y_C and z_C.

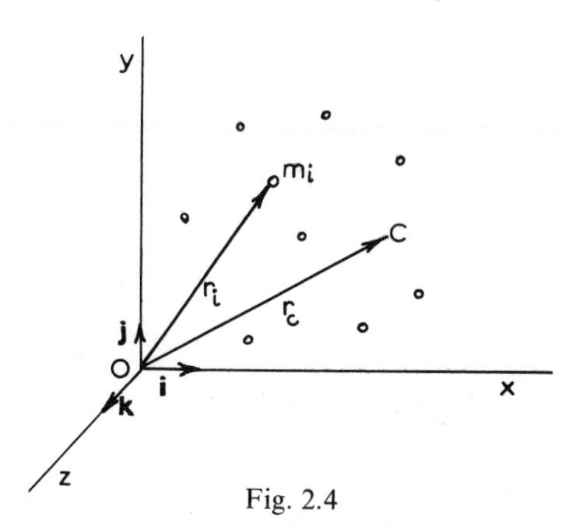

Fig. 2.4

For a *rigid body* these expressions take the form

$$x_C = \frac{\int x\,dm}{M} \qquad y_C = \frac{\int y\,dm}{M} \qquad z_C = \frac{\int z\,dm}{M}$$

where M is the total mass of the body; *these expressions are the same as those for the centre of gravity*; this means that the *centre of gravity* and the *centre of mass* are the same point for a rigid body in a uniform parallel gravitational field.

If the origin O of the coordinate system is taken at C, then

$$\int x\,dm = \int y\,dm = \int z\,dm = 0 \tag{2.7}$$

This holds for axes through the centre of gravity; since C is also determined by (2.6), we have $\mathbf{r}_C = \mathbf{0}$. Therefore

$$\sum m_i \mathbf{r}_i = \mathbf{0} \tag{2.8}$$

for position vectors $\mathbf{r}_i$ taken from C. We also find

$$\sum m_i \dot{\mathbf{r}}_i = \sum m_i d\mathbf{r}_i/dt = \sum d(m_i \mathbf{r}_i)/dt = d(\sum m_i \mathbf{r}_i)/dt = \mathbf{0} \tag{2.9}$$

These two vector formulae will be found useful in future developments of dynamics.

2.4 Differential equations of motion

Newton's second law, $\mathbf{F}=m\ddot{\mathbf{r}}$, for the motion of a particle or the centre of mass of a rigid body, is a vector equation and we may, therefore, take components of the vectors in the equation along any line, to find the scalar equation of the projected motion along that line. Generally coordinate axes are introduced to give the simplest possible expressions for the scalar equations of motion; these equations are of the form $F_x=m\ddot{x}$, and may be used to solve two types of problem.

In the first type the position x is given as a function of time or may be established as a function of time. If this function can be differentiated twice, we can directly determine the velocity $\dot{x}$ and acceleration $\ddot{x}$ along the line, and direct substitution in $F_x=m\ddot{x}$ gives the force as a function of time. In the second type of problem, the acting forces are given and the problem is to determine the resultant motion. This involves integration and can be done directly only in some special cases.

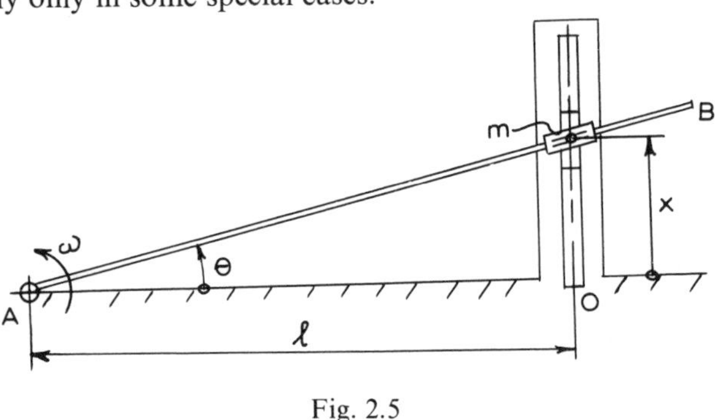

Fig. 2.5

2.4.1 Displacements given as functions of time

When the displacements are given, or may be determined as functions of time, the forces necessary to produce the motion may be determined directly from Newton's second law: the procedure is best illustrated by some examples.

Example 2.1. Figure 2.5 shows a radial arm AB which is rotating in a horizontal plane about A with constant angular velocity ω. The arm is connected through a sliding sleeve and a pin connection to a block of mass m which slides in a fixed slot in the plane as shown. Neglecting friction, find the total force on m in the direction of the groove, the force from the arm on m, and the reaction from the slot.

Solution. Taking the angle of rotation of the arm as θ and the x-axis along the slot, we have $\dot{\theta}=\omega$ (constant) and $\theta=\omega t$. From the geometry of the figure, $x=l\tan\theta=l\tan\omega t$; hence

$$\dot{x} = \frac{l\omega}{\cos^2 \omega t} \qquad \text{and} \qquad \ddot{x} = 2l\omega^2 \frac{\sin \omega t}{\cos^3 \omega t}$$

so that

$$F_x = m\ddot{x} = 2ml\omega^2 \frac{\sin \omega t}{\cos^3 \omega t}$$

From the condition of no friction, the force F from the sleeve on the block is perpendicular to the arm AB, and the reaction from the slot N is perpendicular to the x-axis, so that $F\cos\theta=F_x$; therefore $F=F_x/\cos\theta=F_x/\cos\omega t$, and $N-F\sin\theta=0$. Therefore $N=F\sin\theta=F_x\tan\omega t$.

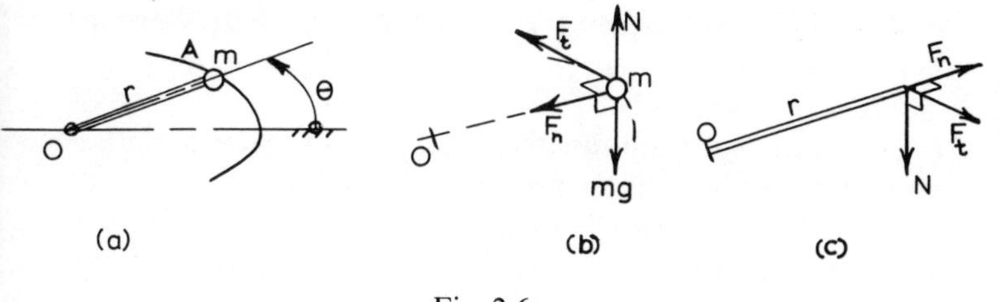

Fig. 2.6

Example 2.2. Figure 2.6(a) shows a rigid radial arm OA of length r, which is rotating in a horizontal plane about O. A particle of mass m is attached to the end of the arm. The angle of rotation θ is given by the function of t, $\theta=2t^2-5t$ rad. Determine all the forces acting on the particle and on the end A of the arm, as functions of t, m and r.

Solution. With $\theta=2t^2-5t$ rad we find the angular velocity and acceleration of the arm $\omega=\dot{\theta}=4t-5$ rad/s and $\dot{\omega}=\ddot{\theta}=4$ rad/s^2. Taking normal and tangential components of the acceleration, these are $r\omega^2$ towards O and $r\dot{\omega}$ perpendicular to OA, while there is no acceleration in the vertical direction. The forces on the mass are then $F_n=mr\omega^2=mr(4t-5)^2$, and $F_t=mr\dot{\omega}=4mr$, directed as shown in Fig. 2.6(b); the gravity force mg also acts on the particle, and is balanced by a vertical force $N=mg$ from the arm. Figure 2.6(c) shows the forces acting on the arm at A.

Example 2.3. Figure 2.7 shows a horizontal turntable which is rotating about a vertical axis through O at a constant angular velocity $\omega=0.5$ rad/s. A man of mass 85 kg walks from O towards A along a radial line drawn on the disc; the velocity of the man relative to the disc is $V=2$ m/s constant.

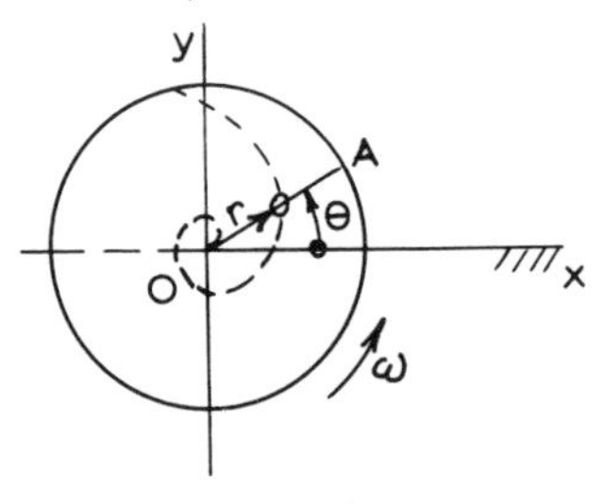

Fig. 2.7

Assuming that the man starts from O when $t=0$, determine the horizontal force on his feet at the instant when $t=2$ s.

Solution. Fixing a coordinate system (x, y) as shown, and determining the position of the man by the polar coordinates (r, θ), we have $r=Vt=2t$m and $\theta=\omega t=0{\cdot}5t$ rad. The path of the man in the horizontal plane is then the spiral $r=4\theta$, as indicated on the figure.

The components of acceleration in polar coordinates are of magnitude $a_r=\ddot{r}-r\dot{\theta}^2$ and $a_t=r\ddot{\theta}+2\dot{r}\dot{\theta}$; we have now $\dot{r}=2$, $\ddot{r}=0$, $\dot{\theta}=0{\cdot}5$ and $\ddot{\theta}=0$, so that $a_r=-r/4=-t/2$ m/s^2, and $a_t=2\times 2\times 0{\cdot}5=2$ m/s^2.

When $t=2$ s, the position of the man is found by $r=2t=4$ m, and $\theta=0{\cdot}5t=1$ rad; the acceleration components are $a_r=-1$ m/s^2, and $a_t=2$ m/s^2; the resultant acceleration is $a=\sqrt{(a_r^2+a_t^2)}=\sqrt{5}$ m/s^2, and the force on the man is $F=ma=85\sqrt{5}=190$ N.

The coefficient of friction for leather on metal may be taken as $\mu=0{\cdot}4$; the maximum friction force is then $\mu N=0{\cdot}4\times 85\times 9{\cdot}81 = 334$ N, so that the man will not be sliding.

2.4.2 Motion under a constant force

The simplest case of motion occurs when the resultant force is constant in magnitude and in a fixed direction; if the initial velocity of the particle is zero, or in the direction of the force, the motion is rectilinear, along the line of action of the force; this situation is shown in Fig. 2.8, where the x-axis has been taken along the line of action of the resultant force F.

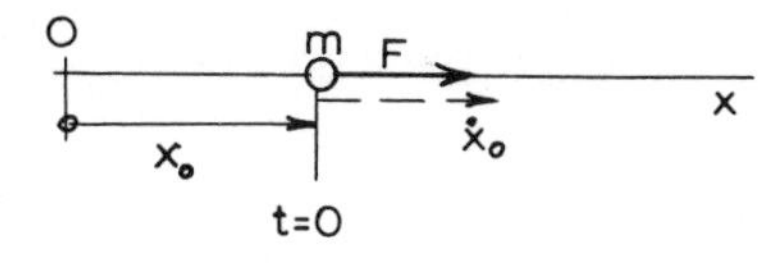

Fig. 2.8

The equation of motion is $m\ddot{x}=F$ (constant), so that $\ddot{x}=F/m = a =$ constant acceleration; two successive integrations give as functions of time the velocity $\dot{x}=at+C_1$ and the displacement

$$x = \tfrac{1}{2}at^2+C_1t+C_2$$

The integration constants C_1 and C_2 must be determined from the initial conditions, if the displacement and velocity at the time $t=0$ are $x=x_0$ and $\dot{x}=\dot{x}_0$, we find $C_1=\dot{x}_0$ and $C_2=x_0$; the complete solution of the problem is then

$$\begin{aligned} x &= \tfrac{1}{2}at^2+\dot{x}_0t+x_0 \\ \dot{x} &= at+\dot{x}_0 \\ \ddot{x} &= a = F/m \text{ (constant)} \end{aligned}$$

Example 2.4. A body is dropped in a free fall starting from rest, determine the displacement as a function of time.

Solution. Assuming that the height about the earth's surface is relatively small, we can take g as a constant; if the body has a mass that is large compared to the same volume of air and is of a compact shape, we can neglect the air resistance for small velocities; the only force acting is then the constant weight force $F=W=mg$.

Taking the positive x-direction to be vertically downwards, we have $m\ddot{x}=mg$, or $\ddot{x}=g=$constant. With $x=0$ and $\dot{x}=0$ at $t=0$, we find $\dot{x}=gt$ and $x=\frac{1}{2}gt^2$ for a free fall starting from rest; these expressions were first established by Galileo during his investigations of falling bodies about the year 1600.

Example 2.5. A body of mass m (Fig. 2.9) is released from rest on a plane inclined at α to the horizontal. The coefficient of friction between the mass and the plane is μ; determine the equation of motion of the body, and the distance moved as a function of time.

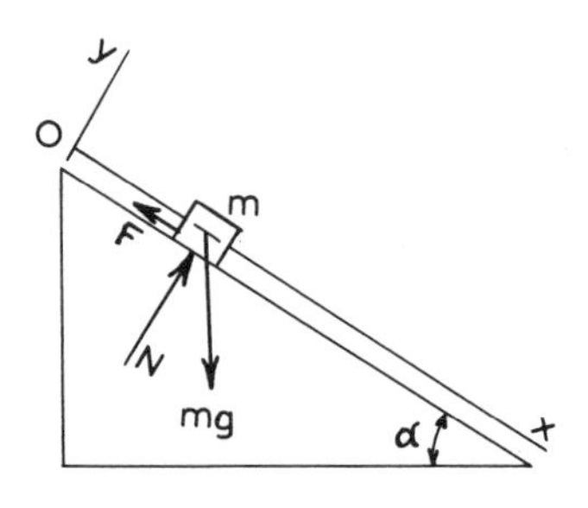

Fig. 2.9

Solution. Introducing a coordinate system (x, y) as shown, the body is in a rectilinear motion along the x-axis. The acting forces are the gravity force mg and a reaction from the plane, which is resolved into a normal pressure N and a force F along the plane; F is called the friction force and is of magnitude $F=\mu N$.

Since there is no motion in the y-direction, $\ddot{y}=0$, so that the forces in this direction are balanced. We have then $N-mg\cos\alpha=0$, or $N=mg\cos\alpha$, so that $F=\mu mg\cos\alpha$.

The equation of motion in the x-direction is $m\ddot{x}=mg\sin\alpha-F=mg\sin\alpha-\mu mg\cos\alpha$, so that $\ddot{x}=g\sin\alpha-\mu\cos\alpha=$constant. To enable the mass to slide down, we must have $\ddot{x}>0$:

$$g\sin\alpha-\mu\cos\alpha>0 \qquad \text{or} \qquad \mu<g\tan\alpha$$

If $\mu>g\tan\alpha$, the mass will not start to move; if $\mu<g\tan\alpha$, we find $\dot{x}=g(\sin\alpha-\mu\cos\alpha)t$, and $x=\frac{1}{2}g(\sin\alpha-\mu\cos\alpha)t^2$, where the integration constants have been taken as zero since the body starts from rest.

As an example of motion under a constant force, where the initial velocity is not in the direction of the force, consider the case of *projectile motion*:

Example2.6. Figure 2.10 shows a projectile of mass m which starts with an initial velocity V_0 inclined at an angle α to the horizontal x-axis; neglecting air resistance, determine the velocity and displacement of the projectile as functions of time. Determine also the path of the projectile, the range and the maximum height obtained.

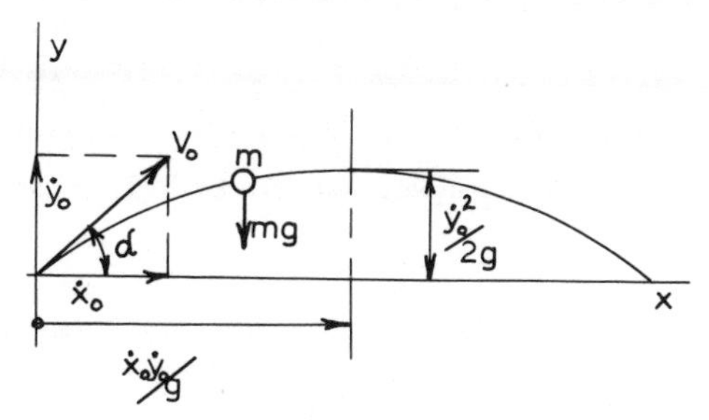

Fig. 2.10

Solution. The equations of motion are

$$m\ddot{x}=0 \qquad m\ddot{y}=-mg$$

Integrating these equations and taking $\dot{x}=\dot{x}_0$, $\dot{y}=\dot{y}_0$ when $t=0$, we find the *velocity components*:

$$\dot{x}=\dot{x}_0 \qquad \dot{y}=-gt+\dot{y}_0$$

The component of velocity in the x-direction is *constant*; seen from above, the projectile moves along the x-axis with constant velocity $\dot{x}_0$ until it suddenly stops.

Integrating the velocity equations and taking $x=y=0$ when $t=0$ gives the *displacements*:

$$x=\dot{x}_0t \qquad y=-\tfrac{1}{2}gt^2+\dot{y}_0t$$

Eliminating t gives the *equation of the path*

$$y=\frac{x}{\dot{x}_0}\left(\dot{y}_0-\frac{gx}{2\dot{x}_0}\right)$$

This is a parabola; substituting $\dot{x}_0 = V_0 \cos\alpha$, $\dot{y}_0 = V_0 \sin\alpha$ gives a second form for the path

$$y = x\left(\tan\alpha - \frac{gx}{2V_0^2\cos^2\alpha}\right)$$

The vertex is found when

$$\frac{dy}{dx} = \frac{\dot{y}_0}{\dot{x}_0} - \frac{gx}{\dot{x}_0^2} = 0 \qquad \text{or} \qquad x = \frac{\dot{x}_0\dot{y}_0}{g}$$

This value gives $y = \dot{y}_0^2/2g$ as the maximum height; this result may also be found from $\dot{y} = 0$.

The range, from symmetry, is $r = 2\dot{x}_0\dot{y}_0/g$, this may also be found by setting $y = 0$ in the equation for the parabola. The time to impact is $t = x/\dot{x}_0$, where $x = 2\dot{x}_0\dot{y}_0/g$, so that $t = 2\dot{y}_0/g$ s. The range may also be given as

$$r = \frac{2\dot{x}_0\dot{y}_0}{g} = \frac{2}{g}V_0\cos\alpha\, V_0\sin\alpha = \frac{V_0^2}{g}\sin 2\alpha$$

for a given velocity V_0, the maximum range is found for $\sin 2\alpha = 1$, or $\alpha = 45°$.

The solution given is quite unrealistic, since the air resistance is of considerable importance at the velocities of normal projectiles. Introducing air resistance results in equations of motion that can only be solved numerically.

2.4.3 Force as a function of time

If the resultant force is given as a function of time that can be integrated twice, the motion produced by the force may be found directly by integration.

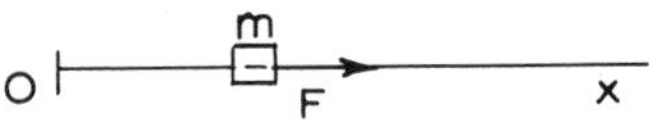

Fig. 2.11

Example 2.7. Figure 2.11 shows a mass m in rectilinear motion under the action of a force $F = F_0 \cos\omega t$; where F_0 and ω are given constants. Determine the equation of motion and the displacement–time function, if the displacement and velocity at $t = 0$ is $x = x_0$ and $\dot{x} = \dot{x}_0$.

Solution. The equation of motion, from Newton's second law, is $m\ddot{x}=F_0 \cos \omega t$, so that $d\dot{x}=(F_0/m) \cos \omega t\, dt$, or

$$\dot{x} = \frac{F_0}{m}\int \cos \omega t\, dt + C_1 = \frac{F_0}{m\omega} \sin \omega t + C_1$$

when $t=0$, $\dot{x}=\dot{x}_0$, so $C_1=\dot{x}_0$; we have now

$$dx = \frac{F_0}{m\omega} \sin \omega t\, dt + \dot{x}_0\, dt$$

and integration gives

$$x = -\frac{F_0}{m\omega^2} \cos \omega t + \dot{x}_0 t + C_2$$

When $t=0$, $x=x_0$, or $x_0=-F_0/m\omega^2+C_2$, so that $C_2=x_0+F_0/m\omega^2$. This finally gives

$$x = -\frac{F_0}{m\omega^2} \cos \omega t + \dot{x}_0 t + \left(x_0 + \frac{F_0}{m\omega^2}\right)$$

2.4.4 Resistant force proportional to the displacement

Consider the system shown in Fig. 2.12; it consists of a mass m which slides without friction on a horizontal plane; the mass is connected by a spring to a vertical wall. The mass is shown in the static-equilibrium position where there is no force in the spring, and the displacements x are measured from this position, positive to the right.

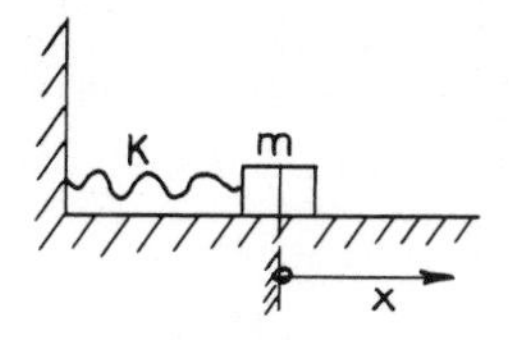

Fig. 2.12

The force in the spring is determined by the force law $F=Kx$, where K is a constant, called the *spring constant*, which is the force necessary to extend or compress the spring one unit of length. The units of K are N/m or N/cm. If the mass is displaced a distance x

to the right, the spring pulls back on it with a force of magnitude Kx to the left, the force on the mass must then be stated as $F = -Kx$. Note that if x is negative, the force is to the right, so that this expression holds for x positive or negative.

The equation of motion of the mass is now

$$m\ddot{x} = -Kx \qquad \text{or} \qquad \ddot{x} + (K/m)x = 0$$

Introducing a new constant $\omega_0 = \sqrt{(K/m)}$ rad/s, we have

$$\ddot{x} + \omega_0^2 x = 0 \tag{2.10}$$

A solution to this equation is $x = A \cos \omega_0 t + B \sin \omega_0 t$, where A and B are arbitrary constants; this may be seen by direct substitution. It is in fact the general solution since it contains *two* arbitrary constants A and B corresponding to the order of the equation.

From the general mathematical solution to eq. (2.10), we must now find the one solution that fits the physical situation; the constants A and B must be found from the starting conditions of the motion. Let us assume that the mass is started in motion by giving it a displacement x_0 followed by a blow to give it an initial velocity $\dot{x}_0$; counting time t from this instant, we have the starting conditions at $t=0$: $x = x_0$ and $\dot{x} = \dot{x}_0$; substituting $x = x_0$ and $t = 0$ in the general solution gives $A = x_0$; differentiating gives $\dot{x} = -A\omega_0 \sin \omega_0 t + B\omega_0 \cos \omega_0 t$, so that $\dot{x}_0 = B\omega_0$; for these starting conditions the solution is then

$$x = x_0 \cos \omega_0 t + \frac{\dot{x}_0}{\omega_0} \sin \omega_0 t$$

Writing this as

$$x = A_0 \cos (\omega_0 t - \varphi) = A_0 \cos \omega_0 t \cos \varphi + A_0 \sin \omega_0 t \sin \varphi$$

we have $A_0 \cos \varphi = x_0$, and $A_0 \sin \varphi = \dot{x}_0/\omega_0$, so that

$$A_0 = \sqrt{\left(x_0^2 + \frac{\dot{x}_0^2}{\omega_0^2}\right)} \qquad \text{and} \qquad \tan \varphi = \frac{\dot{x}_0}{x_0 \omega_0}$$

the solution may then be given as

$$x = \left[\sqrt{\left(x_0^2 + \frac{\dot{x}_0^2}{\omega_0^2}\right)}\right] \cos (\omega_0 t - \varphi)$$

The maximum displacement is then A_0, and this called the *amplitude* of the motion. The angle φ is called the *phase angle*.

The acceleration is $\ddot{x} = -A_0\omega_0^2 \cos(\omega_0 t - \varphi) = -\omega_0^2 x$, so the acceleration is proportional to the displacement, but in the opposite direction.

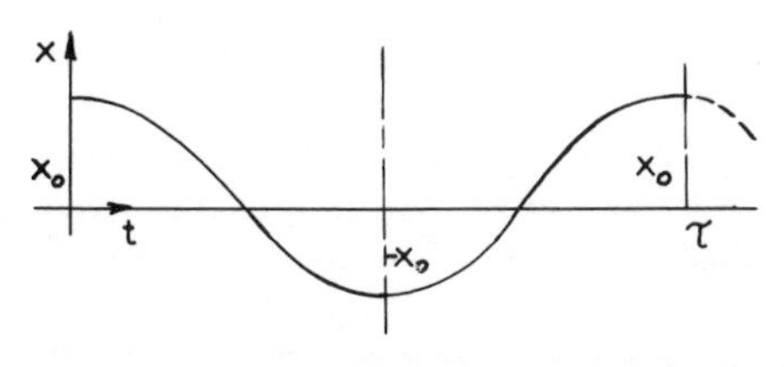

Fig. 2.13

The simplest solution is found for $\dot{x}_0 = 0$, which gives $A_0 = x_0$ and $\varphi = 0$, so that $x = x_0 \cos \omega_0 t$; this motion is shown in Fig. 2.13; it is clearly a *vibratory motion* along the x-axis, which completely repeats itself after a certain time τ; the motion is called a *periodic motion* with period $\tau = 2\pi/\omega_0$ s. This is the time for one complete vibration; the constant $\omega_0 = \sqrt{(K/m)}$ rad/s is called the *natural circular frequency* of the motion. The *frequency* f is the number of complete vibrations in one second:

$$f = \frac{1}{\tau} = \frac{1}{2\pi}\omega_0 = \frac{1}{2\pi}\sqrt{\left(\frac{K}{m}\right)} \text{ cycles/s}$$

The motion just described is called *simple harmonic motion*, and is of great importance in physics and engineering. The so-called *restoring force* $-Kx$ may be found in many different forms in various systems. Because of its importance, a general discussion of vibratory motion will be deferred to a special chapter, Chapter 10, where further examples may be found.

2.4.5 Force proportional to the velocity, but in the opposite direction

For motion, with small velocities, in a resisting fluid medium, it is found that the resisting force may be taken proportional to the velocity. The force may then be expressed by $-cv$, where the constant c depends on the size and shape of the body, and must be found experimentally.

Taking a simple case of this motion of a body moving down

through a fluid, the *driving force* is the gravity force mg. Taking the x-axis vertical, the equation of motion is

$$m\frac{dv}{dt} = mg - cv \qquad \text{or} \qquad \frac{dv}{dt} = g - \frac{c}{m}v = g - \beta v$$

where $\beta = c/m$. The resisting force increases with the velocity and eventually reaches the same magnitude as the constant driving force. When this happens, the forces on the mass are balanced, so that the body continues with a constant velocity and zero acceleration. We have then $dv/dt = 0$, or $g - \beta v = 0$, so that $v = g/\beta = v_0$, and this is the *limiting* or *terminal* velocity.

Introducing v_0 in the equation of motion gives

$$\frac{dv}{dt} = \beta v_0 - \beta v = \beta(v_0 - v) \qquad \text{or} \qquad dt = \frac{dv}{\beta(v_0 - v)}$$

$$t = \frac{1}{\beta}\int\frac{dv}{v_0 - v} + C_1 = -\frac{1}{\beta}\int\frac{d(1 - v/v_0)}{1 - v/v_0} + C_1$$

$$= -\frac{1}{\beta}\log\left(1 - \frac{v}{v_0}\right) + C_1$$

If the body starts at rest, we have $v = 0$ at $t = 0$, so that $C_1 = 0$, and

$$\log\left(1 - \frac{v}{v_0}\right) = -\beta t$$

$$1 - \frac{v}{v_0} = e^{-\beta t} \qquad \text{or} \qquad v = v_0(1 - e^{-\beta t})$$

We find from this that

$$dx = v_0(1 - e^{-\beta t})\, dt$$

$$x = v_0\int(1 - e^{-\beta t})\, dt + C_2$$

$$= v_0 t - \frac{v_0}{\beta}\int e^{-\beta t}\, d\beta t + C_2 = v_0\left[\frac{1}{\beta}e^{-\beta t} + t\right] + C_2$$

If $x = 0$ when $t = 0$, we get $C_2 = -v_0/\beta$, so that $x = (v_0/\beta)\,[e^{-\beta t} + \beta t - 1]$. An example of this motion is small particles settling in a fluid, for instance silt settling in water.

2.4.6 Resistance proportional to the square of the velocity

For higher velocities in a resisting medium, the resistance is found to be proportional to v^2; taking the same case as in Section 2.4.5, the equation of motion is

$$m\frac{dv}{dt} = mg - cv^2 \qquad \text{or} \qquad \frac{dv}{dt} = g - \frac{c}{m}v^2 = g - \beta v^2$$

where $\beta = c/m$; when the resisting force is equal to the driving force, $dv/dt = 0$, or $v = \sqrt{(g/\beta)} =$ limiting velocity v_0, so that

$$\frac{dv}{dt} = \beta(v_0^2 - v^2)$$

$$dt = \frac{dv}{\beta(v_0^2 - v^2)}$$

$$t = \frac{1}{\beta}\int\frac{dv}{v_0^2 - v^2} + C_1 = \frac{1}{\beta v_0^2}\int\frac{dv}{1-(v/v_0)^2} + C_1$$

$$= \frac{v_0}{g}\int\frac{d(v/v_0)}{1-(v/v_0)^2} + C_1 = \frac{v_0}{g}\tanh^{-1}\left(\frac{v}{v_0}\right) + C_1$$

If at $t=0$, $v=0$, we find $C_1 = 0$ and

$$t = \frac{v_0}{g}\tanh^{-1}\left(\frac{v}{v_0}\right)$$

so that $v = v_0 \tanh(gt/v_0) = \dot{x}$,

$$x = v_0\int\tanh\left(\frac{gt}{v_0}\right)dt + C_2 = \frac{v_0^2}{g}\log\cosh\left(\frac{gt}{v_0}\right) + C_2$$

If $x=0$ when $t=0$, we get $C_2 = 0$, so that

$$x = \frac{v_0^2}{g}\log\cosh\left(\frac{gt}{v_0}\right) = \frac{1}{\beta}\log\cosh\left(\frac{gt}{v_0}\right)$$

A series of examples has been given in which the solution to the equation of motion can be found by differentiation or integration. In many more complicated problems, the solution to the equation of motion cannot be found in this way and we must then find solutions by graphical or numerical differentiation or integration. This type of problem is, however, outside the scope of this book. In some cases the equation of motion may be solved if displacements are small; a simple example of this is the following:

Example 2.8. Figure 2.14 shows a *simple pendulum* consisting of a concentrated mass m suspended on a string of length l. The pendulum is swinging in a vertical plane under the action of gravity. Air resistance may be neglected. Determine the equation of motion; simplify the equation for small angular displacements and solve the equation.

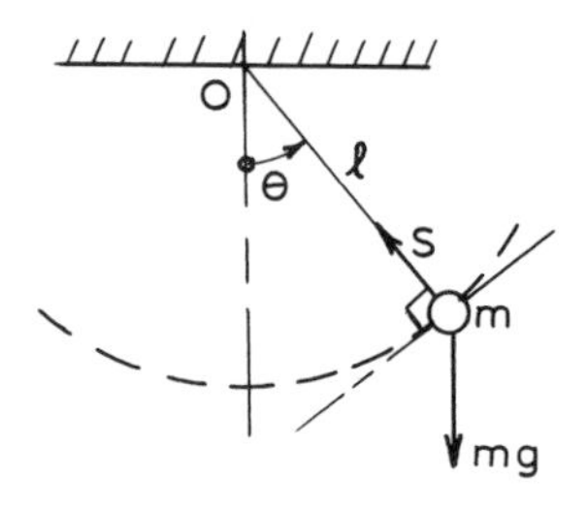

Fig. 2.14

Solution. The mass m is in a circular motion with centre O and radius l; the normal and tangential acceleration components are $l\omega^2$ and $l\dot{\omega}$, so that the corresponding forces are $F_n = ml\omega^2 = ml\dot{\theta}^2$ and $F_t = ml\ddot{\theta}$.

The forces acting are the string tension S and the gravity force mg; projecting these forces on the radial line and the tangent at m, we find $F_n = S - mg\cos\theta$ and $F_t = -mg\sin\theta$; the minus sign for F_t is due to the fact that the force is in the opposite direction to θ. The equations of motion are now

$$S - mg\cos\theta = ml\dot{\theta}^2 \tag{a}$$

$$-mg\sin\theta = ml\ddot{\theta} \tag{b}$$

Equation (a) gives $S = mg\cos\theta + ml\dot{\theta}^2$, which determines S if θ can be found as a function of time. The equation (b) gives the equation of motion:

$$\ddot{\theta} + \frac{g}{l}\sin\theta = 0$$

This is a *non-linear* differential equation; the time t of the swing depends on the angle θ and can only be stated as a function of θ in terms of elliptic integrals, which require numerical methods for solution.

The equation may be given a much simpler form if we restrict the motion to small angles θ; if $\theta=20°=0{\cdot}34907$ rad, sin $\theta=0{\cdot}34202$, so that θ is about 2% bigger than sin θ; for $\theta=10°=0{\cdot}17453$ rad, sin $\theta=0{\cdot}17365$, so that θ is only about $\frac{1}{2}$% bigger than θ; up to about 20° of swing, we can, therefore, expect reasonable accuracy if we take sin $\theta=\theta$; the equation then takes the form $\ddot{\theta}+(g/l)\,\theta=0$, this is a *linear* differential equation. Comparing this with the equation $\ddot{x}+\omega_0^2x=0$ shows that the equation gives simple harmonic motion of the pendulum, with a frequency

$$f = \frac{\omega_0}{2\pi} = \frac{1}{2\pi}\sqrt{\frac{g}{l}}$$

2.5 Alternative form of Newton's second law: impulse–momentum equation

Newton's second law for the motion of a particle or the centre of mass of a rigid body: $\mathbf{F}=m\,d\mathbf{v}/dt$ may be integrated directly to give

$$\int_1^2 \mathbf{F}\,dt = \int_1^2 m\,d\mathbf{v} = m(\mathbf{v}_2-\mathbf{v}_1) \tag{2.11}$$

the term $\mathbf{F}\,dt$ or $\int_1^2 \mathbf{F}\,dt$ is called *the impulse* of the force $\mathbf{F}$ in the time interval, and the vector $m\mathbf{v}$ is called *the momentum* of the particle. Equation (2.11) then states that the impulse is equal to the total change in momentum in the same time; the equation is called *the impulse–momentum equation* and is a vector equation; scalar component equations may be taken along any line. In the x-axis direction, the equation is

$$\int_1^2 F_x\,dt = m(\dot{x}_2-\dot{x}_1)$$

If the force is *constant* (2.11) takes the form

$$\mathbf{F}(t_2-t_1) = m(\mathbf{v}_2-\mathbf{v}_1) \tag{2.12}$$

The unit of impulse is the newton second$=$Kg m/s; that for momentum is, of course, the same.

If there is no resultant force acting on a particle in a certain time interval, the impulse is zero, and (2.11) states that there is no change in the momentum in that time interval, so that the velocity is constant in magnitude and direction. In that case the momentum is conserved, and this is a special case of the so-called *principle of conservation of momentum*.

In the case of two interacting particles the forces between the particles are equal and opposite and act for the same time; their impulses are then equal and opposite and make no contribution to the total impulse on the system. If there are no external forces acting, the total momentum of the system is conserved.

The impulse–momentum equation is based directly on Newton's second law, and contains exactly the same information. It is, however, convenient in certain problems where the impulse of a force is known and the variation of the force with time is not.

Example 2.9. Figure 2.15 shows the approximate variation of the gas pressure in a rifle barrel after firing. The mass of the bullet is m and that of the rifle M; the muzzle velocity of the bullet is v; neglecting friction, determine the recoil velocity of the rifle at the moment when the bullet leaves the barrel.

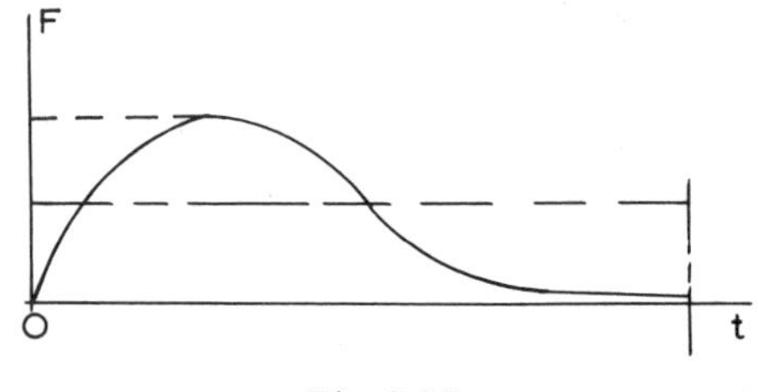

Fig. 2.15

Solution. The total impulse on the bullet is $\int_0^t F\,dt = mv$, which is the area under the curve. If a curve is available, the impulse may be found by measuring the area with a planimeter; in the present case the curve is not known in detail and is very difficult to find experimentally due to the high pressure and extremely short time. Incidentally, the same impulse is obtained from a constant force indicated by the horizontal line, if the area under the line is made the same as that under the curve.

The gas pressure on the rifle is equal and opposite to the pressure on the bullet and acts for the same time. The impulse on the rifle is then $\int_0^t F\,dt = Mv_r$, where v_r is the recoil velocity of the rifle. We have now $Mv_r = mv$, or $v_r = (m/M)v$, so that v_r has been determined without knowledge of the forces acting.

Example 2.10. In order to reduce the effect of the recoil forces, a gun barrel of mass 363 kg is arranged to slide axially on guides against the action of a compression spring with spring constant $K=35\times10^4$ N/m. The spring is initially compressed to hold the barrel against its forward stop, with a force of 26 700 N. The gun fires a shell of mass 8·16 kg with a muzzle velocity of 549 m/s.

Determine the distance through which the barrel recoils, if friction and the time taken for the shell to traverse the barrel are neglected.

Solution. The momentum of the system is conserved, so that $MV-mv=0$, where $M=363$ kg, $m=8{\cdot}26$ kg and $v=549$ m/s. This gives $V=mv/M=12{\cdot}35$ m/s.

The barrel is now in rectilinear motion, with initial velocity 12·35 m/s against a resisting force $F_x=26\,700+Kx$; taking $x=0$ and $\dot{x}=12{\cdot}35$ when $t=0$ at the initial position, we have the equation of motion

$$m\ddot{x} = -F_x$$

so that

$$\ddot{x}+965x = -73{\cdot}6$$

A particular solution to this equation is $x=C=\text{constant}$; by substitution, the value of C is found to be $C=-0{\cdot}0761$. The solution to the homogeneous part of the equation $\ddot{x}+965x=0$ is $x=A\sin\omega t+B\cos\omega t$, where $\omega=\sqrt{965}=31{\cdot}1$ rad/s. The total solution is then

$$x = A\sin\omega t+B\cos\omega t-0{\cdot}0761$$

$$\dot{x} = A\omega\cos\omega t-B\omega\sin\omega t$$

When $t=0$, $x=0$, so $B=0{\cdot}0761$. Since $\dot{x}_0=12{\cdot}35$, $A\omega=12{\cdot}35$, or $A=12{\cdot}35/31{\cdot}1=0{\cdot}397$. When the barrel stops $\dot{x}=0$, so that $A\omega\cos\omega t-B\omega\sin\omega t=0$, or $\tan\omega t=A/B=5{\cdot}21$, $\omega t=79{\cdot}14°$. Substituting this gives $x=0{\cdot}397\cdot0{\cdot}982+0{\cdot}0761\cdot0{\cdot}1884-0{\cdot}0761=0{\cdot}328$ m recoil.

The impulse–momentum equation is particularly useful in problems with large forces acting for very short times, as shown in the examples. The same situation occurs in impact problems to be considered later and a similar type of problem occurs when a continuous stream of particles changes direction abruptly; the forces between the particles are then equal and opposite, and only external forces need be considered.

Example 2.11. Figure 2.16 shows a horizontal water jet, which is stopped by a vertical wall. The jet is of cross-section A, the density is ρ and the constant velocity of the water is V. Determine the force on the wall.

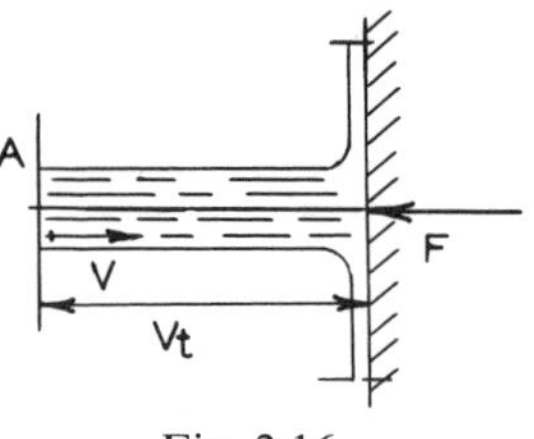

Fig. 2.16

Solution. The volume of water passing in t s is VtA, the mass is $M = VtA\rho$ and the momentum is $MV = V^2tA\rho$. The force on the wall is F and on the water $-F$; the impulse–momentum equation gives the result

$$-Ft = M(V_2 - V_1) = M(0 - V) = -MV$$

so that

$$F = \frac{MV}{t} = V^2 A\rho$$

2.6 Work and power

It is common knowledge that to move or lift a body, certain forces must be applied and act through the distances moved, that is a certain amount of *work* must be done on the body; to use this important concept in dynamics, we need a concise definition that can be used in the simplest possible way, so we define the work done by a force $\mathbf{F}$ (Fig. 2.17), with a point of application that moves through a distance $d\mathbf{s}$, as the product $|\mathbf{F}||d\mathbf{s}| \cos \beta = d$ (work done).

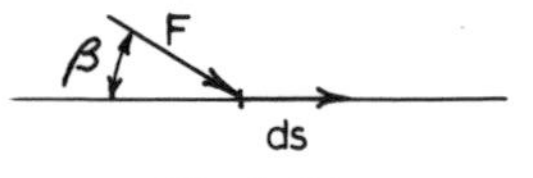

Fig. 2.17

The work done is thus defined as the product of the magnitude of the force *component* in the direction of the displacement, and the magnitude of the displacement; work is therefore a *scalar quantity*. In vector language, the work is $d(\text{work done}) = \mathbf{F} \cdot d\mathbf{s}$, or the dot product of the force and the displacement.

It follows from the definition that the component of the force perpendicular to the displacement does not do any work.

If the force moves from a position 1 to a position 2, the work done is

$$\text{W.D.} = \int_1^2 \mathbf{F} \cdot d\mathbf{s} = \int_1^2 (F_x\,dx + F_y\,dy + F_z\,dz)$$

the unit for work is the newton metre (Nm)$=$kg m^2/s^2, and this unit is called the *joule* (J). It is the work done when the point of application of a constant force of 1 N is displaced through a distance of 1 m in the direction of the force.

An important concept associated with work is *power*, which is the *rate* of work done or *the instantaneous work dW/dt*. This is

$\mathbf{F} \cdot \dfrac{d\mathbf{s}}{dt} = \mathbf{F} \cdot \mathbf{V}$; the unit for power is the *watt* (W) $= 1$ J/s $= 1$ Nm/s

The *mechanical efficiency* of a system or machine is defined by the ratio:

$$\frac{\text{work output}}{\text{work input}} \quad \text{or} \quad \frac{\text{useful work}}{\text{expended work}}$$

2.7 Kinetic energy

If a particle of mass m is acted upon by a resultant force $\mathbf{F}$ of constant magnitude and in a fixed x-direction, Newton's second law gives the equation of motion: $\ddot{x} = F/m$ (constant). If we count time from the instant of application of the force and assume that the particle is initially at rest at position $x=0$, the velocity V at time t is

$$V = \dot{x} = \int_0^t (F/m)\,dt = (F/m)t$$

The distance moved is

$$x = \int_0^t (F/m)t\,dt = \tfrac{1}{2}(F/m)t^2$$

the work done by the force is $Fx = \frac{1}{2}(F^2/m)t^2 = \frac{1}{2}mV^2 = T$; the expression $\frac{1}{2}mV^2$ is called the *kinetic energy* of the particle at the time t; this is the work required to increase its velocity from rest to a velocity V.

The unit for kinetic energy is the same as for work, and is the kg m^2/s^2 = Nm = J. The concept of kinetic energy is of fundamental importance in dynamics.

For a rigid body moving in such a way that all particles have the same velocity V_c at any instant, we find the total kinetic energy T by summing up for all the particles: $T = \frac{1}{2}V_c^2 \sum m = \frac{1}{2}MV_c^2$, where M is the total mass of the body.

2.8 Alternative form of Newton's second law: work–energy equation

If the resultant instantaneous force on a particle of mass m is $\mathbf{F}$ and the instantaneous velocity is $\mathbf{V}$, the displacement of the particle in the time dt is $\mathbf{V}\,dt$; the work done on the particle is then d(W.D.) $= \mathbf{F}\cdot\mathbf{V}\,dt$, and the total work in moving from a position 1 to a position 2 is W.D. $= \int_1^2 \mathbf{F}\cdot\mathbf{V}\,dt$. Introducing Newton's second law $\mathbf{F} = d\mathbf{V}/dt$, we obtain

$$\begin{aligned} \text{W.D.} &= \int_1^2 m\frac{d\mathbf{V}}{dt}\cdot\mathbf{V}\,dt = m\int_1^2 \mathbf{V}\cdot d\mathbf{V} \\ &= \tfrac{1}{2}m\int_1^2 d(\mathbf{V}\cdot\mathbf{V}) = \tfrac{1}{2}m\,[V^2]_1^2 \\ &= \tfrac{1}{2}m\,(V_2^2 - V_1^2) = T_2 - T_1 \end{aligned} \tag{2.13}$$

This shows that the total work done, in moving from a position 1 to a position 2, is equal to the total *change* in kinetic energy between the two positions; this equation is called *the work–energy equation*, and is an alternative form of Newton's second law. Although no new information is introduced by this equation, it is in a form that is convenient in many problems. It is particularly useful in problems where time is not involved in the solution.

In terms of its components, for instance in the x-direction, the equation takes the form

$$\int_{x_1}^{x_2} F_x \, dx = \tfrac{1}{2} m (\dot{x}_2^2 - \dot{x}_1^2) \tag{2.14}$$

Example 2.12. A body (Fig. 2.18) of mass m is released from rest and falls through a distance h to the ground. Determine its impact velocity.

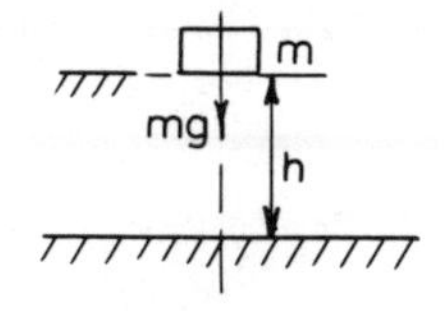

Fig. 2.18

Solution. The work done on the body is mgh and the change in its kinetic energy is $\frac{1}{2} mV^2$; the work–energy equation now states that $mgh = \frac{1}{2} mV^2$, so that $V = \sqrt{(2gh)}$.

Example 2.13. Figure 2.19 shows a body of mass m, which is sliding down a plane inclined at an angle α to the horizontal direction; the body starts from rest, and the coefficient of friction between the body and the plane is μ.

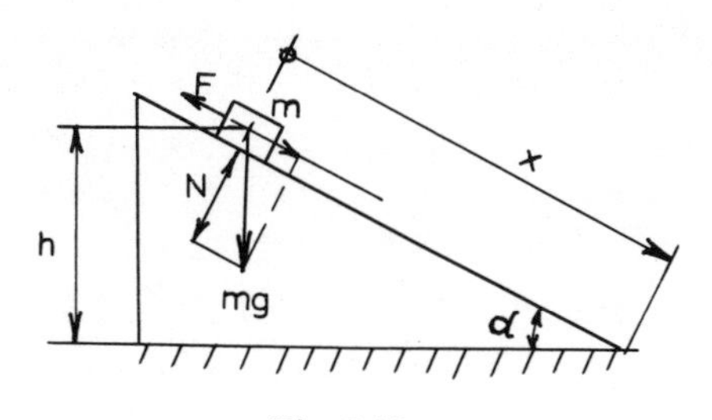

Fig. 2.19

Determine the velocity of the body as a function of the distance x that it has moved along the plane.

Solution. The resultant force in the x-direction is $F_x = mg \sin \alpha - \mu mg \cos \alpha$; this is a constant force, and the work done in moving through a distance x is W.D. $=F_x x$; the component $mg \cos \alpha$ of the weight force is perpendicular to the motion and does no work.

The change in kinetic energy is $T=\frac{1}{2}mV^2$; so that $\frac{1}{2}mV^2=F_x x$, and

$$V = \sqrt{\left(\frac{2}{m} F_x x\right)} = \sqrt{\left[2g\,(\sin\alpha-\mu\cos\alpha)x\right]}$$

For motion down the plane, we must have $\sin\alpha-\mu\cos\alpha>0$; therefore $\tan\alpha>\mu$.

If $\mu=0$, $V=\sqrt{[2g(\sin\alpha)x]}=\sqrt{(2gh)}$; this is the same velocity that the body would reach in a free fall through the distance h.

Example 2.14. Determine the recoil in Example 2.10, by using the work–energy equation.

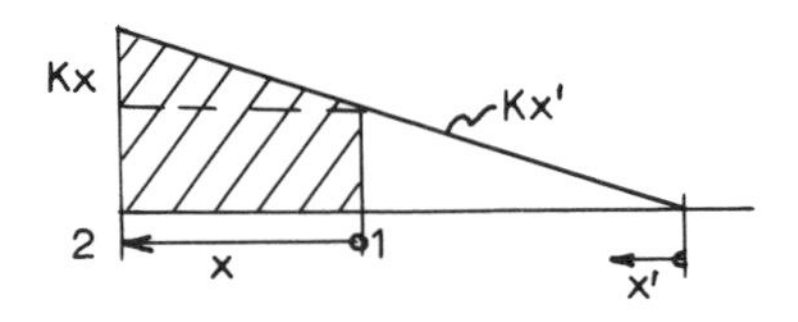

Fig. 2.20

Solution. The initial velocity of the barrel is $V=12{\cdot}35$ m/s, and the mass $M=363$ kg. The initial kinetic energy is then $T_1=\frac{1}{2}MV^2 =\frac{1}{2}\times 363\times 12{\cdot}35^2=27\,650$ Nm.

When the barrel stops in position 2, $V_2=0$ so that $T_2=0$. The work done (Fig. 2.20) is $\int_1^2 F\,dx$; this is the cross-hatched area, so that

$$\text{W.D.} = 26\,700x+\tfrac{1}{2}\times 35\times 10^4 x^2$$

The work–energy equation gives now

$$-\text{W.D.} = T_2-T_1 = -27\,650 \text{ Nm}$$

the result is

$$x^2+0{\cdot}1523x-0{\cdot}1579 = 0$$

so that $x=0{\cdot}328$ m.

Example 2.15. An aircraft is to be launched by catapult at a speed of 241 km/h, at a constant acceleration of $3g$. Determine the minimum distance for take-off.

When landing, the aircraft is arrested by a heavy rubber band acting as a spring with spring constant K N/m. The aircraft weighs 80 000 N and has a landing speed of 194 km/h; during landing, a constant braking force from the wheels and propellers of 22 000 N is applied.

Determine the value of K, if the plane must be stopped in a distance of 45·7 m.

Solution. The launching speed is $V_1=241$ km/h$=66{\cdot}9$ m/s; the change in kinetic energy is $\frac{1}{2}MV_1^2=\frac{1}{2}(W/g)V_1^2$. For a constant acceleration of $3g$, the force on the plane is $F=(W/g)3g=3W$; the work done through a distance x is $Fx=3Wx$, so that $3Wx=\frac{1}{2}(W/g)V_1^2$, or $x=V_1^2/6g=66{\cdot}9^2/69{\cdot}81=75{\cdot}8$ m, which is the minimum distance for take-off.

The landing speed $V_2=193$ km/h$=53{\cdot}6$ m/s; the kinetic energy at landing is

$$\tfrac{1}{2}MV_2^2 = \frac{1}{2}\times\frac{80\,000}{9{\cdot}81}\times 53{\cdot}6^2 = 10^6\times 11{\cdot}71 \text{ Nm}$$

W.D. by the constant braking force

$$= 22\,000\times 45{\cdot}6 = 10^6\times 1{\cdot}014 \text{ Nm}$$

W.D. by the rubber band $= \frac{1}{2}Kx^2 = \frac{1}{2}K\times 45{\cdot}7^2 = 1043K$ Nm

The work–energy equation now states that $1043K+10^6\times 1{\cdot}014 =10^6\times 11{\cdot}71$, from which $K=10\,230$ N/m.

Example 2.16. Figure 2.21 shows a block of mass M which starts from rest at A and slides in a vertical plane along AB on a smooth circular cylinder; the block leaves the cylinder at B. Determine the distance a defining the point of impact, if the radius of the cylinder is $r=2m$.

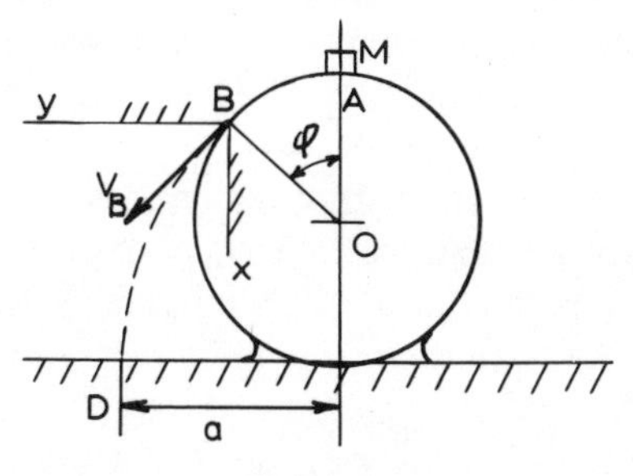

Fig. 2.21

Solution. The equation of motion in the normal direction for motion between A and B is

$$Mg \cos \theta - N = Mr\omega^2 = M\frac{V^2}{r}$$

where N is the normal force on the block and $0 \leqslant \theta \leqslant \varphi$. The block leaves the cylinder at B, where $\theta = \varphi$ and $N = 0$, so that $Mg \cos \varphi = MV_B^2/r$, therefore $\cos \varphi = V_B^2/rg$.

During the motion from A to B, the normal force does no work, and

$$\text{W.D.} = Mgr(1 - \cos \varphi)$$

The change in kinetic energy is $\frac{1}{2}MV_B^2$, so that $Mgr(1 - \cos \varphi) = \frac{1}{2}MV_B^2$. Substituting $\cos \varphi = V_B^2/rg$ gives the result $V_B^2 = 2rg/3$, so that $V_0 = 3{\cdot}62$ m/s, and $\cos \varphi = \frac{2}{3}$.

Introducing a coordinate system (x, y) as shown, we have the equations of motion $M\ddot{x} = Mg$, and $M\ddot{y} = 0$, so that

$$\dot{x} = gt + V_0 \sin \varphi \qquad x = \tfrac{1}{2}gt^2 + V_0 (\sin \varphi)t$$

$$\ddot{y} = 0 \qquad \dot{y} = V_0 \cos \varphi \qquad y = (V_0 \cos \varphi)t$$

Substituting the values for V_0 and φ gives $x = 4{\cdot}905t^2 + 2{\cdot}70t$ and $y = 2{\cdot}41t$.

The impact occurs when $x = r + r \cos \varphi = 3{\cdot}334$ m, so that $4{\cdot}905t^2 + 2{\cdot}70t - 3{\cdot}334 = 0$, or $t = 0{\cdot}594$ s, which gives $y = 1{\cdot}43$ m, and

$$a = r \sin \varphi + y = 2{\cdot}92 \text{ m}$$

2.9 Force function; conservative forces

The work done by a force $\mathbf{F}$, moving from a position 1 to a new position 2, was defined by the line integral $\int_1^2 \mathbf{F} \cdot d\mathbf{s}$; in general this amount of work depends on the path from 1 to 2, but it is possible to show that for certain forces the work done in moving from 1 to 2 is always the same, independent of the path.

Suppose that the force in question can be expressed as the partial derivatives of a function Φ (x, y, z), so that $F_x = \partial\Phi/\partial x$, $F_y = \partial\Phi/\partial y$ and $F_z = \partial\Phi/\partial z$. We have then

$$d\Phi = \frac{\partial\Phi}{\partial x}dx + \frac{\partial\Phi}{\partial y}dy + \frac{\partial\Phi}{\partial z}dz = F_x\,dx + F_y\,dy + F_z\,dz = \mathbf{F} \cdot d\mathbf{s}$$

The function Φ is called a *potential* or a *force function*. The work integral now becomes

$$\int_1^2 \mathbf{F}\cdot d\mathbf{s} = \int_1^2 d\Phi = \Phi_2 - \Phi_1$$

so that the work depends only on the starting and finishing position, and not on the path. Forces of this nature are called *potential* or *conservative forces*. Since conservative forces are derivatives of Φ, the magnitude and direction of the forces are given by the force function and only the point of application needs to be stated to completely define the force.

In the case of dry friction or viscous friction forces, the work clearly depends on the length of the path, and for air-resistance forces the work also depends on the velocity of motion; these forces are clearly non-conservative.

The most important conservative forces in mechanics are gravity forces and elastic forces, to be discussed shortly.

2.10 Potential energy

The *difference* $V_2 - V_1$ in *potential energy* between two points of a *system* acted upon by a *conservative force* $\mathbf{F}$ is defined, for convenience, as follows:

$$V_2 - V_1 = -\int_1^2 \mathbf{F}\cdot d\mathbf{s}$$

This definition shows that the difference in potential energy is the negative of the work done by the force in moving from the first position to the second. The work integral has a unique value only if the force is conservative, so that the integral is independent of the path from position 1 to 2; a potential energy can, therefore, only be defined for conservative forces.

The potential energy V is now given by $V = -\int \mathbf{F}\cdot d\mathbf{s} = -\int d\Phi = -\Phi + C$, where C is an arbitrary constant; since we shall always use the difference in potential energy between two positions, or the derivative of potential energy, the constant C is of no consequence and may be taken as zero; this means that the potential energy at a particularly arbitrarily chosen datum position may be taken as zero.

If V_2 is taken as zero, we have $V_1 = \int_1^2 \mathbf{F}\cdot d\mathbf{s}$, or the potential energy in position 1 is the work done by $\mathbf{F}$ in moving from position

1 *back* to the datum position; this position is chosen for convenience in any particular problem.

Since $V = -\Phi + C$, we have

$$F_x = \frac{\partial \Phi}{\partial x} = -\frac{\partial V}{\partial x} \qquad F_y = \frac{\partial \Phi}{\partial y} = -\frac{\partial V}{\partial y} \qquad F_z = \frac{\partial \Phi}{\partial z} = -\frac{\partial V}{\partial z}$$

the components of a conservative force may thus be found as the partial derivatives of the potential-energy function V.

A force with a *fixed direction* and *constant magnitude* is a conservative force, the force function is $\Phi = ax + by + cz$, and the potential energy function is $V = -\Phi + C = -ax - by - cz + C$, so that $F_x = a$, $F_y = b$ and $F_z = c$.

In the special case where the constant force is the *gravity force* mg on a particle of mass m, we find, for the y-axis in the vertical direction, the force function $\Phi = -mgy$ and $V = mgy + C$. Hence

$$F_x = 0 \qquad F_z = 0 \qquad F_y = -\frac{\partial V}{\partial y} = -mg$$

If the force is *proportional to the displacement*, and in the opposite direction, we find, by taking the x-axis in the direction of the displacement, that $F_x = -Kx$, where K is the proportionality factor and $F_x = 0$ when $x = 0$; now

$$F_x = -\frac{\partial V}{\partial x} = -Kx \qquad F_y = F_z = 0 \qquad \frac{\partial V}{\partial x} = Kx$$

so that $V = \frac{1}{2}Kx^2 + C_1$. This is the elastic force in a spring or elastic band, assuming that all losses and the mass of the spring may be neglected; K is the spring constant.

The gravitational force $F = \gamma\, m_1 m_2 / r^2$ is also conservative. In electricity, the electrostatic field is conservative, while the magnetic field is non-conservative, since the forces on a charged particle depend on both the position and the velocity of the particle.

Example 2.17. Determine the potential-energy function for the systems shown in Fig. 2.22 (a) and (b).

Solution. (a) The two forces acting on the mass m are the string tension S and the gravity force mg. The string tension is always perpendicular to the motion and so produces no work, the gravity force moves through a vertical distance $l - l\cos\theta$; taking the vertical position of the pendulum as datum position with $V=0$, we find $V=mgl(1-\cos\theta)$.

(b) In system (b), it is convenient to take the static equilibrium position as datum position with $V=0$. The unstressed length of the spring is l_0; in the static position the spring is extended a length Δ, and the spring tension is mg to balance the gravity force. From the spring force diagram in Fig. 2.22(b) it follows that the potential energy at position x is $mgx+\frac{1}{2}Kx^2$ for the spring, and $-mgx$ for the gravity force, so that $V=\frac{1}{2}Kx^2$.

Using the static equilibrium position as datum position, the gravity force does not appear in the final expression for V.

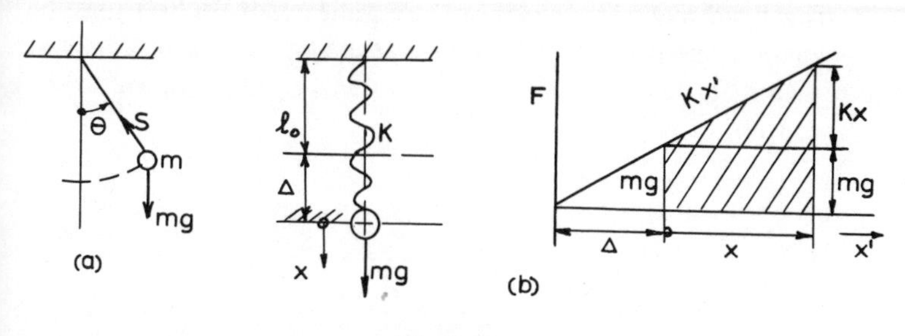

Fig. 2.22

2.11 Principle of conservation of mechanical energy

If a particle is acted upon by a system of conservative forces with resultant **F**, the difference in potential energy between two positions 1 and 2 is by definition

$$V_2-V_1 = -\int_1^2 \mathbf{F}\cdot d\mathbf{s}$$

From the work–energy equation we find that the work done is

$$\text{W.D.} = \int_1^2 \mathbf{F}\cdot d\mathbf{s} = T_2-T_1$$

Introducing the potential energy gives $-(V_2-V_1)=T_2-T_1$, or $T_1+V_1=T_2+V_2$, so that

$$V+T=\text{constant} \qquad (2.15)$$

This means that the sum of the potential and kinetic energy (the mechanical energy) is constant for a particle moving under the action of conservative forces; this is the *principle of conservation of mechanical energy* for a particle.

The principle is sometimes useful in establishing an equation of motion by differentiation with respect to time, since $dV/dt + dT/dt = 0$.

If friction forces or other non-conservative forces are acting, the more powerful work–energy equation may be employed instead of the principle of conservation of mechanical energy.

Example 2.18. Figure 2.23 shows two bodies of mass M_1 and M_2, which are connected by a string over a pulley which can rotate about a fixed axis. The inertia of the pulley and friction may be neglected; given that $M_1 > M_2$ and that the system starts from rest in position A, determine the velocity of M_1 in position B, and the acceleration of M_1.

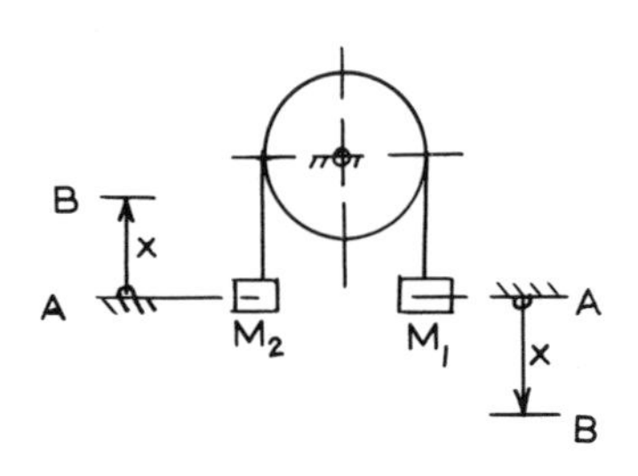

Fig. 2.23

Solution. In position A we have the kinetic energy of the system $T_A = 0$; taking position A as our datum position, we also have $V_A = 0$. In position B the kinetic energy is $T_B = \frac{1}{2}(M_1 + M_2)\dot{x}^2$, and $V_B = (M_2 - M_1)gx$. The principle of conservation of mechanical energy now gives the equation

$$T_B + V_B = T_A + V_A$$

therefore

$$\tfrac{1}{2}(M_1 + M_2)\dot{x}^2 + (M_2 - M_1)gx = 0$$

so that

$$\dot{x} = \sqrt{\left[\frac{2(M_1 - M_2)}{M_1 + M_2} gx\right]}$$

differentiating the equation leads to

$$(M_1+M_2)\dot{x}\ddot{x}+(M_2-M_1)g\dot{x} = 0 \qquad \text{or} \qquad \ddot{x} = \frac{M_1-M_2}{M_1+M_2}g$$

This result may be seen to be correct dimensionally. An investigation of the result for the so-called '*logical* extremes', shows that as $M_1 \to \infty$, $\ddot{x} \to g$, so that M_1 is in a free fall; if $M_2=0$, we find again $\ddot{x}=g$, or M_1 in a free fall as it should be.

Example 2.19. Figure 2.24 shows a vertical glass U-tube with a uniform bore of cross-sectional area A. The tube is open at both ends, and contains a column of liquid of total length l and density ρ. Neglecting friction and the motion of air, determine the equation of motion of the liquid after a displacement x from the equilibrium position.

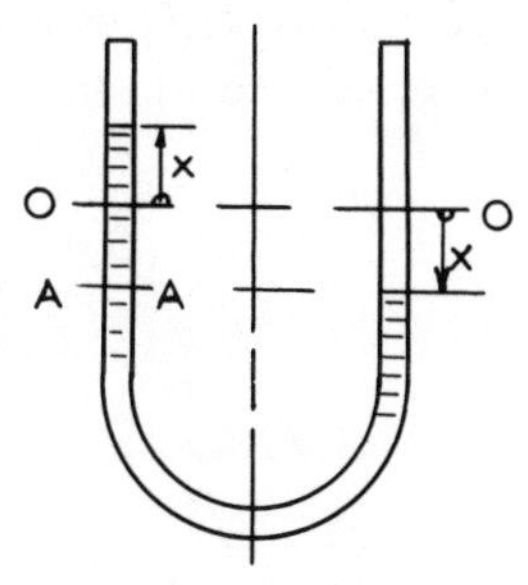

Fig. 2.24

Solution. All the particles of the liquid move with the same displacement x and velocity $\dot{x}$. The total mass is $M=Al\rho$, and the kinetic energy is $T=\frac{1}{2}M\dot{x}^2=\frac{1}{2}(Al\rho)\dot{x}^2$.

For a displacement x, the unbalanced force is $2A\rho gx$, which is proportional to the displacement. The action is, therefore, the same as for a spring of spring constant $K=2A\rho g$, so that the potential energy is

$$V = \tfrac{1}{2}Kx^2 = A\rho gx^2$$

We have now

$$\begin{aligned}\tfrac{1}{2}Al\rho\dot{x}^2+\rho Agx^2 &= \text{constant}\\ Al\rho\dot{x}\ddot{x}+2\rho Agx\dot{x} &= 0\end{aligned}$$

or $\ddot{x}+2(g/l)x=0$ as equation of motion; this is simple harmonic motion with frequency

$$f=\frac{1}{2\pi}\sqrt{\frac{2g}{l}}\text{ cycles/s}$$

2.12 Impact

An impact may be defined as a sudden contact between two bodies involving large contact forces acting for a short time. This definition implies that at least one of the bodies must be moving before the impact; the large forces occur because of the big change in velocities in a very short time.

If a light steel hammer is used to strike a blow on a large piece of steel an impression is made, indicating that large forces act at the impact; pressing the hammer against the steel piece produces no impression on the steel.

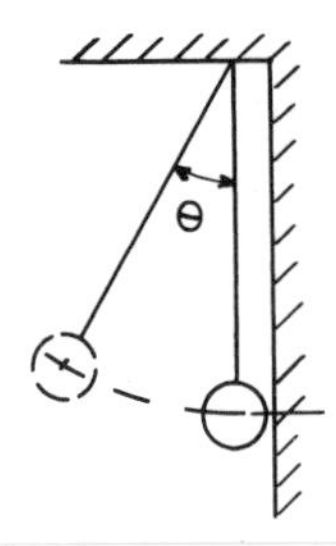

Fig. 2.25

To investigate this further, experiments may be performed as shown in Fig. 2.25. A pendulum with a 3 cm diameter brass sphere as a bob is released at an angle θ from the vertical position. It will be noticed that the pendulum, after impact against the wall, swings out to an angle slightly smaller than θ, because its starting velocity after impact is slightly less than its impact velocity. The time of the impact may be measured electrically, and is found to be of the order $1{\cdot}5\times10^{-4}$ s. In this extremely short time the bob has stopped, and regained an equal velocity in the opposite direction. The maximum value of the impact force may be found by measuring the diameter of the impact area on the bob, after covering the wall with soot or dye, and a static compression test, giving the same compression area, enables us to measure the maximum force; some idea of the magnitude of the force may be found by calculating the *constant* force necessary to give the same impulse: the mass of the sphere is $M=0{\cdot}122$ kg, the impact velocity $V=0{\cdot}3$ m/s, and the impulse–momentum equation states that $\int_0^t F\,dt = M(V_2 - V_1)$. Taking force and velocities positive to the right, this gives

$$-Ft = M(-V-V) = -2MV$$

so that

$$F = \frac{2MV}{t} = \frac{2\times 0{\cdot}122\times 0{\cdot}3}{1{\cdot}5\times 10^{-4}} = 488 \text{ N}$$

The weight W of the sphere is $W=1{\cdot}20$ N, so that $F\sim 400W$; the actual maximum force may be 2–3 times bigger. These observations enables us to deal with a series of practical problems in an *approximate way* by the following two rules for impact of this nature:

1. The impact forces are so great that other forces like gravity, friction etc. may be *neglected* during impact.
2. The time of impact is so short that no appreciable motion can take place; the configuration just after impact may therefore be assumed to be the same as the configuration just before impact.

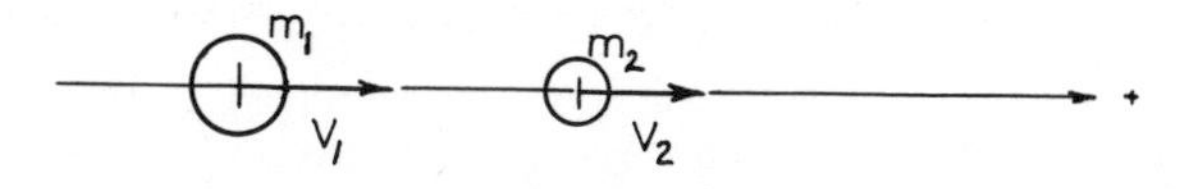

Fig. 2.26

Consider now two spheres (Fig. 2.26) of mass m_1 and m_2 and in rectilinear motion with velocities V_1 and V_2; with $V_1 > V_2$ an impact will occur; this is the simplest form of impact and is called a *direct central impact*. Neglecting all external forces during impact, taking velocities as positive to the right, and calling the velocities after impact V_1' and V_2', we can apply the *law of conservation of momentum*, since the impact forces are equal and opposite and act through the same time. This gives the equation

$$m_1V_1 + m_2V_2 = m_1V_1' + m_2V_2'$$

which holds for all impacts, independent of the elasticity of the bodies.

If we now assume that the bodies are *perfectly elastic*, they will return to the same shape after impact. The work done on the deformation during impact will be completely recovered, the acting forces are conservative elastic forces, and the *principle of conservation of mechanical energy* may be applied, so that the sum of the kinetic and potential energy just before impact is equal to the mechanical energy just after impact: $T+V=T'+V'$, however, $V=V'=0$, so that $T=T'$, or

$$\tfrac{1}{2}m_1 V_1^2+\tfrac{1}{2}m_2 V_2^2 = \tfrac{1}{2}m_1 V_1'^2+\tfrac{1}{2}m_2 V_2'^2$$

the two equations may be stated as follows:

$$\begin{aligned} m_1(V_1-V_1') &= m_2(V_2'-V_2) \\ m_1(V_1^2-V_1'^2) &= m_2(V_2'^2-V_2^2) \end{aligned}$$

the ratio of these equations gives $V_1+V_1'=V_2'+V_2$, or $V_1'-V_2' = -(V_1-V_2)$, so that the relative velocity after impact is the negative of the relative velocity before impact.

In any practical case there is a loss in kinetic energy, since the material is not perfectly elastic. This may be taken into account by writing the equation in the form $V_1'-V_2'=-e(V_1-V_2)$, where the factor e is called the *coefficient of restitution* of the material, and $0\leqslant e\leqslant 1$. In this form the equation is called *Newton's empirical law of impact*, since it is based on experiments.

We now have the following two equations for direct central impact:

$$\begin{aligned} m_1 V_1+m_2 V_2 &= m_1 V_1'+m_2 V_2' \\ V_1'-V_2' &= -e(V_1-V_2) \end{aligned} \tag{2.16}$$

assuming that V_1, V_2 and e are known, the velocities V_1' and V_2' may be determined from (2.16).

If $e=0$, we find $V_1'=V_2'$, and the first equation gives the solution; the two bodies continue as one body and in a distorted shape, there is no tendency to rebound, this is called a *plastic impact*; a material like putty has a value of e close to zero. If $e=1$, we have the same relative velocity before and after impact. This is called a *perfectly elastic impact*, and the total energy is constant during impact. For $0<e<1$, the impact is called *semi-elastic*. For glass or polished hardened steel the value of e is about 0·9, for ivory $e\sim 0{\cdot}8$, and for lead $e\sim 0{\cdot}1$.

In the experiment in Fig. 2.25, we have $V_2 = V_2' = 0$; the second equation in (2.16) then gives $V_1' = -V_1$ if we take $e = 1$, so that the pendulum returns with the same speed as the impact speed.

Example 2.20. Figure 2.27 shows a pile of mass M_2 which is driven into the ground by blows from a hammer of mass M_1 which falls freely through a height of h. The impact may be assumed plastic, and one blow moves the pile a distance b into the ground. Determine the resistance R from the ground if it is assumed constant.

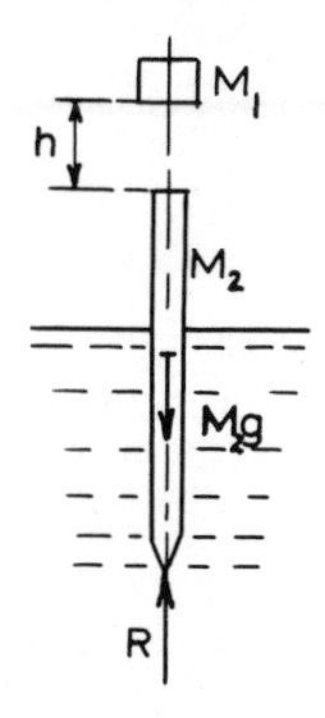

Fig. 2.27

Solution. The impact velocity of the hammer is $v_1 = \sqrt{(2gh)}$, while the initial velocity of the pile is $v_2 = 0$; taking $e = 0$ in eq. (2.16) gives $v_2' = v_1' = v'$, so that M_1 and M_2 move as one body after impact; the first of the equations (2.16) now gives $m_1\sqrt{(2gh)} = (m_1 + m_2)v'$, or $v' = m_1\sqrt{(2gh)}/(m_1 + m_2)$. The kinetic energy right after impact is $T_1 = \frac{1}{2}(m_1 + m_2)v'^2 = ghm_1^2/(m_1 + m_2)$; the pile and hammer move through a distance b and stop; the work done during this motion is $(m_1 + m_2)gb - Rb$; the work–energy equation states that W.D. $= T_2 - T_1$, which gives

$$(m_1 + m_2)gb - Rb = 0 - \frac{ghm_1^2}{m_1 + m_2}$$

so that

$$R = \frac{ghm_1^2}{b(m_1 + m_2)} + (m_1 + m_2)g$$

Example 2.21. Figure 2.28 shows a *ballistic pendulum* consisting of a bag of sand of mass M suspended from a string of length l. A bullet of mass m is fired into the sand in a horizontal direction at a speed v, while the pendulum is at rest; the pendulum swings out an angle α. Assuming that the bullet is retained by the sand, develop a formula for the velocity v of the bullet in terms of the ratio M/m, l and α. Determine v if $M/m = 402{\cdot}5$, $l = 0{\cdot}954$ m and $\alpha = 28{\cdot}4°$.

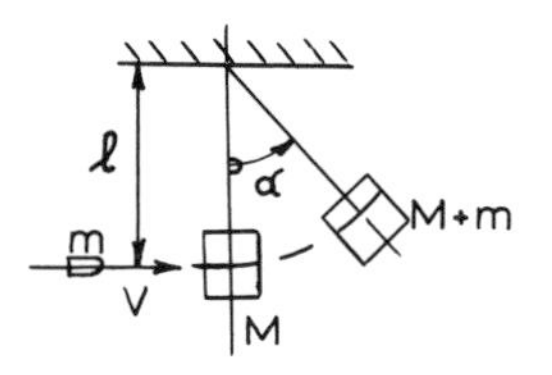

Fig. 2.28

Solution. The velocities before impact are $v_1 = v$ and $v_2 = 0$; after impact the velocity is v' for the sand and the bullet. Conservation of momentum, eq. (2.16), gives

$$mv = (M+m)v' \qquad \text{or} \qquad v = \frac{M+m}{m}v'$$

The kinetic energy just after impact is $T_1 = \frac{1}{2}(M+m)v'^2$, while $T_2 = 0$; the work done is W.D. $= -(M+m)gl(1-\cos\alpha) = T_2 - T_1 = -\frac{1}{2}(M+m)v'^2$, so that $v' = \sqrt{[2gl(1-\cos\alpha)]}$; the final result is

$$v = \left(\frac{M}{m}+1\right)\sqrt{[2gl(1-\cos\alpha)]}$$

With the numerical values given,

$$v = (402{\cdot}5+1)\sqrt{[2\times 9{\cdot}81\times 0{\cdot}954(1-\cos 28°{\cdot}4)]} = 604 \text{ m/s}$$

2.13 D'Alembert's principle

Newton's second law for a particle, $\mathbf{F} = m\ddot{\mathbf{r}}$, may be stated in the form

$$\mathbf{F} + (-m\ddot{\mathbf{r}}) = \mathbf{0} \tag{2.17}$$

The term $-m\ddot{\mathbf{r}}$ has the dimension of a force, the minus sign indicating a direction always opposite to the acceleration $\ddot{\mathbf{r}}$. The expression $-m\ddot{\mathbf{r}}$ may be considered as a fictitious force, convenient in certain applications, although it is not a force in the usual meaning of the word, and cannot be applied by external means. It has been given the name *inertia force*.

Writing eq. (2.17) obviously corresponds to writing an equation in dynamics in the same form as a static equilibrium equation; this form of the equation of motion was first given by the French mathematician D'Alembert in his book *Traité de Dynamique* in 1743. Using D'Alembert's principle, we apply a force $-m\ddot{\mathbf{r}}$ to the particle in question, so that the forces may be considered to be in equilibrium. This situation is called *dynamic equilibrium*.

Once the inertia forces have been applied, the equilibrium equations of *statics* may be applied, and this may sometimes result in considerable simplifications in the solution.

Example 2.22. Figure 2.29 shows a light frictionless pulley of radius r, which carries two masses M_1 and M_2 connected by a string; assuming that $M_1 > M_2$, determine the acceleration of the masses, by applying D'Alembert's principle.

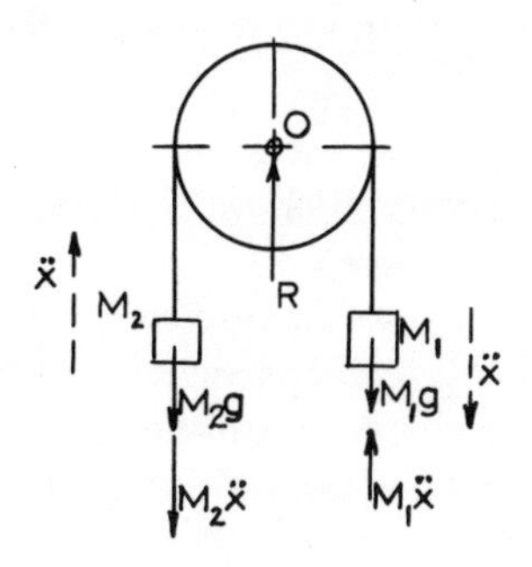

Fig. 2.29

Solution. The external forces R, $M_1 g$ and $M_2 g$ are first applied to the system. The acceleration of M_1 is $\ddot{x}$ downwards, the inertia force is $-M_1\ddot{x}$ and this force is applied, as shown to M_1; M_2 has an acceleration $\ddot{x}$ upwards, so the inertia force $-M_2\ddot{x}$ is then applied downwards on M_2. The system is now in dynamic equilibrium, and equations of statics may be used. Taking moments about point O, to eliminate the unknown reaction R, gives the result

$$M_2gr + M_2r\ddot{x} + M_1r\ddot{x} - M_1gr = 0$$

from which

$$\ddot{x} = \frac{M_1 - M_2}{M_1 + M_2} g$$

The same result was found in Example 2.18. The advantage of the present method is that the fixed reaction and the string tension need not be considered.

2.14 Principle of virtual work

A virtual displacement $\delta\mathbf{r}$ is an assumed infinitesimal displacement of a particle, in which the forces and constraints are unchanged; the notation $\delta\mathbf{r}$ is used to distinguish a virtual displacement from a real displacement $d\mathbf{r}$, which takes place in a time dt in which forces and constraints may change.

We may write $\delta\mathbf{r}$ in terms of its components as $\delta x\mathbf{i} + \delta y\mathbf{j} + \delta z\mathbf{k}$. Since $\delta\mathbf{r}$ is infinitesimal, it follows the same rules as $d\mathbf{r} = dx\mathbf{i} + dy\mathbf{j} + dz\mathbf{k}$: for instance, if a displacement y at a point of a mechanism is related to the displacement x of another point by $y = \sin x$, we get $\delta y = \cos x\ \delta x$; the determination of virtual displacements then becomes a problem in geometry.

Any static force system acting on a particle may be reduced to a single resultant force $\mathbf{F}$; if the particle is given a virtual displacement $\delta\mathbf{r}$, the work done by $\mathbf{F}$ is the *virtual work* $\delta(\text{W.D.}) = \mathbf{F} \cdot \delta\mathbf{r}$. If the force system is in equilibrium, it will also be in equilibrium during this virtual displacement, following the definition of virtual displacements, and we have $\delta(\text{W.D.}) = 0$, for any virtual displacement. This is the *principle of virtual work* in statics.

In dynamics, the acting forces are generally not in equilibrium, but if D'Alembert's principle is applied, we have dynamic equilibrium, and the principle of virtual work may then be applied as in a static case.

The combination of D'Alembert's principle and the principle of virtual work is a powerful method in many dynamics problems.

The principle of virtual work in dynamics may now be stated in the form:

$$(\mathbf{F}-m\ddot{\mathbf{r}})\cdot\delta\mathbf{r} = 0 \tag{2.18}$$

Equation (2.18) is called *D'Alembert's equation*, and is of fundamental importance in further developments in dynamics; it was used by Lagrange, about 50 years later, to develop his famous equations in dynamics, which will be discussed in Chapter 9. For more complicated problems Lagrange's equations largely supersede D'Alembert's equation.

Example 2.23. Determine the acceleration of the masses in Example 2.22, by using the principle of virtual work.

Solution. Applying all external forces and inertia forces as shown in Fig. 2.29, we have created dynamic equilibrium. Giving the mass M_1 a virtual displacement δx, the principle of virtual work states that $\delta(\text{W.D.})=0$.

Now $\delta(\text{W.D.})=M_1g\ \delta x-M_1\ddot{x}\ \delta x-M_2g\ \delta x-M_2\ddot{x}\ \delta x=0$, cancelling δx, we find

$$\ddot{x} = \frac{M_1-M_2}{M_1+M_2}g$$

as in Example 2.22.

The great advantage of the principle is that internal forces, occurring in equal and opposite pairs, produce no work in the work summation; fixed reactions and normal forces also produce no work. All of these forces may therefore be omitted in the calculation of the work done.

2.15 Degrees of freedom and equations of constraint

The number of *degrees of freedom n* of a particle is defined as the number of *independent* coordinates that are necessary to determine the position of the particle.

A *fixed* particle may be said to have $n=0$; a particle moving in *rectilinear motion* has $n=1$, since one coordinate is sufficient to define the position of the particle; if the particle is in *plane curvi-*

linear motion, we also find $n=1$, the two coordinates x and y, to specify the position of the particle in the plane, are connected by an equation $y=f(x)$, which gives the path of the particle. In this case x and y are therefore not independent; the equation $y=f(x)$ is called an *equation of constraint*, and each equation of constraint lowers the number of degrees of freedom by one.

If the particle is *free to move in a plane*, $n=2$. A particle moving on a *space curve* has $n=1$, for a *surface in space* $n=2$. A particle *free to move in space* has $n=3$; this is then the maximum possible number of degrees of freedom of a particle.

The concepts of degrees of freedom and equations of constraint are of fundamental importance in more advanced dynamics.

2.16 Moment of momentum. Central force motion

The concept of moment of a force or a vector was widely used in statics, and found to be of great importance, since the moment gave the 'turning action' of a force about a line. The moment of any vector **F** about a *point O* is defined as the vector cross product $\mathbf{M}_O=\mathbf{r}\times\mathbf{F}$, where **r** is a vector from O to *any point* on the line of action of **F**. The *scalar magnitude* of the component of M_O along *any line* through O is defined as the moment of F about that line.

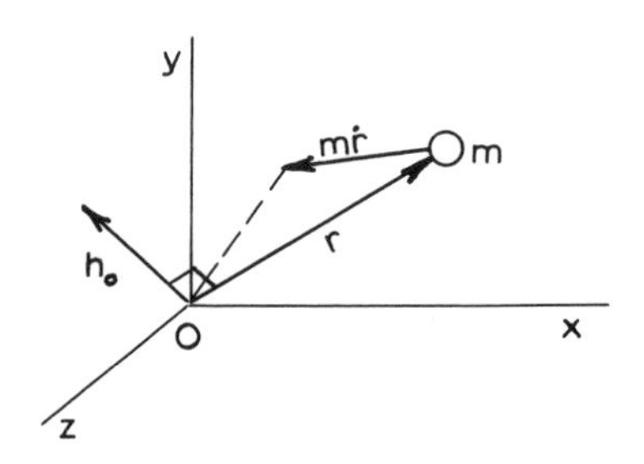

Fig. 2.30

To investigate the moment of forces acting on a moving particle, consider the particle in Fig. 2.30, which is moving under the action of a resultant force **F**. Newton's second law applied to the particle gives $\mathbf{F}=m\ddot{\mathbf{r}}$, where **r** is the position vector of the particle in the inertial xyz-system shown, with the *fixed point O* as origin.

The moment of $\mathbf{F}$ about O is $\mathbf{M}_O = \mathbf{r} \times \mathbf{F}$; introducing $\mathbf{F} = m\ddot{\mathbf{r}}$, gives

$$\mathbf{M}_O = \mathbf{r} \times m\ddot{\mathbf{r}} = \frac{d}{dt}(\mathbf{r} \times m\dot{\mathbf{r}}) - \dot{\mathbf{r}} \times m\dot{\mathbf{r}} = \frac{d}{dt}(\mathbf{r} \times m\dot{\mathbf{r}})$$

since the cross product of parallel vectors vanishes.

The vector $m\dot{\mathbf{r}}$ is the *momentum* of the particle, the vector $\mathbf{r} \times m\dot{\mathbf{r}}$ is *the moment* of the *momentum vector* in point O, so that the *moment of momentum* of the particle in O is $\mathbf{h}_O = \mathbf{r} \times m\dot{\mathbf{r}}$, and this vector is perpendicular to the plane determined by $\mathbf{r}$ and $m\dot{\mathbf{r}}$ as shown in the figure. We have now

$$\mathbf{M}_O = \dot{\mathbf{h}}_O \tag{2.19}$$

or the moment of the resultant force about O is equal to the time derivative of the moment of momentum about the same point.

If $\mathbf{M}_O = \mathbf{0}$, we have $\dot{\mathbf{h}}_O = \mathbf{0}$, so that $\mathbf{h}_O$ is constant in magnitude and direction; this is the case if $\mathbf{r} \times \mathbf{F} = \mathbf{0}$, or if (1) $\mathbf{r} = \mathbf{0}$, (2) $\mathbf{F} = \mathbf{0}$, or (3) $\mathbf{r}$ and $\mathbf{F}$ are parallel; only case (3) is of further interest, since in case (1) the particle is fixed at O, and in case (2) it is moving in a straight line with constant velocity. Case (3) occurs if the force is always directed along a line through O; such a force is called a *central force*, and the motion of the particle under the action of a central force is called *central-force motion.*

For central force motion, we have now $\mathbf{M}_O = \mathbf{0}$, or $\mathbf{h}_O$ constant in magnitude and direction; since $\mathbf{h}_O$ is normal to the plane defined by $\mathbf{r}$ and $m\dot{\mathbf{r}}$, it follows that *all central-force motion is plane motion.*

2.17 Planetary motion

In the motion of the planets in our solar system, we may consider the planets as particles, because of the great mass of the sun and the great distances involved.

The gravitational attraction of the sun is given by Newton's universal law of gravitation, eq (2.3), $F = \gamma\, Mm/r^2$, where M is the mass of the sun, m the mass of the planet in question, and r the distance from the sun to the planet; γ is the universal gravitational constant.

Introducing a new constant $K = \gamma M$, we have $F = Km/r^2$. A given planet attracts the sun with an equal and opposite force, but because of the great difference in mass, we may take the sun as a fixed point, and also neglect the gravitational attraction from the other planets; the movement of each planet is then a case of *central-force motion*, and the orbit of the planet is a *plane curve.*

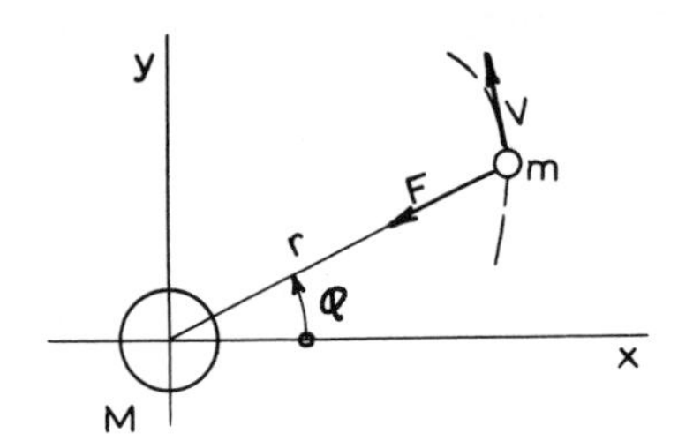

Fig. 2.31

Figure 2.31 shows the general situation in planetary motion. The equations of motion of the planet are expressed, in the most convenient form, in the polar coordinates (r,φ). The equations of motion are

$$m(\ddot{r}-r\dot{\varphi}^2) = -F = -\frac{Km}{r^2} \tag{a}$$

$$m(r\ddot{\varphi}+2\dot{r}\dot{\varphi}) = 0 \tag{b}$$

$$\ddot{r}-r\dot{\varphi}^2 = -\frac{K}{r^2} \tag{a$'$}$$

$$r\ddot{\varphi}+2\dot{r}\dot{\varphi} = 0 \tag{b$'$}$$

Equation (b′) may be given the form

$$\frac{1}{r}\frac{d}{dt}(r^2\dot{\varphi}) = 0$$

which may be shown by working out the time derivative; this means that $r^2\dot{\varphi}$ is constant, or

$$r^2\dot{\varphi} = C \tag{2.20}$$

Equation (2.20) states that $d\varphi/dt = C/r^2$; we have now

$$\dot{r} = \frac{dr}{dt} = \frac{dr}{d\varphi}\frac{d\varphi}{dt} = \frac{C}{r^2}\frac{dr}{d\varphi} = -C\frac{dr^{-1}}{d\varphi}$$

Introducing a new variable $u = 1/r$, we have $\dot{r} = -C\, du/d\varphi$ and $\dot{\varphi} = Cu^2$, so that

$$\ddot{r} = \frac{d^2r}{dt^2} = \frac{d}{dt}\left(\frac{dr}{dt}\right) = \frac{d}{dt}\left(-C\frac{du}{d\varphi}\right) = -C\frac{d}{d\varphi}\left(\frac{du}{d\varphi}\right)\frac{d\varphi}{dt}$$

$$= -C\frac{d^2u}{d\varphi^2}(Cu^2) = -C^2u^2\frac{d^2u}{d\varphi^2}$$

Substituting these expressions in eq. (a) gives

$$-C^2u^2\frac{d^2u}{d\varphi^2} - \frac{1}{u}(C^2u^4) = -Ku^2$$

$$\frac{d^2u}{d\varphi^2} + u = \frac{K}{C^2}$$

The corresponding homogeneous equation, $d^2u/d\varphi^2 + u = 0$, has the solution $u = A \sin \varphi + B \cos \varphi$, where A and B are arbitrary constants. A particular solution of the complete equation is $u = K/C^2$, so that the general solution is

$$u = A \sin \varphi + B \cos \varphi + \frac{K}{C^2} = D \cos(\varphi - \varphi_0) + \frac{K}{C^2}$$

where D and φ_0 are new constants depending on the initial conditions of the motion. We now have the trajectory of the planet in the form

$$\frac{1}{r} = \frac{K}{C^2} + D \cos(\varphi - \varphi_0) \tag{2.21}$$

This equation is a standard form for a *general conic section in polar coordinates*, as may be seen from the following:

A conic section is defined as a curve for which all points are such that the ratio of their distances from a fixed point O and a fixed line L is a constant e. The fixed point is called the *focus*, the fixed line the *directrix*, and the ratio e *the eccentricity*. Figure 2.32 shows the general situation. P is a point on the curve; we have the curve determined by $e = OP/PB$. We draw a line OA perpendicular to the directrix, and note that the point P', the same distance from OA as P, is clearly also a point on the curve, so the curve has then an axis of symmetry OA. We find now that

$$e = \frac{r}{p - r \cos(\varphi - \varphi_0)}$$

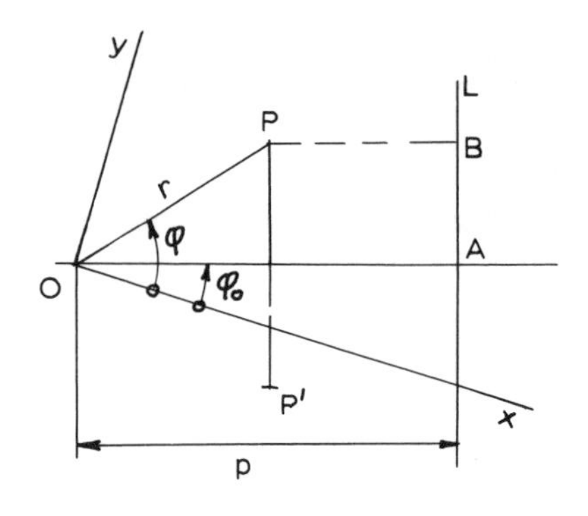

Fig. 2.32

which leads to

$$\frac{1}{r} = \frac{1}{pe} + \frac{1}{p}\cos(\varphi - \varphi_0) \tag{2.22}$$

Equation (2.22) is of the same form as (2.21); by measuring the angle φ from the axis of symmetry, we find $\varphi_0 = 0$, and the equation becomes

$$\frac{1}{r} = \frac{1}{pe} + \frac{1}{p}\cos\varphi \tag{2.23}$$

Comparing this with (2.21) we have

$$\frac{1}{pe} = \frac{K}{C^2} \quad \text{and} \quad \frac{1}{p} = D$$

so that

$$e = \frac{C^2}{Kp} = \frac{DC^2}{K} \tag{2.24}$$

The situation is now as shown in Fig. 2.33. In *rectangular coordinates* $r = \sqrt{(x^2+y^2)}$ and $\cos\varphi = x/\sqrt{(x^2+y^2)}$; substituting this in (2.23) gives the result

$$x^2(1-e^2) + y^2 + 2pe^2x - p^2e^2 = 0 \tag{2.25}$$

To investigate the various types of curve given by (2.25), consider first the case when $e=0$. Since $pe=r+er\cos\varphi$ from (2.23), we find for $e\to 0$, that $pe\to r$; for $e=0$, we have then from (2.25) $x^2+y^2=r^2$, the equation of a circle. If $e=1$, we find $y^2+2px-p^2=0$ which is a *parabola*.

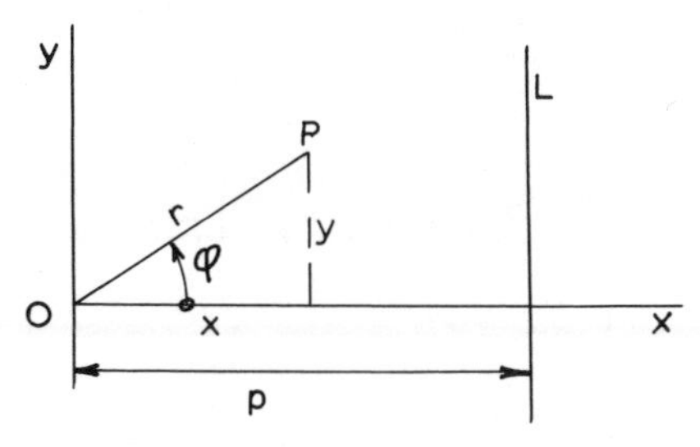

Fig. 2.33

If $e>1$, we can displace the xy-system a distance d to the right, so that the new coordinates (x_1, y_1) are related to (x, y) by the expressions $y=y_1$ and $x=x_1+d$; substituting this in (2.25), we find the coefficient $[2pe^2-(e^2-1)2d]$ for x_1. Equating this to zero gives $d=pe^2/(e^2-1)$, and the equation takes the final form

$$\frac{x_1^2}{[pe/(e^2-1)]^2}-\frac{y_1^2}{[pe/\sqrt{(e^2-1)}]^2}=1$$

which is the standard form for a *hyperbola*.

If $e<1$, we can displace the coordinate system a distance d to the left, so that $y=y_1$ and $x=x_1-d$. Substituting in (2.25) gives the coefficient $[2pe^2-(1-e^2)2d]$ for x_1. Equating this to zero gives $d=pe^2/(1-e^2)$, and the equation takes the final form

$$\frac{x_1^2}{[pe/(e^2-1)]^2}+\frac{y_1^2}{[pe/\sqrt{(e^2-1)}]^2}=1$$

which is the standard form of an *ellipse* with semi-major *axis* $a=pe/(1-e^2)$ and semi-minor axis $b=pe/\sqrt{(1-e^2)}=a\sqrt{(1-e^2)}$.

If any planets were originally in orbit with $e\geqslant 1$, they would soon have vanished into space; the remaining planets are, therefore, in *elliptical orbits with the sun at one focal point*. (In the case of the earth, $e=0{\cdot}017$, so that the orbit of the earth is very nearly circular.) This proves *Kepler's first law*, that the orbits of the planets are ellipses with the sun at one focal point. This law was given in 1609.

From eq. (2.20), $r^2\dot{\varphi}=C$, we find (Fig. 2.34), that the area swept by the radius vector in a time dt is $dA=\frac{1}{2}r(r\,d\varphi)=\frac{1}{2}r^2\,d\varphi$, so that

$$\frac{dA}{dt} = \tfrac{1}{2}r^2\frac{d\varphi}{dt} = \frac{C}{2} = \text{constant}$$

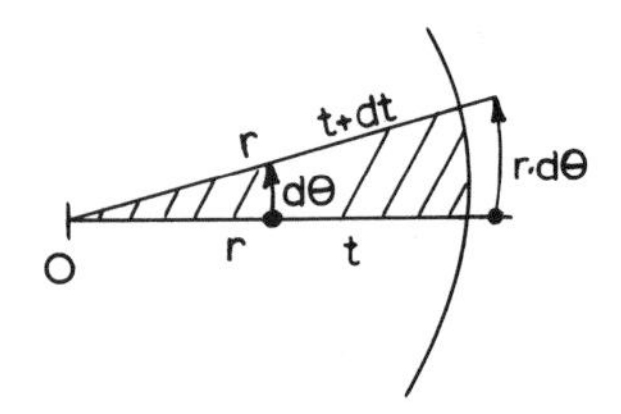

Fig. 2.34

This gives

$$A = \int_0^t \tfrac{1}{2}C\,dt = \tfrac{1}{2}Ct$$

This proves *Kepler's second law*, which states that the radius vector from the sun to the planet sweeps out equal areas in equal times. This law was also given in 1609.

Since the orbit of a planet is an ellipse of total area $A=\pi ab$, we find the *total orbiting time*

$$T = \frac{2A}{C} = \frac{2\pi ab}{C} = \frac{2\pi a^2}{C}\sqrt{(1-e^2)} \tag{2.26}$$

From eq. (2.24) we have $C^2=Ke/D=Kep$, and since $a=pe/(1-e^2)$ we find $C^2=Ka(1-e^2)$. Substituting this in (2.26) gives

$$T^2 = \frac{4\pi^2a^4(1-e^2)}{C^2} = \left(\frac{4\pi^2}{K}\right)a^3$$

This is *Kepler's third law*, which states that the square of the period is proportional to the cube of the semi-major axis of the orbit of any planet. This law was first given in 1619.

2.18 Orbiting space vehicles

Space vehicles orbiting the earth are a relatively recent accomplishment; the first Russian satellite was launched in 1957. The space vehicle is lifted through the earth's atmosphere by multi-stage rockets to its launching height, and is given a launching velocity in a certain direction. From then on it moves under the action of the gravitational force from the earth in *central-force motion.*

The general starting conditions are illustrated in Fig. 2.35. The launching velocity is V_0 at a distance r_0 from the earth's centre, and V_0 is inclined at an angle α, the launching angle, to the direction perpendicular to r_0. The angle φ is measured from the axis of symmetry, the x-axis; the position of this axis is determined by the angle β, which is as yet unknown.

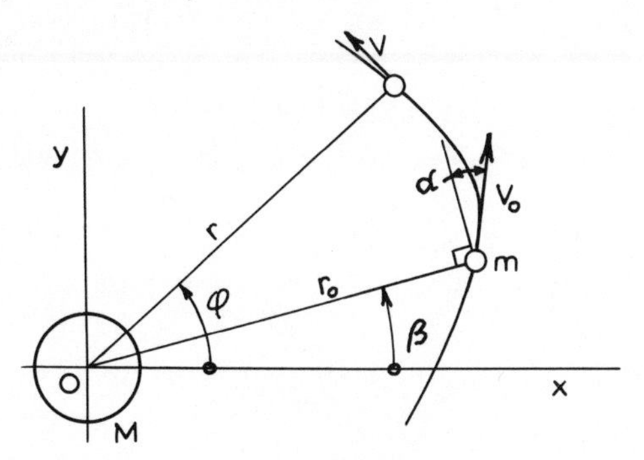

Fig. 2.35

The coordinate system (x, y) is fixed at the earth's centre and translates with the earth. It is taken as an inertial system, in which the earth rotates once in 24 hours; the plane of the orbit is the xy-plane, which keeps a fixed orientation in space while the earth rotates under it.

In polar coordinates, the equation for the vehicle trajectory is, from (2.23),

$$\frac{1}{r} = \frac{K}{C^2} + D \cos \varphi \tag{2.27}$$

with eccentricity $e = DC^2/K$. With the given r_0, V_0 and α, we have to determine C, D and β. The constant $K = \gamma M$, where M is the mass of the earth and we may determine this from the consideration that at the earth's surface we have

$$mg = \frac{\gamma Mm}{R^2} = \frac{Km}{R^2}$$

where R is the earth's radius, so that $K=gR^2$; with $g=9{\cdot}81$ m/s^2, and taking the mean radius of the earth $R=6370$ km, we find that

$$K = \frac{9{\cdot}81}{10^3}(3600)^2\,(6370)^2 = 5{\cdot}15\times10^{12}\ \text{km}^3/\text{h}^2$$

In polar coordinates, the velocity is $\mathbf{V}=\mathbf{V}_r+\mathbf{V}_\varphi$, where $V_r=\dot{r}$ and $V_\varphi=r\dot{\varphi}$. At launching we have $V_{r0}=V_0\sin\alpha$ and $V_{\varphi 0}=V_0\cos\alpha$, so that

$$\dot{r}_0 = V_0\sin\alpha \qquad \text{and} \qquad r_0\dot{\varphi}_0 = V_0\cos\alpha$$

Equation (2.20), $r^2\dot{\varphi}=C$, now gives us $C=r_0^2\dot{\varphi}_0=r_0(r_0\dot{\varphi}_0)$, so that

$$C=r_0V_0\cos\alpha \tag{2.28}$$

To determine D and β, we differentiate (2.27) with respect to time:

$$-\frac{1}{r^2}\dot{r} = -D\sin\varphi\,\dot{\varphi} \qquad \text{or} \qquad \dot{r} = r^2\dot{\varphi}\,D\sin\varphi = CD\sin\varphi$$

When $\varphi=\beta$, we have

$$\dot{r} = \dot{r}_0 = V_0\sin\alpha \qquad \text{or} \qquad \frac{V_0\sin\alpha}{C} = D\sin\beta$$

Equation (2.27), for $r=r_0$ and $\varphi=\beta$, gives

$$\frac{1}{r_0}-\frac{K}{C^2} = D\cos\beta$$

By squaring and adding these two expressions, we find that

$$D = \left[\left(\frac{V_0\sin\alpha}{C}\right)^2+\left(\frac{1}{r_0}-\frac{K}{C^2}\right)^2\right]^{1/2} \tag{2.29}$$

The ratio of the expressions gives

$$\tan\beta = \frac{r_0V_0C\sin\alpha}{C^2-r_0K} \tag{2.30}$$

These formulae completely determine the trajectory for given launching conditions.

If the eccentricity $e>1$, the trajectory is a *hyperbola*, and if $e=1$ it is a *parabola*; the limiting magnitude of V_0 which gives a parabolic trajectory is called the *escape velocity* V_E. If $e = DC^2/K = 1$, we have $D^2 = K^2/C^4$; substituting D from (2.29) gives

$$\left(\frac{V_0 \sin \alpha}{C}\right)^2 + \left(\frac{1}{r_0} - \frac{K}{C^2}\right)^2 = \frac{K^2}{C^4}$$

$$\left(\frac{V_0 \sin \alpha}{C}\right)^2 = \frac{K^2}{C^4} - \frac{1}{r_0^2} + \frac{2K}{r_0 C^2} - \frac{K^2}{C^4}$$

so that

$$V_0^2 \sin^2 \alpha = \frac{2K}{r_0} - \frac{C^2}{r_0^2} = \frac{2K}{r_0} - V_0^2 \cos^2 \alpha$$

This gives $V_0^2 = 2K/r_0$, or

$$V_E = \sqrt{\left(\frac{2K}{r_0}\right)} \tag{2.31}$$

The escape velocity is thus independent of the launching angle α. The *orbiting time* T is, from eq. (2.26): $T = 2\pi ab/C$, where

$$a = \frac{e}{D(1-e^2)} \qquad \text{and} \qquad b = a\sqrt{(1-e^2)}$$

For launching conditions with the *launching angle* $\alpha = 0$, we have V_0 perpendicular to r_0, in which case the formulae simplify to the following:

$$C = r_0 V_0 \qquad D = \frac{1}{r_0} - \frac{K}{C^2}$$

and $\beta = 0$ or $\beta = \pi$; under these conditions the eccentricity $e = DC^2/K$ depends only on the magnitude of V_0 for a given r_0.

If V_0 is so large that $e > 1$, the trajectory is a *hyperbola*, and the vehicle will soon vanish into space. If V_0 is of a magnitude such that $e = 1$, the trajectory is a *parabola*, and the vehicle will again escape from the earth. This value of V_0 is the escape velocity $V_E = \sqrt{(2K/r_0)}$.

If V_0 has such a magnitude that $0 < e < 1$, the trajectory is an *ellipse*, so that we have a real case of orbiting. This happens when $V_0 < \sqrt{(2K/r_0)}$, as long as V_0 is large enough to take the satellite into an orbit outside the earth's atmosphere; the lower limit for V_0 in this case depends on r_0.

To get a *circular orbit* with radius r_0, we must have $e=0$. As shown before, if $e \to 0$ we get $pe \to r$, or $e \to rD$, so that $D=0=1/r_0 - K/C^2$. This gives $C=\sqrt{(r_0 K)}$; the launching velocity for a circular orbit is then $V_{oc}=C/r_0=\sqrt{(K/r_0)}$, or $V_{oc}=V_E/\sqrt{2}$.

All this assumes, of course, that r_0 and V_0 are sufficiently large for the orbit to be outside the earth's atmosphere, otherwise the drag forces of the air will slow the vehicle down and it will spiral in and burn up in the atmosphere or crash on the earth's surface. If r_0 and V_0 are too small for orbiting, the vehicle will only move through a curve in space, enter the atmosphere (if it starts outside the atmosphere), and burn up or crash.

Example 2.24. An earth satellite is to be launched at an altitude of 500 km above the earth with a launching speed of 32 000 km/h at zero launching angle. Determine: (a) the eccentricity of the orbit, (b) the trajectory, (c) the maximum and minimum distance from the earth's surface, (d) the maximum and minimum velocity in orbit, and (e) the orbiting time.

Determine also the velocity for a circular orbit, and the escape velocity.

Solution. For a mean earth radius of 6370 km, we have $r_0=6370+500=6870$ km. For $\alpha=0$, we have $C=r_0 V_0=6870 \times 32\,000 = 22 \times 10^7$ km^2/h. Using $D=1/r_0 - K/C^2$, with $K=5{\cdot}15 \times 10^{12}$ km^3/h^2, we obtain

$$D = \frac{1}{6870} - \frac{5{\cdot}15 \times 10^{12}}{22^2 \times 10^{14}} = \frac{0{\cdot}392}{10^4} \text{ km}^{-1}$$

(a) the eccentricity

$$e = \frac{DC^2}{K} = \frac{0{\cdot}392}{10^4} \times \frac{22^2 \times 10^{14}}{5{\cdot}15 \times 10^{12}} = 0{\cdot}369 < 1$$

so that the satellite goes into an *elliptic orbit*.
(b) the trajectory is

$$\frac{1}{r} = \frac{K}{C^2} + D \cos \varphi$$

$$\frac{K}{C^2} = \frac{D}{e} = \frac{0{\cdot}392}{10^4 \times 0{\cdot}369} = 1{\cdot}062 \times 10^{-4}$$

The trajectory is determined by

$$\frac{1}{r} = 1{\cdot}062 \times 10^{-4} + 0{\cdot}392 \times 10^{-4} \cos \varphi$$

(c) Figure 2.36 shows the general situation; the position of minimum distance r_0 from the earth is called *perigee*. The distance from the earth's centre is r_0, so that the minimum height of the satellite is 500 km. The position of maximum distance r_{max} from the earth is called *apogee*; it may be found from the trajectory for $\varphi = \pi$, so that

$$\frac{1}{r_{max}} = 1{\cdot}062 \times 10^{-4} - 0{\cdot}392 \times 10^{-4} = 0{\cdot}670 \times 10^{-4} \quad \text{or}$$

$$r_{max} = 14\,900 \text{ km}$$

The apogee is thus at a height of 8030 km.

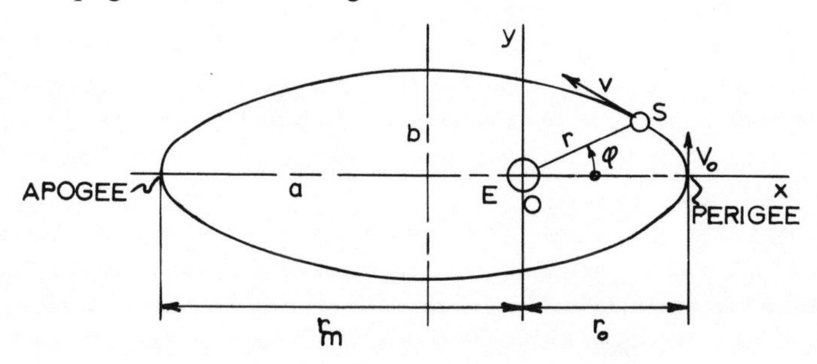

Fig. 2.36

(d) From Kepler's second law, the maximum speed is at perigee and the minimum speed at apogee, so that $V_{max} = 32\,000 \text{ km/h} = V_0$.

$$V_{min} = r_{max}\,\dot{\varphi}_\pi = \frac{r_{max}^2\,\dot{\varphi}_\pi}{r_{max}} = \frac{C}{r_{max}} = \frac{22 \times 10^7}{14\,900} = 14\,760 \text{ km/h}$$

(e) The orbiting time is $T = 2\pi ab/C$

$$a = \frac{e}{D(1-e^2)} = \frac{0{\cdot}369 \times 10^4}{0{\cdot}392(1-0{\cdot}369^2)} = 10\,900 \text{ km}$$

$b = a\sqrt{(1-e^2)} = 10\,130$ km, so that

$$T = \frac{2\pi \times 10\,900 \times 10\,130}{22 \times 10^7} = 3{\cdot}16 \text{ h} \simeq 190 \text{ minutes}$$

the semi-major axis a may also be found from $a = \frac{1}{2}(r_0 + r_{max})$.

To obtain a *circular orbit* for the *same* launching conditions, we must have

$$V_{oc} = \sqrt{\left(\frac{K}{r_0}\right)} = \sqrt{\left(\frac{5{\cdot}15 \times 10^{12}}{6870}\right)} = 27\,400 \text{ km/h}$$

The escape velocity is

$$V_E = \sqrt{2}(V_{oc}) = 38\,750 \text{ km/h}$$

Problems

2.1 In Fig. 2.37, M_1 represents a mass which slides on a plane which is inclined at the angle α to the horizontal. A string attached to M_1 passes over a fixed pulley at the top of the slope, and then around a free pulley before being anchored to 'ground'. The free pulley carries a mass M_2.

Neglecting the inertia and friction of the pulleys, derive a formula for the tension in the string, for the case when M_1 is moving up the plane.

Calculate the magnitude of the tension when $\alpha = 45°$, $M_2 = 200$ kg, $M_1 = 100$ kg and the coefficient of friction between M_1 and the plane is 0·2.

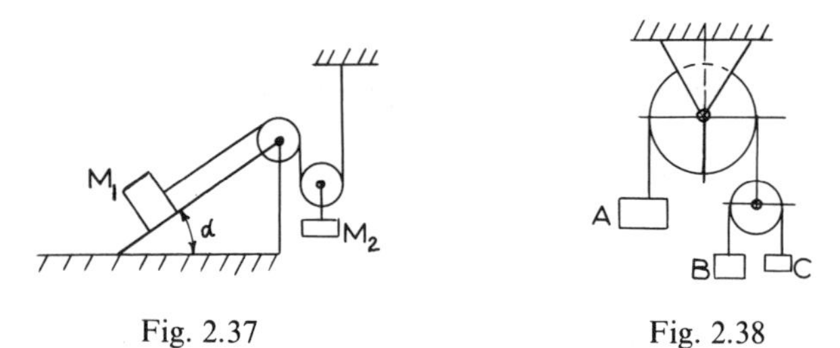

Fig. 2.37 Fig. 2.38

2.2 A system of weights and pulleys is arranged in a vertical plane as shown in Fig 2.38. Neglecting friction, inertia and weight of the pulleys find the acceleration of each weight if their magnitudes are in the ratio $A:B:C = 3:2:1$.

2.3 The mass m (Fig. 2.39) is supported by wires subjected to tensile force T. Assuming small amplitudes, determine the natural frequency of vibration in a plane perpendicular to the wire. Show that the period of vibration is greatest when $a=b$. Assume that T is large and unchanged for small displacements.

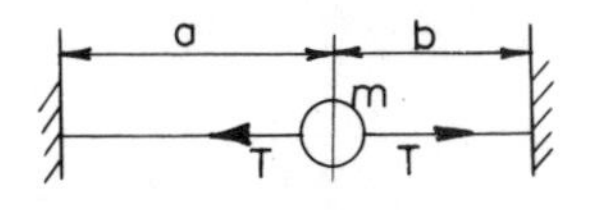

Fig. 2.39

2.4 A particle of mass m is falling vertically through a resisting medium with resistance $F=-mA\dot{x}$, where A is a constant. The starting conditions are at $t=0$, $x=\dot{x}=0$.

Find the velocity function of the particle and the terminal velocity for $t\to\infty$. Determine the displacement–time function.

2.5 Find the velocity of escape for a projectile fired from the surface of the moon in a vertical direction. Assume that the moon has the same density as the earth and a mass equal to 1/75 of the earth.

The radius of the earth is 6370 km.

2.6 In Fig. 2.40 a small ball of weight $W=5$ N starts from rest at O and rolls down the smooth track OCD under the influence of gravity. Find the reaction R on the ball at C if the curve OCD is defined by the equation $y=h\sin(\pi x/l)$, and $h=l/3$.

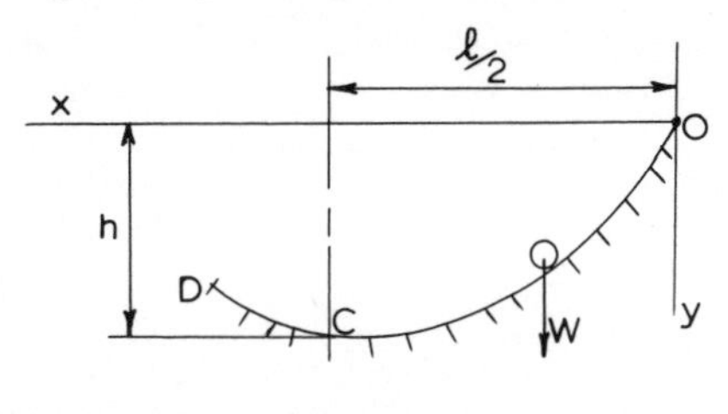

Fig. 2.40

2.7 Show that the central-force field due to the gravitational attraction of a sphere is conservative, and determine the potential-energy function.

2.8 A weight W_1 falls through a height h onto a block W_2 supported by a spring of constant K (Fig. 2.41). Assuming plastic impact, find the maximum compression Δ of the spring from the equilibrium position shown. $W_1 = W_2 = 10$ N, $K = 10$ N/cm, $h = 3$ cm.

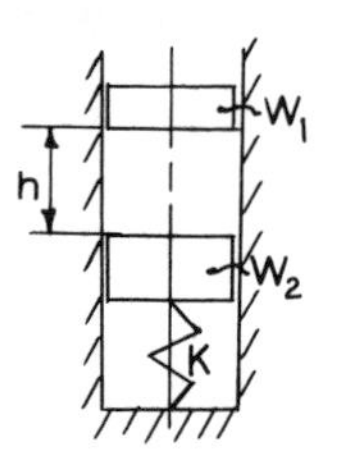

Fig. 2.41

2.9 For the pile and piledriver shown in Fig. 2.42, $W_1 = 2000$ N, $W_2 = 1000$ N, $h = 10$ m and $e = \frac{1}{4}$. If the resistance to penetration is constant $= 60\,000$ N, find the minimum number of blows necessary to drive the pile to a depth of more than 1 m.

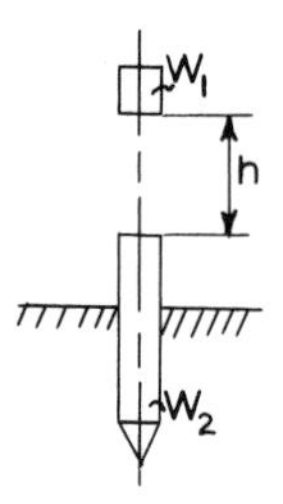

Fig. 2.42

CHAPTER THREE

Kinematics of a Rigid Body

3.1 Types of motion. Angular velocity

Some of the various types of motion of a rigid body are illustrated by the mechanism in Fig. 3.1. As the crank arms O_1A and O_2B, of equal length, rotate in the plane of the mechanism about O_1 and O_2, the piston P moves in the cylinder in a *translation*, in this case a rectilinear translation. The crank arms rotate in pure rotation in plane motion, and they are *rotating about fixed axes* through O_1 and O_2 perpendicular to the figure.

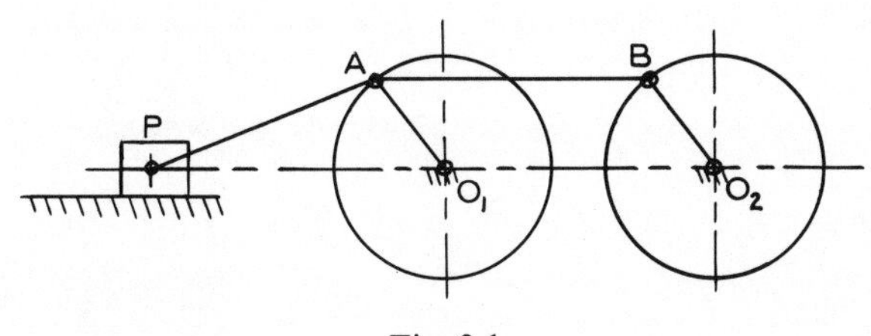

Fig. 3.1

The side rod AB is moving in the plane, and is always parallel to O_1O_2; it is in plane *curvilinear translation*, while the connecting rod PA is in *general plane motion*.

For the purpose of kinematic and dynamic analysis, the motion of a rigid body may be classified under only six different headings; these are, in increasing order of complexity: (1) rest, (2) translation, (3) rotation about a fixed axis, (4) plane motion, (5) rotation about a fixed point, and (6) general three-dimensional motion.

The first five are all special cases of (6), in fact case (3) is a special case of (5) and (4).

Case (1) is dealt with in statics, while case (5) in general is so-called gyroscopic motion; cases (5) and (6) are much more complicated than the others, and are outside the scope of this book. We shall therefore concentrate on cases (2), (3) and (4); the great majority of engineering problems belong in these categories.

In Fig. 3.1, only the radial arms and the connecting rod change their angular position with respect to O_1O_2. Before the motion of a rigid body can be discussed, the concept of change of angular relationship which we can call rotation must be discussed in more detail.

In rotation of a rigid body about a *fixed axis*, all points on the axis remain stationary, while all other points in the body move in circles in planes perpendicular to the axis, and with centres on the axis of rotation.

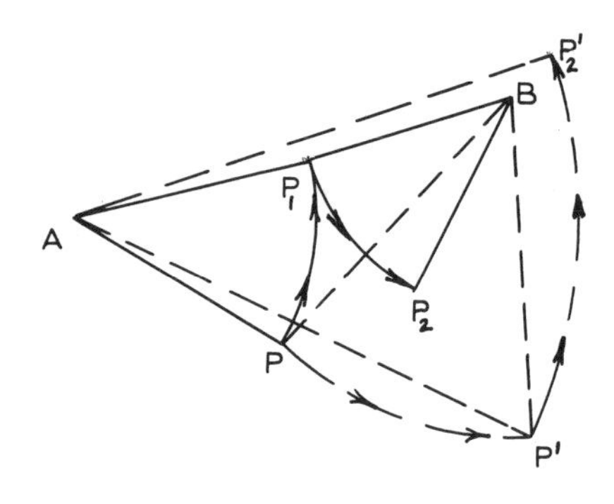

Fig. 3.2

Because of the power of the vector concept, it seems reasonable to try and define a rotation as a vector on the axis of rotation, of length equal to the magnitude of the angle of rotation and with sense following the right-hand screw rule. For rotations to be vectors, they must obey the commutative addition law; to investigate whether this is the case, consider first rotation in a plane as shown in Fig. 3.2: a point P is rotated about an axis perpendicular to the paper through A, the angle of rotation is θ_1, and P moves to position P_1. Then P_1 is rotated about an axis through B to position P_2, the angle of rotation being θ_2; if the order of rotation is reversed,

P moves first to P_1' and then to P_2'; clearly the final positions P_2 and P_2' are quite different, in fact for angles $\theta_1 = 43{\cdot}5°$ and $\theta_2 = 47°$, the linear distance P_2P_2' is about 39 mm on the figure.

Finite angular rotations cannot, therefore, be considered vectors, since the order of rotations determines the final position, and this violates the commutative addition law of vectors.

If we now repeat Fig. 3.2, but this time with angles $\theta_1 = \theta_2 = 5°$, we will find that P_2 and P_2' practically coincide, in fact P_2P_2' is well under 1 mm; clearly if we take infinitesimal angles $\Delta\theta_1$ and $\Delta\theta_2$ and let $\Delta\theta_1 \to 0$ and $\Delta\theta_2 \to 0$, the final positions of P_2 and P_2' are identical, so that we have $\Delta\boldsymbol{\theta}_1 + \Delta\boldsymbol{\theta}_2 = \Delta\boldsymbol{\theta}_2 + \Delta\boldsymbol{\theta}_1$. Infinitesimal rotations thus are vectors.

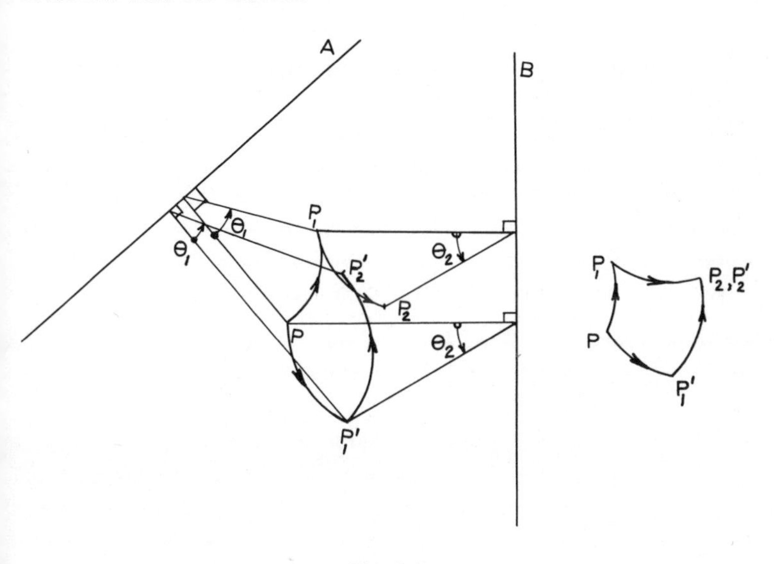

Fig. 3.3

In three-dimensional motion, the situation is as shown in Fig. 3.3, with rotations about axis A and axis B as shown. Again the positions of P_2 and P_2' are different for finite angles, while for infinitesimal angles $\Delta\theta_1 \to 0$ and $\Delta\theta_2 \to 0$, the radial distances of P and P_1 from axis B approach each other, so that the arcs P_1P_2 and PP_1' become parallel; the same happens to arcs PP_1 and $P_1'P_2'$, so that in the limit, we end up with a figure as shown, where P_2 and P_2' coincide.

The *instantaneous angular velocity* ω, which was introduced in Section 1.2, may now be defined as a *vector*

$$\boldsymbol{\omega} = \lim_{\Delta t \to 0} \frac{\Delta \boldsymbol{\theta}}{\Delta t} = \frac{d\boldsymbol{\theta}}{dt} = \dot{\boldsymbol{\theta}}$$

This is through O and perpendicular to the rotating radial line, of magnitude equal to the instantaneous angular velocity ω and with sense following the right-hand screw rule. The vector $\boldsymbol{\omega}$ then determines the instantaneous axis of rotation.

The *instantaneous angular acceleration* $\dot{\omega}$ introduced in Section 1.2 is now the *vector*

$$\dot{\boldsymbol{\omega}} = \lim_{\Delta t \to 0} \frac{\Delta \boldsymbol{\omega}}{\Delta t} = \frac{d\boldsymbol{\omega}}{dt} = \ddot{\boldsymbol{\theta}}$$

The velocity and the components of the acceleration of a point in circular motion, discussed in Example 1.6, may now be given the following vector formulation (Fig. 3.4): the velocity is $\mathbf{V} = \boldsymbol{\omega} \times \mathbf{r}$; the *normal acceleration* component $\mathbf{A}_\mathrm{N} = \boldsymbol{\omega} \times (\boldsymbol{\omega} \times \mathbf{r}) = \boldsymbol{\omega} \times \mathbf{V}$; the *tangential acceleration component* is $\mathbf{A}_\mathrm{T} = \dot{\boldsymbol{\omega}} \times \mathbf{r}$. It may be seen that the magnitudes and directions of these components agree with previously found results in Example 1.6. This vector formulation is very convenient in the kinematics of rigid bodies.

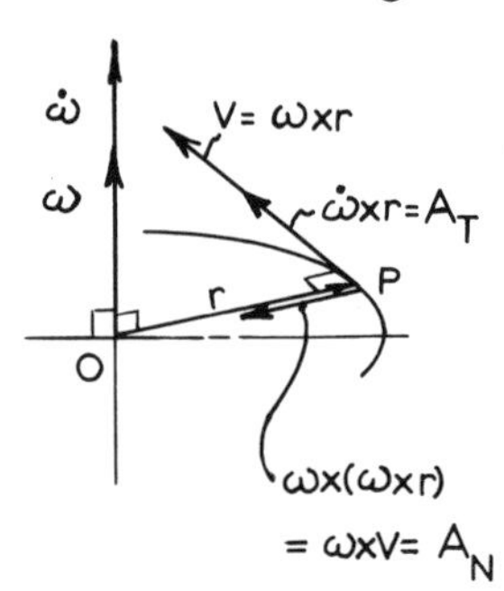

Fig. 3.4

3.2 Chasle's theorem in plane motion

The position of a rigid body in plane motion is completely determined by the position of a line A_1B_1 (Fig. 3.5) in the body and in the plane of motion.

If the body is moved to a second position A_2B_2, we may accomplish this motion by first moving the line in translation so that A_1 moves to A_2. The body must then be rotated through an angle θ about A_2 to reach the final position A_2B_2. It is also possible to translate the line to position B_2A_2', and follow this by a rotation about B_2 as shown; finally the motion may be determined by a *pure rotation* about a point C found as the intersection of the perpendicular bisectors to A_1A_2 and B_1B_2; in all cases the angle of rotation is the same.

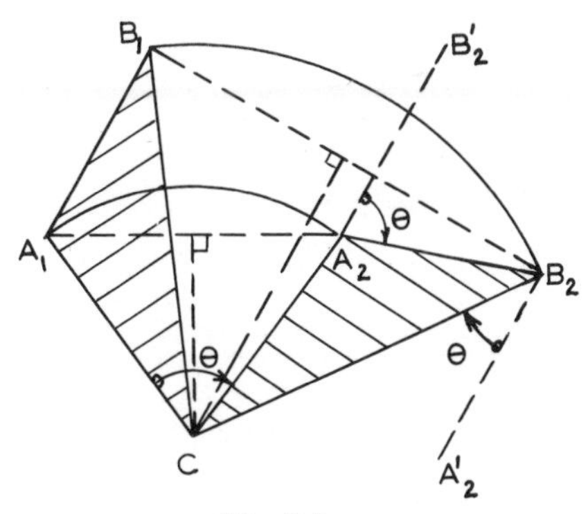

Fig. 3.5

The motion of a plane figure may be considered as made up of a series of infinitesimal displacements from one instantaneous position to the next, each displacement being a superposition of a pure rotation about a point called *the instantaneous centre of rotation* for that particular position, and a translational motion.

Figure 3.6 shows the infinitesimal displacements from a time t to a time $t+\Delta t$. The displacements $\Delta\mathbf{A}$ and $\Delta\mathbf{B}$ are different, but $\Delta\theta$ is the same for all points chosen for the translation; the ratios $\Delta\mathbf{A}/\Delta t$ and $\Delta\boldsymbol{\theta}/\Delta t$ are the average translational velocity and rotational speed; if $\Delta t \to 0$ we have the *instantaneous velocity* and angular velocity $\mathbf{V}_A$ and $\boldsymbol{\omega}$, which gives the instantaneous motion by superposition. We find that $\Delta\mathbf{B} = \Delta\mathbf{A} + \Delta\mathbf{B}'\mathbf{B}_2 = \Delta\mathbf{A} + \Delta\boldsymbol{\theta} \times \mathbf{A}_2\mathbf{B}'$, so that

$$\frac{\Delta\mathbf{B}}{\Delta t} = \frac{\Delta\mathbf{A}}{\Delta t} + \frac{\Delta\boldsymbol{\theta}}{\Delta t} \times \mathbf{A}_2\mathbf{B}'$$

when $\Delta t \to 0$ we have $\mathbf{V}_B = \mathbf{V}_A + \boldsymbol{\omega} \times \mathbf{AB}$. We may now state Chasle's theorem as follows: Any point P in the body with instantaneous velocity $\mathbf{V}_P$ may be selected, and $\mathbf{V}_P$ may be taken as the instantaneous velocity of translation of the whole body. A pure rotational velocity $\boldsymbol{\omega}$ about an axis through P may be superimposed. This will give the actual instantaneous motion of the body. The vector $\boldsymbol{\omega}$ will be the same for any point P chosen as a 'pole' for the motion, and only the translational velocity will change if another pole is chosen. The actual instantaneous axis of rotation is the axis through the points of the body having zero velocity at the instant considered.

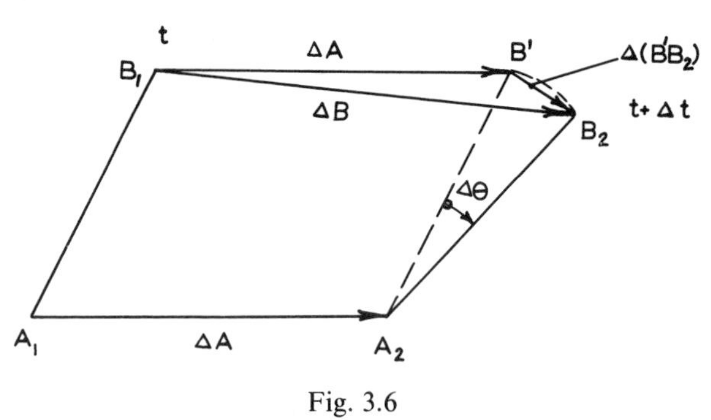

Fig. 3.6

3.3 Euler's and Chasle's theorems in three-dimensional motion

Euler's theorem states that any finite motion of a rigid body about a *fixed point* may be accomplished by a pure rotation about a particular axis through the point.

The position of a rigid body with one point O fixed is completely determined by a triangle OA_1B_1 (Fig. 3.7) fixed in the body. The points A_1 and B_1 have been taken at the same distance from O, so that $OA_1 = OB_1 = R$; A_1 and B_1 then move on a sphere with centre O and radius R as shown. If A_1 is moved along some curve on the sphere to a position A_2, B_1 will move to a position B_2 as shown. If we draw the great circular arcs A_1A_2 and B_1B_2 and introducing two planes OCN and ODN passing through O and bisecting the arcs A_1A_2 and B_1B_2, the line of intersection of the planes is ON.

The length A_1B_1 is equal to the length A_2B_2, and since both lines are chords in a great circle of radius R, the length of the arc

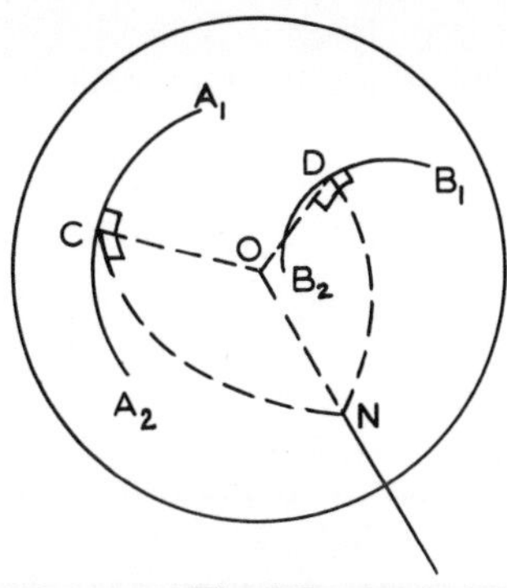

Fig. 3.7

A_1B_1 is equal to the length of the arc A_2B_2; similarly the arc B_1N is equal to the arc B_2N and A_1N is equal to A_2N; the two spherical triangles A_1B_1N and A_2B_2N are then equal. If now a rotation about the axis ON brings the side B_1N into the position B_2N, the side A_1N will coincide with A_2N; a rotation about the axis ON therefore brings the body from position A_1B_1 to position A_2B_2.

Chasle's theorem extends Euler's theorem to a general displacement in space. Consider Fig. 3.8, which shows a triangle PAB in a certain rigid body; the body is moved in space so that PAB takes the final position $P'A'B'$. The triangle PAB may be moved to its final position by first moving the body in a parallel translation to bring P to its final position P', which moves A to position A_1 and B to position B_1. According to Euler's theorem, the body may now be moved to its final position by rotation about a unique axis of rotation through P'. If this axis is as shown in the figure, rotating about the axis brings $P'A_1B_1$ into the final position $P'A'B'$.

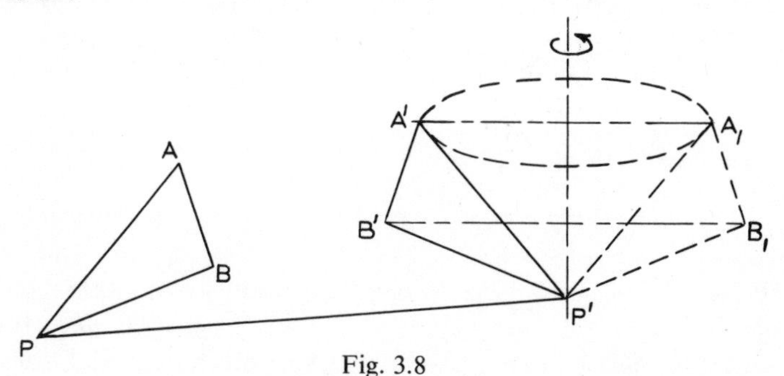

Fig. 3.8

The same change in position may be obtained by moving PAB in a translation so that A finishes in the final position A'; it will be found that to bring the triangle into the final position $P'A'B'$, a rotation must be performed about an axis through A' parallel to the previous axis through P', the angle and direction of the rotation being the same as before.

Suppose now that the displacement PP' is infinitesimal and PP' is $\Delta\mathbf{r}$ and the rotation is $\Delta\boldsymbol{\theta}$, which we can assume to be the case if the displacement happens in an infinitesimal time Δt; if we let $\Delta t \to 0$ we find $\lim \Delta\mathbf{r}/\Delta t = \mathbf{V}_P$ and $\lim \Delta\boldsymbol{\theta}/\Delta t = \boldsymbol{\omega}$, where $\mathbf{V}_P$ is the instantaneous velocity of P, and $\boldsymbol{\omega}$ the instantaneous angular velocity. This may now be given as a vector $\boldsymbol{\omega}$ on an axis through P parallel to the axis of rotation shown in Fig. 3.8; in general the displacement AA' is different from PP', so that $\mathbf{V}_A \neq \mathbf{V}_P$, however $\boldsymbol{\omega}$ is the same for any point in the body chosen, which means that *$\boldsymbol{\omega}$ is a free vector.*

The instantaneous motion of a rigid body may now be determined by selecting any point P in the body, or rigidly connected to the body, as a pole. Assuming that P at the instant considered has a velocity $\mathbf{V}_P$, the velocity of any point may be found by assuming that all points have the velocity $\mathbf{V}_P$, combined with a rotational velocity due to rotation $\boldsymbol{\omega}$ about P. This is Chasle's theorem for three-dimensional motion.

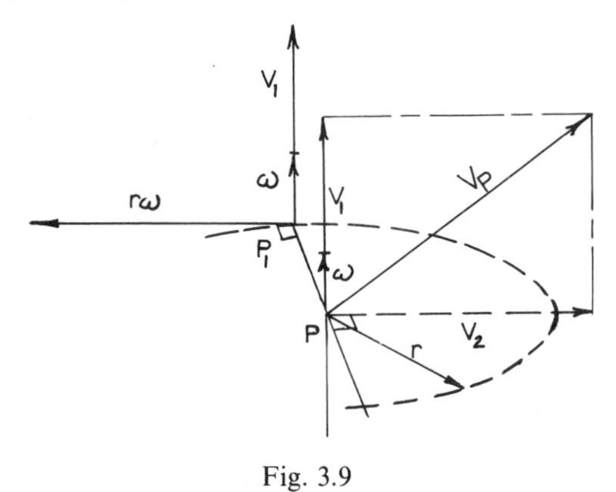

Fig. 3.9

Figure 3.9 shows a pole P in a rigid body with instantaneous velocity $\mathbf{V}_P$ and angular velocity $\boldsymbol{\omega}$. If $\mathbf{V}_P$ is resolved into components $\mathbf{V}_1$ along the $\boldsymbol{\omega}$ axis and $\mathbf{V}_2$ perpendicular to the axis, a radial distance r may be calculated from $r\omega = V_2$, and the circle with radius r is shown on the figure. The point P_1 has only the velocity $\mathbf{V}_1$ parallel to $\boldsymbol{\omega}$; if the ω-axis is drawn through P_1, the motion is reduced to a *screw motion* with angular velocity $\boldsymbol{\omega}$ and velocity $\mathbf{V}_1$ parallel to $\boldsymbol{\omega}$; if $\boldsymbol{\omega} = \mathbf{0}$, the motion is a *translation* with velocity $\mathbf{V}_1$; if $\mathbf{V}_1 = \mathbf{0}$ the motion is a *pure rotation* about the axis and points on the axis through P_1 have no velocity, so this axis is the *instantaneous axis of rotation*. In the case of plane motion $\boldsymbol{\omega}$ is always perpendicular to $\mathbf{V}_P$, so that $\mathbf{V}_1 = \mathbf{0}$, and the point P_1 in the plane of motion is then the *instantaneous centre of rotation*. In plane motion the instantaneous axis of rotation and the instantaneous centre can always be determined, except in the case of translation.

In three-dimensional motion, the *instantaneous screw axis* may always be determined, except in the case of translation. Only if the pole velocity $\mathbf{V}_P$ is *perpendicular* to the angular velocity vector $\boldsymbol{\omega}$ is the instantaneous screw axis the instantaneous axis of rotation. The proper name for the instantaneous screw axis is the *central axis* of the system.

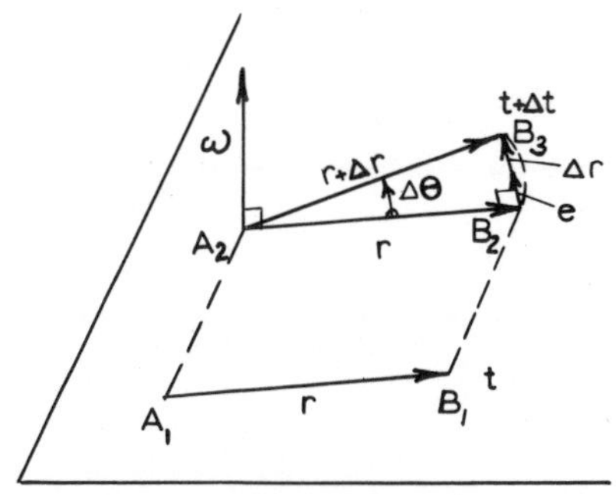

Fig. 3.10

3.4 Derivative of a vector of constant length

Figure 3.10 shows a vector $\mathbf{r}$ in a position A_1B_1 in a plane. After an interval of Δt, the vector is in a position A_2B_3; the length of the vector is r, and this is a *constant length*. The vector may be moved to the final position by first translating it to position A_2B_2 and then rotating it an angle $\Delta\theta$ about A_2. The translation of the vector does not change its magnitude or direction and therefore gives no contribution to $\Delta\mathbf{r}$ and therefore no derivative with respect to time.

The total change in $\mathbf{r}$ is the vector $\Delta\mathbf{r}$ as shown, and we have

$$\lim_{\Delta t\to 0} \frac{\Delta\mathbf{r}}{\Delta t} = \lim r\frac{\Delta\theta}{\Delta t}\mathbf{e} = r\frac{d\theta}{dt}\mathbf{e} = r\omega\mathbf{e}$$

where $\mathbf{e}$ is a unit vector in the direction $\Delta\mathbf{r}$. The vector $\Delta\mathbf{r}$ is a chord in a circular arc, and for $\Delta t\to 0$ we have $\Delta\theta\to 0$, so that in the limit $\Delta\mathbf{r}$ and $\mathbf{e}$ are perpendicular to $\mathbf{r}$. Introducing the instantaneous angular velocity vector $\boldsymbol{\omega}$ as shown, we find

$$\frac{d\mathbf{r}}{dt} = \dot{\mathbf{r}} = r\omega\mathbf{e} = \boldsymbol{\omega}\times\mathbf{r}$$

It may be seen that this vector expression for the derivative $\dot{\mathbf{r}}$ of a *constant length* vector, has the proper magnitude and direction.

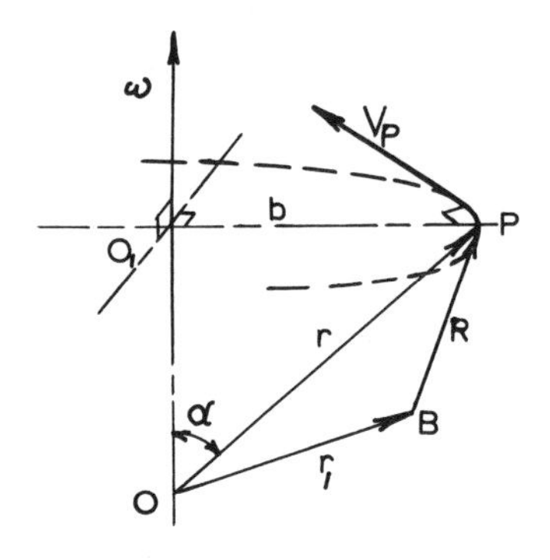

Fig. 3.11

Figure 3.11 shows a point P rotating about a fixed axis. The instantaneous angular velocity is $\boldsymbol{\omega}$; the velocity of P is $\mathbf{V}_P$ as shown, and the speed of P is $V_P = b\omega$.

A position vector $\mathbf{r}$ has been drawn from a point O on the axis of rotation to P; it follows from the definition of velocity that $\dot{\mathbf{r}} = \mathbf{V}_P$. The velocity $\mathbf{V}_P$ may also be expressed by $\mathbf{V}_P = \boldsymbol{\omega}\times\mathbf{r}$. The direction and sense of $\boldsymbol{\omega}\times\mathbf{r}$ may be seen to be in agreement with those of $\mathbf{V}_P$. The magnitude of $\boldsymbol{\omega}\times\mathbf{r}$ is $|\boldsymbol{\omega}|\,|\mathbf{r}|\sin\alpha = r\omega\sin\alpha = b\omega$ which is in agreement with the magnitude of $\mathbf{V}_P$; it follows from this that $\dot{\mathbf{r}} = \boldsymbol{\omega}\times\mathbf{r}$, where r has a *constant length*. The point O

may be taken anywhere on the axis of rotation, the result will be the same.

For any vector $\mathbf{R}$ as shown in Fig. 3.11 of *constant length* and fixed in a rotating rigid body, we find $\mathbf{R}=\mathbf{r}-\mathbf{r}_1$, so that

$$\dot{\mathbf{R}}=\dot{\mathbf{r}}-\dot{\mathbf{r}}_1 = \boldsymbol{\omega}\times\mathbf{r}-\boldsymbol{\omega}\times\mathbf{r}_1 = \boldsymbol{\omega}\times(\mathbf{r}-\mathbf{r}_1) = \boldsymbol{\omega}\times\mathbf{R} \quad (3.1)$$

(3.1) indicates that the derivative of *any constant-length vector* $\mathbf{R}$ fixed in a rotating rigid body with instantaneous angular velocity $\boldsymbol{\omega}$ is $\boldsymbol{\omega}\times\mathbf{R}$.

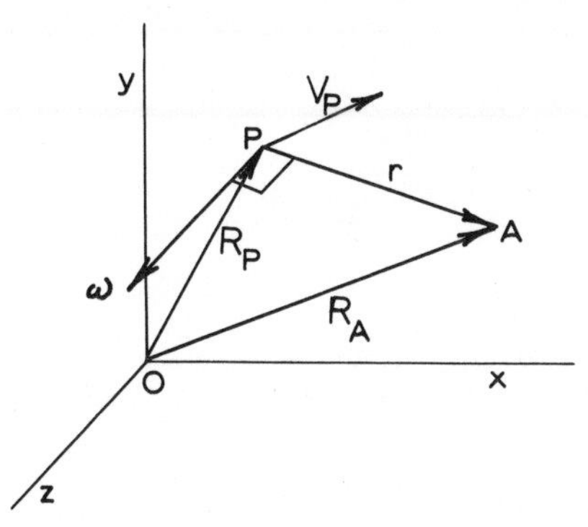

Fig. 3.12

3.5 Kinematics of a rigid link. Velocity equation

Figure 3.12 shows a rigid link PA moving in a plane. Taking P as a pole, we will assume that the instantaneous velocity of P is known and is $\mathbf{V}_P$, and that the instantaneous angular velocity is $\boldsymbol{\omega}$. The vector $\boldsymbol{\omega}$ is perpendicular to the plane of motion and directed outward for the indicated direction of ω. According to Chasle's theorem $\boldsymbol{\omega}$ may be situated at point P.

Taking position vectors $\mathbf{R}_P$ and $\mathbf{R}_A$ from a fixed point O in the plane, and introducing the *fixed length vector* $\mathbf{r}$ from P to A, we find that $\mathbf{R}_A=\mathbf{R}_P+\mathbf{r}$, so that $\dot{\mathbf{R}}_A=\dot{\mathbf{R}}_P+\dot{\mathbf{r}}$. Introducing $\dot{\mathbf{R}}_A=\mathbf{V}_A$, $\dot{\mathbf{R}}_P=\mathbf{V}_P$ and $\dot{\mathbf{r}}=\boldsymbol{\omega}\times\mathbf{r}$, we have $\mathbf{V}_A=\mathbf{V}_P+\boldsymbol{\omega}\times\mathbf{r}$. The term $\boldsymbol{\omega}\times\mathbf{r}$ is called the *relative velocity* of A with respect to P (that is measured with respect to a set of coordinate axes fixed at P) and is denoted $\mathbf{V}_{AP}$, so that we have the *important vector formula*

$$\mathbf{V}_A = \mathbf{V}_P+\mathbf{V}_{AP} \quad (3.2)$$

The motion of A relative to P can only be motion in a circle with P as centre and r as radius, $\mathbf{V}_{AP}=\boldsymbol{\omega}\times\mathbf{r}$ is in agreement with this since it is perpendicular to r and of magnitude $r\omega$; it is determined by fixing the pole, and considering A in a circular motion about the pole.

The formula (3.2) may be extended at once to three-dimensional motion.

Example 3.1. In the system shown in Fig. 3.13 (a), determine the instantaneous velocity of point A in terms of the coordinates x and θ. Assume that the system is moving in the positive directions of the coordinates.

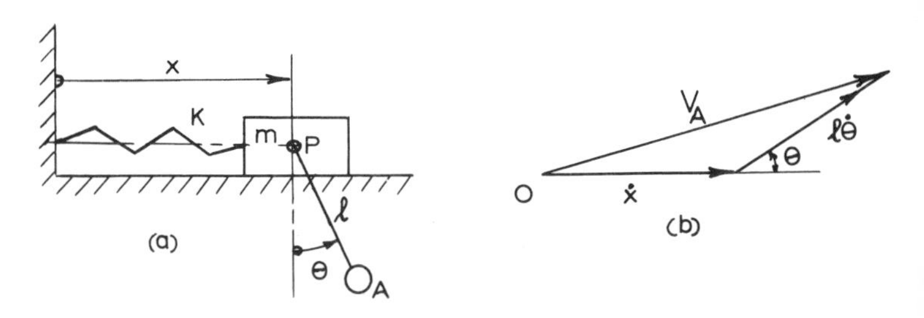

Fig. 3.13

Solution. The vector formula (3.2) states that $\mathbf{V}_A=\mathbf{V}_P+\mathbf{V}_{AP}$; we have $\mathbf{V}_P=\dot{x}\rightarrow$, and $\mathbf{V}_{AP}=l\dot{\theta}$ at $\theta°$.* We may now *sketch* a velocity diagram as shown in Fig. 3.13 (b). Using the cosine rule, we find from the diagram that

$$V_A^2 = \dot{x}^2+l^2\dot{\theta}^2+2\dot{x}l\dot{\theta}\cos\theta$$

To find *numerical* values, it is necessary to know the magnitude of $\dot{x}$ and $\dot{\theta}$ at the position and instant shown in the figure.

Example 3.2. The system shown in Fig. 3.14(a) consists of a bar OP of length a, which is rotating in the plane about O. A second bar PA of length l is hinged to OP at P and is also rotating in the plane of the system. Determine the velocity of point A, in terms of the coordinates φ and θ.

* The direction in which vectors act are indicated here and elsewhere either by arrows or by stating an angle. In the latter case the usual convention is followed, 0° being horizontal to the right and positive rotations being anti-clockwise.

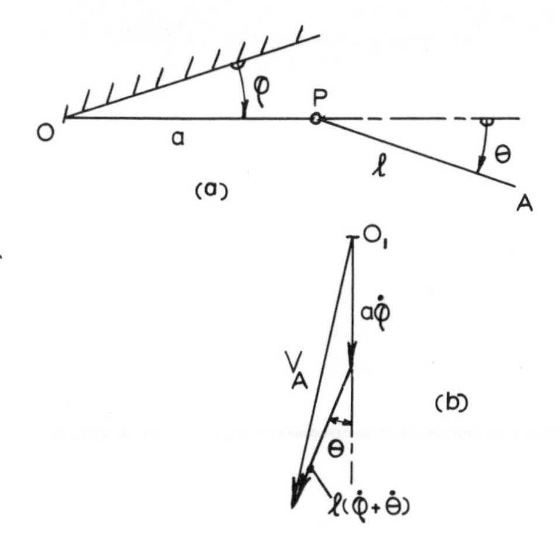

Fig. 3.14

Solution. We have from formula (3.2) that $\mathbf{V}_A = \mathbf{V}_P + \mathbf{V}_{AP}$. Now $\mathbf{V}_P = a\dot{\varphi} \downarrow$; the *total* angular velocity of PA is $\dot{\varphi}+\dot{\theta}$, so that $\mathbf{V}_{AP} = l(\dot{\varphi}+\dot{\theta}) \swarrow$ and $\mathbf{V}_{AP}$ is perpendicular to PA.

The velocity diagram is *sketched* in Fig. 3.14(b). We find

$$V_A^2 = a^2\dot{\varphi}^2 + l^2(\dot{\varphi}+\dot{\theta})^2 + 2a\dot{\varphi}l(\dot{\varphi}+\dot{\theta}) \cos\theta$$

3.6 Velocity diagrams. Graphical solutions

Figure 3.15 shows a slider–crank mechanism. The radial arm OP of length r rotates in the plane about the fixed point O with angular velocity ω and angular acceleration $\dot{\omega}$ as shown; the crank arm OP is pinned to a connecting rod PC of length l at P, and the connecting rod is pinned to a piston at C. The piston is constrained

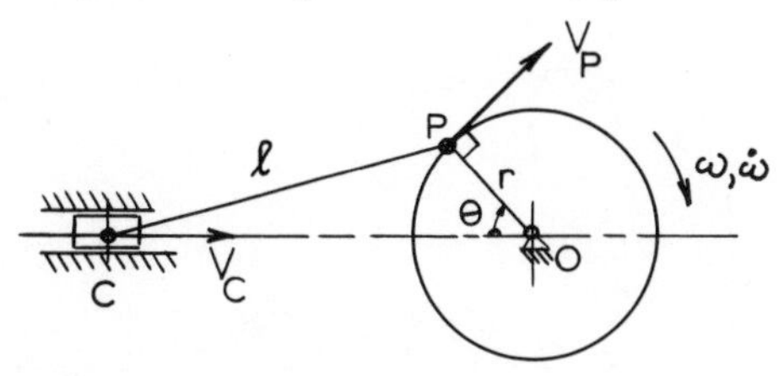

Fig. 3.15

to move in a horizontal cylinder. The angle θ determines the position of the system at the instant considered. In actual cases with the dimensions given, Fig. 3.15 is drawn to a certain scale, and is called a *configuration diagram*.

To determine the piston velocity $\mathbf{V}_C$ at this position, the speed of P may be calculated as $V_P = r\omega$, and the direction of $\mathbf{V}_P$ is as shown perpendicular to OP. Considering P as a point on the connecting rod, P may be used as a pole for the motion of the rod.

Formula (3.2) now gives $\mathbf{V}_C = \mathbf{V}_P + \mathbf{V}_{CP}$; where $\mathbf{V}_{CP} = \boldsymbol{\omega}_1 \times \mathbf{PC}$, which is perpendicular to PC, and $\boldsymbol{\omega}_1$ is the instantaneous angular velocity of PC.

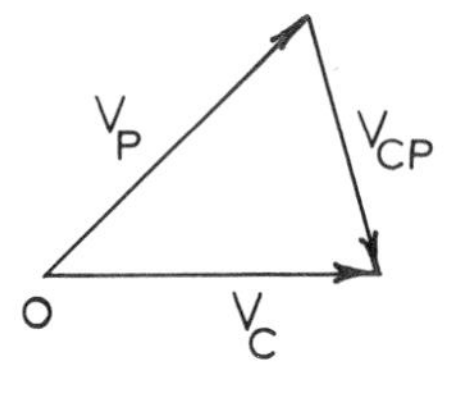

Fig. 3.16

The velocity diagram, which represents the vector equation $\mathbf{V}_C = \mathbf{V}_P + \mathbf{V}_{CP}$, may now be drawn to a suitable scale. This is shown in Fig. 3.16. The piston velocity $\mathbf{V}_C$ may now be measured from the diagram, which also gives the magnitude and direction of $\mathbf{V}_{CP}$. The instantaneous angular velocity $\boldsymbol{\omega}_1$ of PC is determined from $V_{CP} = l\omega_1$, and the direction of ω_1 is found as shown in Fig. 3.17.

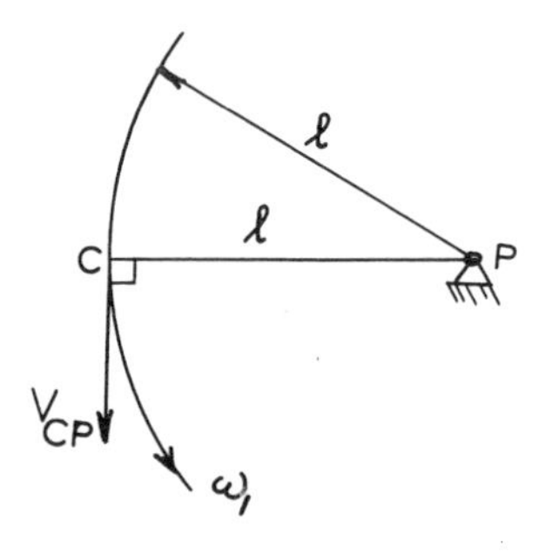

Fig. 3.17

If the velocity diagram is drawn with starting point P, as shown in Fig. 3.18, it may be seen that the components of $\mathbf{V}_P$ and $\mathbf{V}_C$ *along the bar* are of the same magnitude and direction. This must be the case, since PC is assumed rigid, so the distance PC must be constant at all times; the construction in Fig. 3.18 gives a simple way of determining the velocity of one point in a rigid body with known direction of velocity, if the magnitude and direction of the velocity of one other point in the body is known.

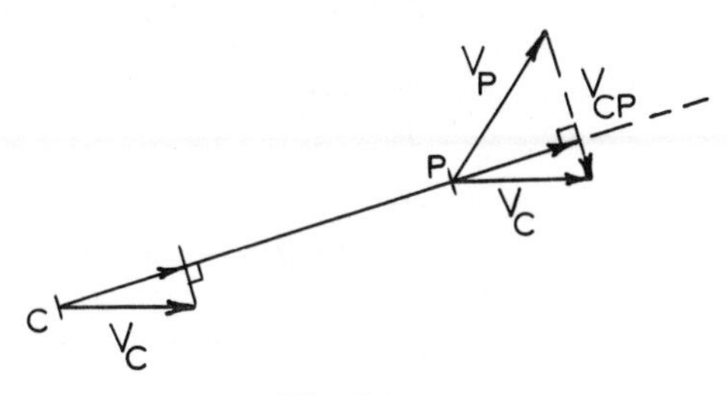

Fig. 3.18

3.7 The use of instantaneous centres to determine velocities

Figure 3.19 shows the mechanism discussed in Section 3.6. The instantaneous axis of rotation through the points instantaneously at rest was discussed in Section 3.2 and 3.3. In plane motion this axis is perpendicular to the plane of motion and intersects the plane

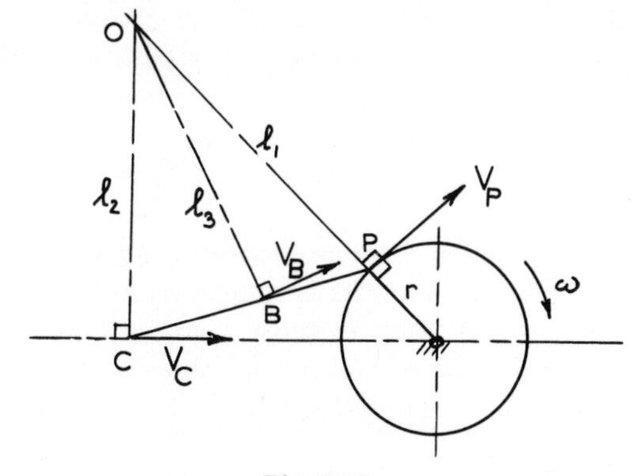

Fig. 3.19

in the *instantaneous centre of* rotation. In Fig. 3.19 the instantaneous centre of rotation O for the connecting rod PC has been found as shown: the centre must be on a line perpendicular to the velocity of any point of the body at that instant; the directions of $\mathbf{V}_P$ and $\mathbf{V}_C$ are known, the intersection O of OC and OP drawn perpendicular to $\mathbf{V}_C$ and $\mathbf{V}_P$ is the instantaneous centre for PC. Measuring l_1, we may find the instant angular velocity ω_1 of PC from $V_P = r\omega = l_1\omega_1$; the velocity of the piston is $V_C = l_2\omega_1$. For the arbitrary point B on PC, $V_B = l_3\omega_1$. The lengths l_2 and l_3 are measured on the diagram.

For a circular disc *rolling without slipping*, it may be shown graphically that the point of contact between the disc and the plane or curved surface on which it is rolling is *the instantaneous centre of rotation* for the disc. This was shown analytically in Example 1.5. It must be remembered that the instantaneous centre refers to the velocities and *not* to the accelerations.

3.8 Analytical determination of velocities

For the simple cases of rectilinear or circular motion, it is often possible to find the displacement as a function of time, and thereby the velocity and acceleration by simple differentiation; in more complicated types of motion this is not possible and the problem may then be solved at a *certain instant in time* by graphical means.

The graphical method employs the vector formula (3.2), and the same formula may be used to determine instantaneous values of the velocity *analytically*. The procedure is best explained by an example.

Example 3.3. Figure 3.20 shows a plane four-bar mechanism, consisting of bars pinned together at the ends; the fixed ground OC is counted as one bar. The mechanism is able to move in the plane of the mechanism. Bar OA is rotated by an external torque about O at a *constant* angular velocity $\omega_1 = 12$ rad/s counter-clockwise. Determine the instantaneous velocity of points A and B and the angular velocities of AB and BC at the given configuration.

Solution. A coordinate system (x, y) is introduced at O and position vectors $\mathbf{R}_1$, $\mathbf{R}_2$ and $\mathbf{R}_3$ drawn along the bars as shown. From the given geometry

$$\mathbf{R}_1 = -0{\cdot}075\mathbf{i}+0{\cdot}13\mathbf{j}\text{ m}$$

$$\mathbf{R}_2 = 0{\cdot}486\mathbf{i}+0{\cdot}177\mathbf{j}\text{ m}$$

$$\mathbf{R}_3 = 0{\cdot}1113\mathbf{i}+0{\cdot}306\mathbf{j}\text{ m}$$

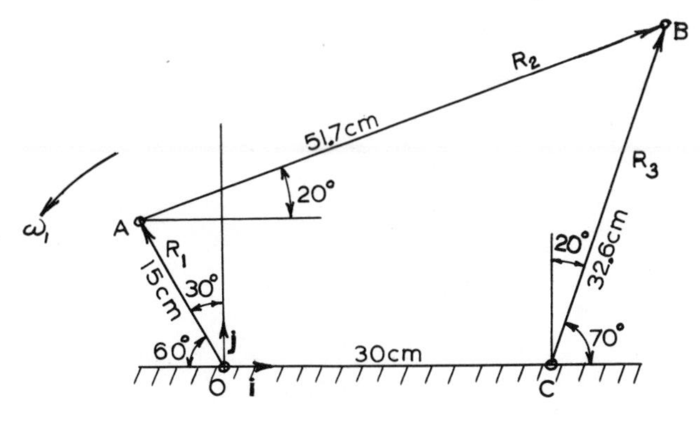

Fig. 3.20

The angular velocities are

$$\boldsymbol{\omega}_1 = 12\mathbf{k}\text{ rad/s}, \boldsymbol{\omega}_2 = \omega_2\mathbf{k}\text{ rad/s and }\boldsymbol{\omega}_3 = \omega_3\mathbf{k}\text{ rad/s}$$

The velocities may now be calculated from formula (3.2):

$$\mathbf{V}_A = \boldsymbol{\omega}_1\times\mathbf{R}_1 = -1{\cdot}56\mathbf{i}-0{\cdot}90\mathbf{j}\text{ m/s} \qquad V_A = 1{\cdot}80\text{ m/s}$$

$$\mathbf{V}_B = \mathbf{V}_A+\mathbf{V}_{BA}$$

$$\mathbf{V}_{BA} = \boldsymbol{\omega}_2\times\mathbf{R}_2 = -0{\cdot}177\omega_2\mathbf{i}+0{\cdot}486\omega_2\mathbf{j}$$

Substituting in $\mathbf{V}_B$ gives

$$\mathbf{V}_B = -(1{\cdot}56+0{\cdot}177\omega_2)\mathbf{i}-(0{\cdot}90-0{\cdot}486\omega_2)\mathbf{j}$$

$\mathbf{V}_B$ may also be found from

$$\mathbf{V}_B = \boldsymbol{\omega}_3\times\mathbf{R}_3 = 0{\cdot}306\omega_3\mathbf{i}+0{\cdot}1113\omega_3\mathbf{j}$$

equating the **i**-terms and the **j**-terms for the two expressions for $\mathbf{V}_B$ gives two equations to determine ω_2 and ω_3; the solution is

$\omega_2 = 3{\cdot}48$ rad/s anti-clockwise, $\omega_3 = 7{\cdot}10$ rad/s. Substituting back gives

$$\mathbf{V}_B = -2{\cdot}17\mathbf{i} + 0{\cdot}791\mathbf{j} \text{ m/s} \qquad V_B = 2{\cdot}31 \text{ m/s}$$
$$\mathbf{V}_{BA} = -0{\cdot}616\mathbf{i} + 1{\cdot}69\mathbf{j} \text{ m/s} \qquad V_{BA} = 1{\cdot}80 \text{ m/s}$$

3.9 Kinematics of a rigid link. Acceleration equation

For the rigid link *PA* in Fig. 3.12, the following equation was established in Section 3.5:

$$\dot{\mathbf{R}}_A = \dot{\mathbf{R}}_P + \boldsymbol{\omega} \times \mathbf{r}$$

Differentiating again, we find

$$\ddot{\mathbf{R}}_A = \ddot{\mathbf{R}}_P + \boldsymbol{\omega} \times \dot{\mathbf{r}} + \dot{\boldsymbol{\omega}} \times \mathbf{r}$$

Since $\mathbf{r}$ is a *constant-length* vector, we have from (3.1) $\dot{\mathbf{r}} = \boldsymbol{\omega} \times \mathbf{r}$, so that

$$\ddot{\mathbf{R}}_A = \ddot{\mathbf{R}}_P + \boldsymbol{\omega} \times (\boldsymbol{\omega} \times \mathbf{r}) + \dot{\boldsymbol{\omega}} \times \mathbf{r}$$

Comparing this expression to the result in Section 3.1, Fig. 3.4, we see that the term $\boldsymbol{\omega} \times (\boldsymbol{\omega} \times \mathbf{r})$ is the normal component of acceleration of A in its relative motion about P, and the term $\dot{\boldsymbol{\omega}} \times \mathbf{r}$ is the tangential component of acceleration of A relative to P.

The magnitudes of the two components for the case of plane motion are $r\omega^2$ and $r\dot{\omega}$ as usual for *circular motion.*

The only relative motion possible for A with respect to P is motion in a circle with P as centre and radius r, and the two relative acceleration components may thus be determined by fixing the pole and considering A in a circular motion about the pole.

Introducing the acceleration of A relative to P as $\mathbf{A}_{AP}$, we have

$$\mathbf{A}_{AP} = \mathbf{A}_{AP}^{\mathrm{N}} + \mathbf{A}_{AP}^{\mathrm{T}} = \boldsymbol{\omega} \times (\boldsymbol{\omega} \times \mathbf{r}) + \dot{\boldsymbol{\omega}} \times \mathbf{r}$$

Since $\ddot{\mathbf{R}}_A = \mathbf{A}_A$ and $\ddot{\mathbf{R}}_P = \mathbf{A}_P$, we have finally

$$\mathbf{A}_A = \mathbf{A}_P + \mathbf{A}_{AP} \tag{3.3}$$

The formula (3.3) may be directly applied to *three-dimensional motion.*

3.10 Graphical determination of acceleration

Consider the mechanism in Fig. 3.15, for which the velocities were determined graphically in paragraph (3.6). The acceleration of C may be determined from formula (3.3):

$$\mathbf{A}_C = \mathbf{A}_P + \boldsymbol{\omega}_1 \times (\boldsymbol{\omega}_1 \times \mathbf{PC}) + \dot{\boldsymbol{\omega}}_1 \times \mathbf{PC} = \mathbf{A}_P + \mathbf{A}_{CP}$$

$$\mathbf{A}_P = \boldsymbol{\omega} \times (\boldsymbol{\omega} \times \mathbf{OP}) + \dot{\boldsymbol{\omega}} \times (\mathbf{OP})$$

In magnitude these two components of $\mathbf{A}_P$ are $r\omega^2$ and $r\dot{\omega}$; the first is directed from P to O and the second is perpendicular to PO in a sense to give the sense of $\dot{\omega}$.

The normal component of $\mathbf{A}_{CP}$ is $\boldsymbol{\omega}_1 \times (\boldsymbol{\omega}_1 \times \mathbf{PC})$; this is of magnitude $l\omega_1^2$ and directed from C to P; the tangential component of $\mathbf{A}_{CP}$ is $\dot{\boldsymbol{\omega}}_1 \times \mathbf{PC}$, which is of unknown magnitude but with line of action perpendicular to $\mathbf{PC}$. We have then four component vectors to determine $\mathbf{A}_C$, and an acceleration diagram may be drawn to scale as shown in Fig. 3.21.

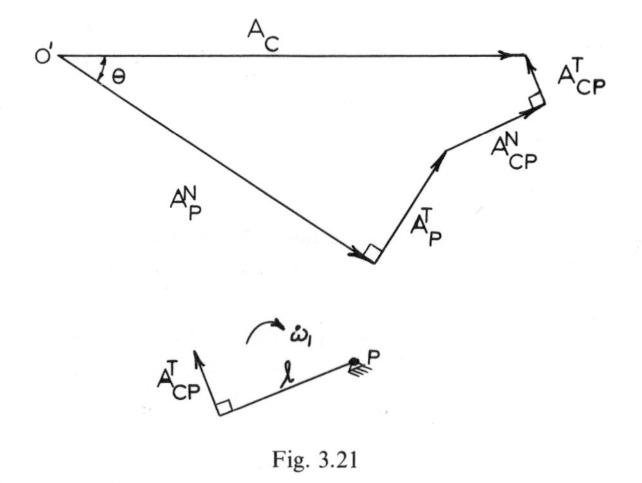

Fig. 3.21

The magnitude and direction of $\mathbf{A}_C$ can now be found from the diagram. The magnitude and direction of the tangential component of $\mathbf{A}_{CP}$ gives the magnitude and direction of $\dot{\omega}_1$ which is $|\mathbf{A}^T_{CP}|/l$ clockwise. Since the order of composition of the vectors is immaterial, a series of acceleration diagrams may be found, all giving the same result for $\mathbf{A}_C$ and $\dot{\omega}_1$.

3.11 Analytical determination of accelerations

The analytical determination of accelerations is best shown by an example.

Example 3.4. For the mechanism in Fig. 3.20, determine the instantaneous accelerations of points A and B, and the instantaneous angular accelerations $\dot{\omega}_2$ and $\dot{\omega}_3$ of AB and BC respectively.

Solution. The linear and angular *velocities* were determined in Example 3.3. The accelerations are determined from formula (3.3). The angular accelerations are $\dot{\boldsymbol{\omega}}_1 = \mathbf{0}$, $\dot{\boldsymbol{\omega}}_2 = \dot{\omega}_2\mathbf{k}$ and $\dot{\boldsymbol{\omega}}_3 = \dot{\omega}_3\mathbf{k}$.

$$\mathbf{A}_A = \boldsymbol{\omega}_1 \times (\boldsymbol{\omega}_1 \times \mathbf{R}_1) + \dot{\boldsymbol{\omega}}_1 \times \mathbf{R}_1 = \boldsymbol{\omega}_1 \times \mathbf{V}_A = 10{\cdot}8\mathbf{i} - 18{\cdot}7\mathbf{j}\ \text{m/s}^2$$

$$\mathbf{A}_A = 21{\cdot}6\ \text{m/s}^2$$

$$\mathbf{A}_B = \mathbf{A}_A + \boldsymbol{\omega}_2 \times (\boldsymbol{\omega}_2 \times \mathbf{R}_2) + \dot{\boldsymbol{\omega}}_2 \times \mathbf{R}_2$$

$$\boldsymbol{\omega}_2 \times (\boldsymbol{\omega}_2 \times \mathbf{R}_2) = \boldsymbol{\omega}_2 \times \mathbf{V}_{BA} = -5{\cdot}88\mathbf{i} - 2{\cdot}14\mathbf{j}\ \text{m/s}^2$$

$$\dot{\boldsymbol{\omega}}_2 \times \mathbf{R}_2 = -0{\cdot}177\dot{\omega}_2\mathbf{i} + 0{\cdot}486\dot{\omega}_2\mathbf{j}\ \text{m/s}^2$$

We have now

$$\mathbf{A}_B = (4{\cdot}92 - 0{\cdot}177\dot{\omega}_2)\mathbf{i} - (20{\cdot}84 - 0{\cdot}486\dot{\omega}_2)\mathbf{j}\ \text{m/s}^2$$

We also have

$$\mathbf{A}_B = \boldsymbol{\omega}_3 \times (\boldsymbol{\omega}_3 \times \mathbf{R}_3) + \dot{\boldsymbol{\omega}}_3 \times \mathbf{R}_3$$

$$\boldsymbol{\omega}_3 \times (\boldsymbol{\omega}_3 \times \mathbf{R}_3) = \boldsymbol{\omega}_3 \times \mathbf{V}_B = -5{\cdot}61\mathbf{i} - 15{\cdot}4\mathbf{j}\ \text{m/s}^2$$

$$\dot{\boldsymbol{\omega}}_3 \times \mathbf{R}_3 = -0{\cdot}306\dot{\omega}_3\mathbf{i} + 0{\cdot}1113\dot{\omega}_3\mathbf{j}\ \text{m/s}^2$$

so that

$$\mathbf{A}_B = -(5{\cdot}61 + 0{\cdot}306\dot{\omega}_3)\mathbf{i} - (15{\cdot}4 - 0{\cdot}1113\dot{\omega}_3)\mathbf{j}\ \text{m/s}^2$$

Equating the **i** and **j** terms from the two expressions for $\mathbf{A}_B$ gives two equations in $\dot{\omega}_2$ and $\dot{\omega}_3$; the solution is

$$\dot{\omega}_2 = 4{\cdot}0\ \text{rad/s}^2\ \text{(anti-clockwise)} \quad \text{and} \quad \dot{\omega}_3 = -32{\cdot}1\ \text{rad/s}^2\ \text{(clockwise)}$$

Substituting back in the previous expression for $\mathbf{A}_B$ gives $\mathbf{A}_B = 4{\cdot}21\mathbf{i} - 18{\cdot}9\mathbf{j}\ \text{m/s}^2$, so that

$$A_B = 19{\cdot}35\ \text{m/s}^2$$

The direction of $\mathbf{A}_B$ is given by the vector expression.

Problems

3.1 Figure 3.22 shows a mechanism consisting of a circular disc with centre C and radius 15·2 cm. The disc rolls on the horizontal plane without slipping; a link AB is attached to the disc at B and rotates freely around B; the other end of the link A is constrained to move in the vertical direction only.

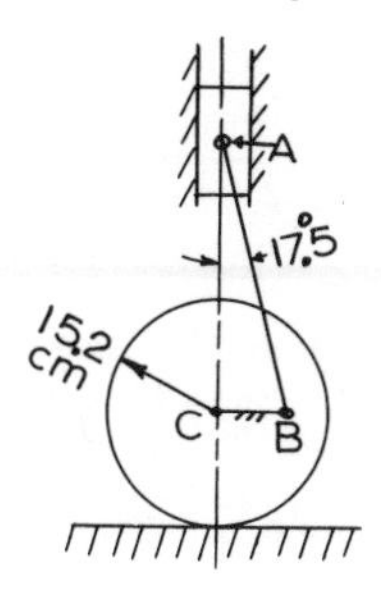

Fig. 3.22

For the position shown, the velocity of C is 0·61 m/s to the right. Find the magnitude and direction of the velocity of A and the angular velocity of AB for the given position, given that $CB = 7{\cdot}62$ cm, $AB = 25{\cdot}4$ cm.

3.2 Figure 3.23 shows an articulated connecting rod used on a two-cylinder V-engine. The crank OA rotates at a constant 2000 rev/min in the clockwise direction. The dimensions are $OA = AC = 5{\cdot}08$ cm, $AB = BC = 15{\cdot}25$ cm, and $CD = 12{\cdot}7$ cm.

Find the velocity of the pistons B and D for the given position in which OA is horizontal.

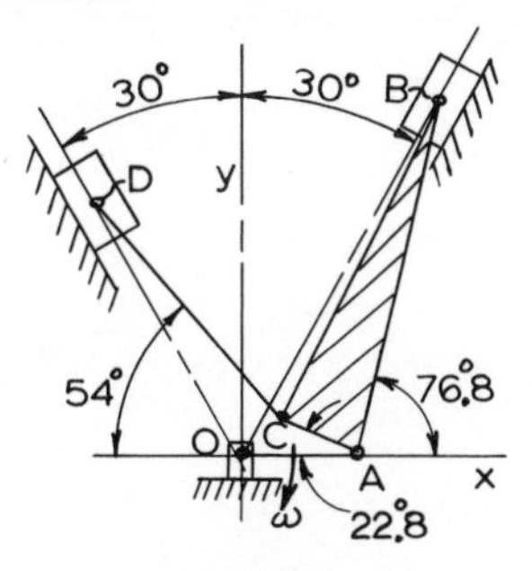

Fig. 3.23

3.3 The four-bar plane-motion mechanism (Fig. 3.24) has a crank angular velocity $\omega_2 = 5$ rad/s. Find the velocity of point C and the angular velocity of link 4, given that $O_2A = 7{\cdot}61$ cm, $O_4C = 10{\cdot}15$ cm, $AB = 25{\cdot}4$ cm, $BC = 17{\cdot}8$ cm, $O_2O_4 = 21{\cdot}6$ cm.

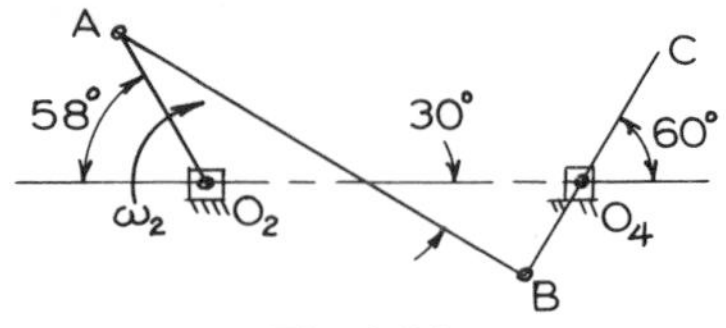

Fig. 3.24

3.4 Figure 3.25 shows an equilateral triangular plate ABC which is moving in the xy-plane. The corners A and B of the triangle are guided by frictionless rollers which move on the x- and y-axes as shown. At the instant when the angle $\theta = 30°$, the velocity of point A is 1·22 m/s to the right and the acceleration of point A is 1·53 m/s^2 to the left.

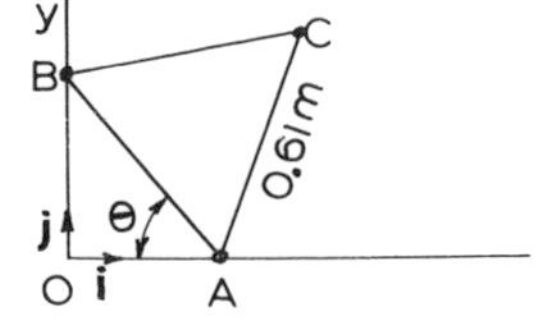

Fig. 3.25

Determine the instantaneous velocity and acceleration of point C and the instantaneous angular velocity and angular acceleration of the triangle when $\theta = 30°$.

3.5 An indicator reducing gear shown in Fig. 3.26 consists of four bars: $AB = 55{\cdot}9$ cm, $BC = ED = 20{\cdot}3$ cm, $DC = EB = 15{\cdot}25$ cm, pinned together to form a pantograph.

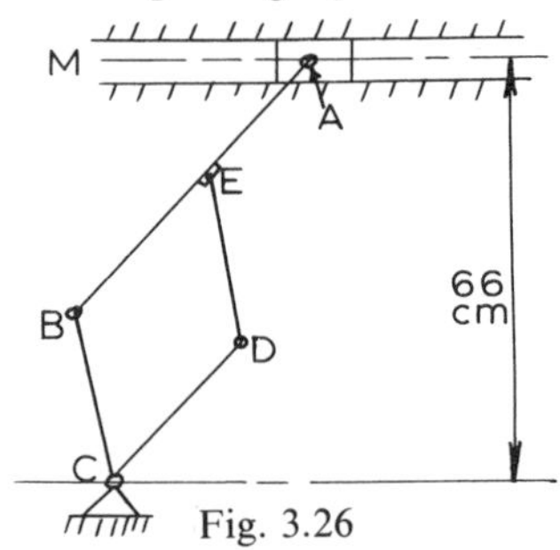

Fig. 3.26

Point A is attached to the crosshead of an engine and C is a fixed point on the engine frame. When angle $MAE = 60°$, A is moving to the left with a velocity of 3·05 m/s and an acceleration of 6·10 m/s^2.

In the given position find the velocity and acceleration of point E and the angular velocity and acceleration of BC.

3.6 In the plane mechanism shown in Fig. 3.27, $CA = 10$ cm and $AB = 20$ cm. The arm CA is rotating about C, and has, at the instant considered, an angular velocity $\omega_2 = 30$ rad/s and angular acceleration $\dot{\omega}_2 = 240$ rad/s^2, directed as shown on the figure. The end B of AB is sliding along a fixed circular guide of 20 cm radius.

Find the velocity and acceleration of point B and the angular velocity and angular acceleration of AB for the position shown.

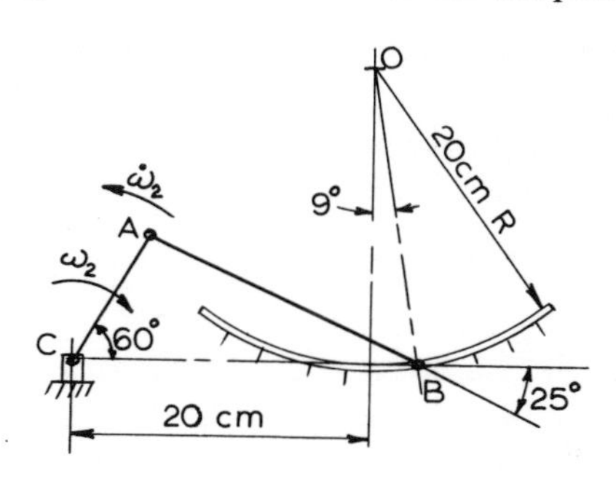

Fig. 3.27

CHAPTER FOUR

Relative Motion

4.1 Relative motion. Rotating coordinate system

To introduce the concept of relative motion, consider the situation shown in Fig. 4.1, which indicates a ship B sailing with velocity $\mathbf{V}_B$ parallel to the shore. A coordinate system (X, Y) has been fixed on the shore with origin O as shown; as usual we consider this a fixed inertial system.

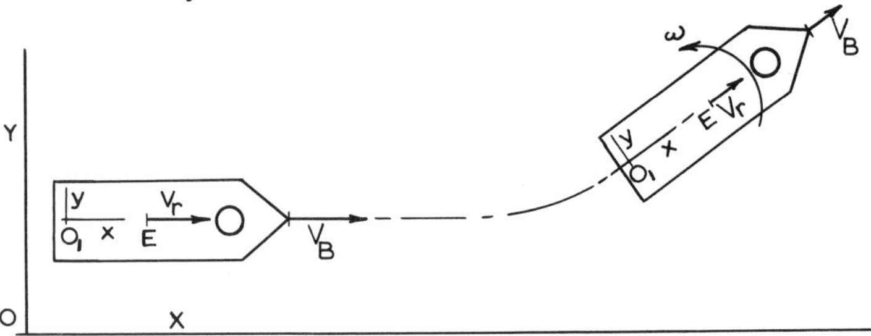

Fig. 4.1

A second coordinate system (x, y) with origin O_1 has been painted on the deck of the ship, and this system is clearly *translating*, with translational velocity $\mathbf{V}_B$, relative to the (X, Y) system; $\mathbf{V}_B$ is called the *base* velocity.

A man E is walking along the deck with a velocity $\mathbf{V}_r$ measured relative to the deck, in the translating (x, y) system; we call $\mathbf{V}_r$ the relative velocity of the man. It is fairly easy to see that the velocity of the man measured in the (X, Y) system is $\mathbf{V}_r + \mathbf{V}_B$, since the man is in rectilinear motion. We call this the *absolute* velocity of the man.

If the man starts to walk on a circle painted on the deck, his motion in the (x, y) system is circular, while his absolute motion in the (X, Y) system becomes much more complicated. Suppose now that the ship turns to port with an angular velocity $\boldsymbol{\omega}$. The coordinate system (x, y) on the deck now becomes a *translating* and *rotating* coordinate system. The relative motion of the man is still circular motion in the (x, y) system, while his absolute motion becomes very complicated.

It will be recalled that Newton's second law is valid only in an inertial system; to write the equation of motion of the man, we shall therefore have to determine his acceleration in the inertial (X, Y) system, the so-called *absolute* acceleration.

The velocity and acceleration of the man in the rotating coordinate system (x, y) are fairly easy to determine in this case of circular motion. The question arises at once of whether it is possible to find a simple relationship between the relative velocity and acceleration measured in the rotating system, and the absolute velocity and acceleration measured in the inertial system (X, Y); it turns out that it is indeed possible to find such a relationship.

To investigate the situation, we shall first consider the relationship between the time derivatives of a vector in a fixed and rotating coordinate system.

4.2 Relationship between the time derivatives of a vector in a fixed and rotating coordinate system

Figure 4.2 shows a body moving in plane motion in an inertial system (X, Y). A coordinate system (x, y) is *fixed in the body* with origin P, and therefore translating and rotating with the body.

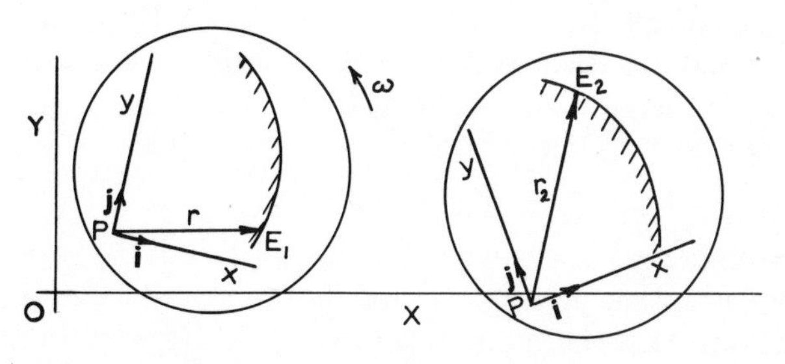

Fig. 4.2

A position vector $\mathbf{r}$ on the body from P to a point E moving on the body is also shown.

The body is shown in two positions, at a time t and at a time $t+\Delta t$. To determine the total change $\Delta\mathbf{r}$ in the vector $\mathbf{r}$, the vectors have been drawn from the same point D as shown in Fig. 4.3.

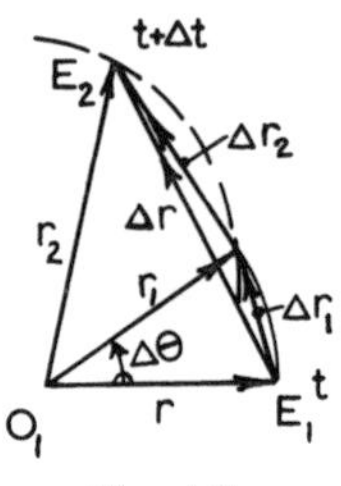

Fig. 4.3

The change $\Delta\mathbf{r}$ in the vector $\mathbf{r}$ may be taken as the sum of two parts $\Delta\mathbf{r}_1$ and $\Delta\mathbf{r}_2$ as shown, so that $\Delta\mathbf{r}=\Delta\mathbf{r}_1+\Delta\mathbf{r}_2$. The vector $\Delta\mathbf{r}_1$ is due to the rotation of the body in the time Δt, so that $|\mathbf{r}|=|\mathbf{r}_1|$, and we have from formula (3.1) for this fixed length vector, that

$$\lim_{\Delta t\to 0}\frac{\Delta\mathbf{r}_1}{\Delta t}=\boldsymbol{\omega}\times\mathbf{r}$$

where $\boldsymbol{\omega}$ is the instantaneous angular velocity of the body.

The vector $\Delta\mathbf{r}_2$ is due to the change in direction and length of the vector $\mathbf{r}$ in the time Δt, as the endpoint of $\mathbf{r}$ moves along the body. $\Delta\mathbf{r}$ is on a chord to the relative path of the point E moving on the body. We have now

$$\lim_{\Delta t\to 0}\frac{\Delta\mathbf{r}_2}{\Delta t}=\left(\frac{d\mathbf{r}_1}{dt}\right)_{\text{rel}}=\left(\frac{d\mathbf{r}}{dt}\right)_{\text{rel}}=\dot{\mathbf{r}}_r$$

where the differentiation is in the rotating system (x, y), or as seen from the body; $\dot{\mathbf{r}}_r$ is along the *tangent* to the path of E across the body, and is the *relative velocity* of E on the body.

We have then the important vector formula:

$$\dot{\mathbf{r}}=\dot{\mathbf{r}}_r+\boldsymbol{\omega}\times\mathbf{r} \tag{4.1}$$

The absolute time derivative $\dot{\mathbf{r}}$ of a vector $\mathbf{r}$ is thus the vector sum of the relative time derivative $\dot{\mathbf{r}}_r$ and the cross product $\boldsymbol{\omega}\times\mathbf{r}$.

If the vector $\mathbf{r}$ is of *constant length*, and *fixed* in the rotating body, it has no change in length and direction in the (x, y) system, and therefore $\dot{\mathbf{r}}_r = \mathbf{0}$, so that $\dot{\mathbf{r}} = \boldsymbol{\omega} \times \mathbf{r}$ as found before, and stated in the formula (3.1).

The formula (4.1) holds for any vector quantity and may also be derived by using the unit vectors $\mathbf{i}$ and $\mathbf{j}$ in the rotating system (x, y). We have then $\mathbf{r} = x\mathbf{i} + y\mathbf{j}$, where x and y are functions of time, and $\mathbf{i}$ and $\mathbf{j}$ are constant length vectors fixed in the rotating body, so that $\dot{\mathbf{i}} = \boldsymbol{\omega} \times \mathbf{i}$ and $\dot{\mathbf{j}} = \boldsymbol{\omega} \times \mathbf{j}$, where the differentiations are absolute.

If we differentiate $\mathbf{r}$ in the rotating (x, y) system, $\mathbf{i}$ and $\mathbf{j}$ are of fixed lengths and directions, and we get $\dot{\mathbf{r}}_r = \dot{x}\mathbf{i} + \dot{y}\mathbf{j}$. Differentiating $\mathbf{r}$ in the absolute (X, Y) system results in $\dot{\mathbf{r}} = \dot{x}\mathbf{i} + x\dot{\mathbf{i}} + \dot{y}\mathbf{j} + y\dot{\mathbf{j}} = (\dot{x}\mathbf{i} + \dot{y}\mathbf{j}) + \boldsymbol{\omega} \times (x\mathbf{i} + y\mathbf{j}) = \dot{\mathbf{r}}_r + \boldsymbol{\omega} \times \mathbf{r}$, as found before.

We may extend this result at once to *three-dimensional motion*, by including a unit vector $\mathbf{k}$ on a z-axis in the rotating system.

4.3 Kinematics of relative motion. Velocity equation

Figure 4.4 shows a body moving in the fixed (X, Y) plane. A coordinate system (x, y) is fixed in the body at a point P. A sliding block with centre A slides in a groove cut in the body; the relative position vector of point A is $\mathbf{r}$ as shown.

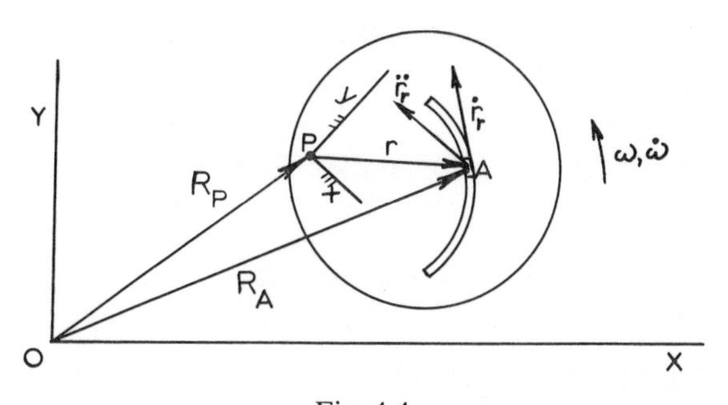

Fig. 4.4

The time derivative of the relative position vector $\mathbf{r}$ in the rotating system is $\dot{\mathbf{r}}_r$ as shown. This is in a direction *along the tangent* to the path of A on the body, and is the sliding velocity of A in the groove, or the velocity of A in the (x, y) system. We call $\dot{\mathbf{r}}_r$ the relative velocity of A with respect to the body or *base*, so that $\dot{\mathbf{r}}_r = \mathbf{V}_r$.

From the figure we have $\mathbf{R}_A = \mathbf{R}_P + \mathbf{r}$; differentiating absolutely with respect to time gives $\dot{\mathbf{R}}_A = \dot{\mathbf{R}}_P + \dot{\mathbf{r}}$. Now from formula (4.1) we have $\dot{\mathbf{r}} = \dot{\mathbf{r}}_r + \boldsymbol{\omega} \times \mathbf{r}$, so that

$$\mathbf{V}_A = [\mathbf{V}_P + \boldsymbol{\omega} \times \mathbf{r}] + \mathbf{V}_r$$

If this formula is used for the special case of point A fixed in the base at a point B, the position vector $\mathbf{r}$ is of constant length and fixed in the body, and therefore $\dot{\mathbf{r}}_r = \mathbf{V}_r = \mathbf{0}$. The velocity of point B is thus $\mathbf{V}_B = [\mathbf{V}_P + \boldsymbol{\omega} \times \mathbf{r}]$. The sum of the first two components of the absolute velocity of the moving point A is therefore equal, at any instant, to the absolute velocity of the point B on the base with which P coincides at that instant.

The formula for $\mathbf{V}_A$ may now be given the simpler form

$$\mathbf{V}_A = \mathbf{V}_B + \mathbf{V}_r \tag{4.2}$$

where $\mathbf{V}_B = \mathbf{V}_P + \boldsymbol{\omega} \times \mathbf{r} = \mathbf{V}_P + \mathbf{V}_{BP}$, in agreement with eq. (3.2). The absolute velocity $\mathbf{V}_A$ of a point moving on a translating and rotating body may thus be found in two steps: first we *fix the point* on the body and determine the velocity of this *base point* $\mathbf{V}_B$, and then to this add geometrically the velocity of the point A in its relative motion across the body $\mathbf{V}_r$, which may be determined by *fixing the body* and considering the point A in its motion across the body.

It will usually be found to be much more convenient to determine these two components of the absolute velocity, than to determine this velocity directly.

From the development of eq. (4.1), it is clear that eq. (4.2) may be applied directly to *three-dimensional motion.*

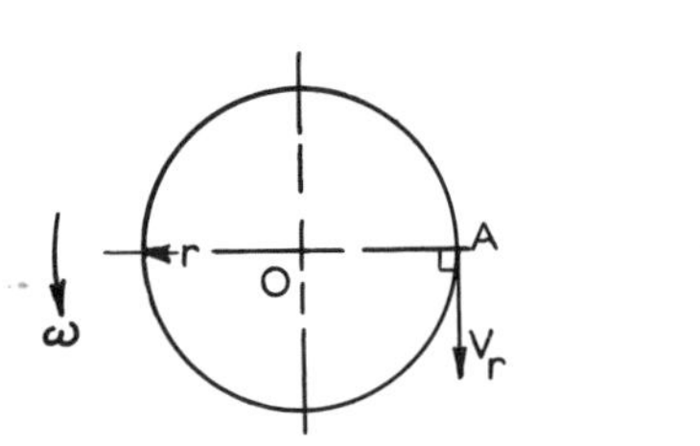

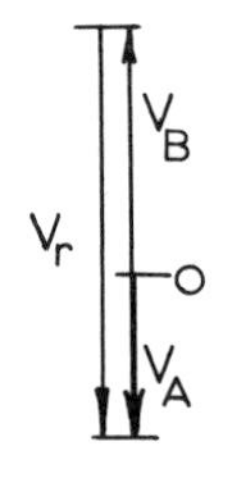

Fig. 4.5

Example 4.1. Figure 4.5 shows a horizontal disc of radius r, which is rotating about a fixed vertical axis through O. The constant angular velocity of the disc is ω as shown. A point A is moving along the rim of the disc, the relative velocity of A being of constant magnitude and equal to $\mathbf{V}_r$. The direction of $\mathbf{V}_r$ is shown on the figure, and $V_r > r\omega$. Determine the absolute velocity of A.

Solution. Equation (4.2) gives $\mathbf{V}_A = \mathbf{V}_B + \mathbf{V}_r$, where $\mathbf{V}_B$ is the absolute velocity of a base point fixed in the disc at the instantaneous position of A, so that $\mathbf{V}_B = r\omega$, acting upwards. The velocity diagram in the figure gives the result

$$\mathbf{V}_A = (V_r - r\omega)$$

which acts downwards, since $V_r > r\omega$.

If $V_r = r\omega$, $\mathbf{V}_A = \mathbf{0}$, so that point A is stationary in space. If $V_r < r\omega$, $\mathbf{V}_A = (r\omega - \mathbf{V}_r)$, acting upwards.

Example 4.2. Figure 4.6 shows a horizontal disc rotating with constant angular velocity ω about a fixed vertical axis through the centre O of the disc. A sliding block with centre A is moving across the disc along a guide fixed on the disc as shown. The relative motion of A is rectilinear motion determined by the coordinate x.

Determine the absolute velocity of A at the position shown.

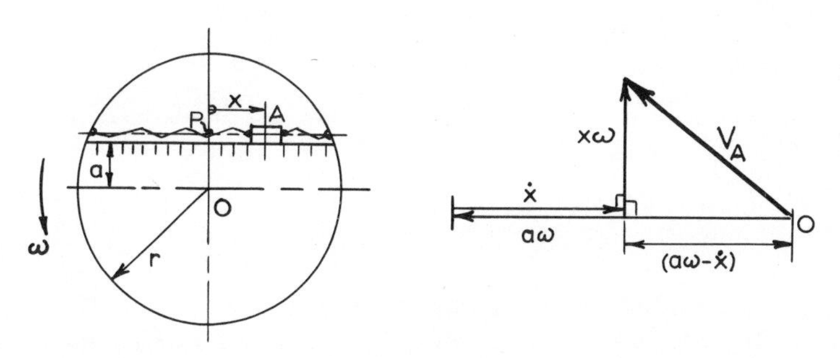

Fig. 4.6

Solution. Taking a base point B in the disc coinciding with the position of A at the instant considered, and a pole P in the disc as shown, we have from eqs. (4.2) and (3.2), that $\mathbf{V}_A = \mathbf{V}_B + \mathbf{V}_r$, and $\mathbf{V}_B = \mathbf{V}_P + \mathbf{V}_{BP}$. Now $\mathbf{V}_P = a\omega$, acting to the left, $\mathbf{V}_{BP} = x\omega$, acting upwards, and $\mathbf{V}_r = \dot{x}$, acting to the right.

The sketch in Fig. 4.6 of the velocity diagram for $\mathbf{V}_A$ gives the solution: $V_A^2 = (a\omega - \dot{x})^2 + x^2\omega^2$. To find the actual magnitude and direction of $\mathbf{V}_A$, it is necessary to know the numerical values of a, ω, x and $\dot{x}$ at the instant considered.

Example 4.3. Figure 4.7 shows a plane mechanism consisting of a radial arm OC rotating with constant angular velocity $\omega = 3$ rad/s about O. At the instant considered OC is inclined at 30° to horizontal. A block A is sliding on OC and is pinned to a block D which is constrained to slide in a vertical slot as shown. By constructing a velocity diagram, determine the absolute velocity of the block D at the instant shown.

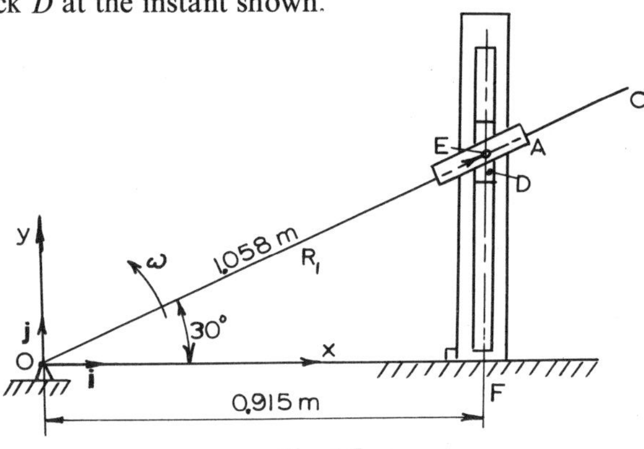

Fig. 4.7

Solution. Taking the centre of the block D as point E, and using the bar OC as a base for E, we consider a point B fixed on OC at the instantaneous position of E as our base point for the motion of E along OC.

Equations (4.2) and (3.2) now state that

$$\mathbf{V}_E = \mathbf{V}_B + \mathbf{V}_r \qquad \text{and} \qquad \mathbf{V}_B = \mathbf{V}_O + \mathbf{V}_{BO}$$

where point O has been taken as pole for the motion of OC. We have $\mathbf{V}_O = \mathbf{0}$, and $\mathbf{V}_{BO} = (OE)\omega\nwarrow$ perpendicular to OE, so that $\mathbf{V}_B = \mathbf{V}_{BO} = 1{\cdot}058 \times 3 = 3{\cdot}174$ m/s at 120°. (See footnote on p. 96) $\mathbf{V}_r$ is the relative velocity of sliding of E along OC, which can only be along the bar OC, so that $\mathbf{V}_r$ acts at 30°.

The velocity diagram is shown in Fig. 4.8, where use has been made of the fact that $\mathbf{V}_E$ is vertical. From the diagram we find $\mathbf{V}_E = 3{\cdot}66$ m/s $\uparrow$, and the velocity of sliding $\mathbf{V}_r = 1{\cdot}83$ m/s at 30°.

The velocity of point E was determined analytically in Example 2.1 as a case of rectilinear motion. The result was $\mathbf{V}_E = l\omega/\cos^2\theta \uparrow$, where l is the length OF, so that $l = 0{\cdot}915$ m, and θ is the angle between OC and the horizontal. $\theta = 30°$.

The result is $V_E = 0{\cdot}915 \times 3/\cos^2 30 = 3{\cdot}66$ m/s, as found above.

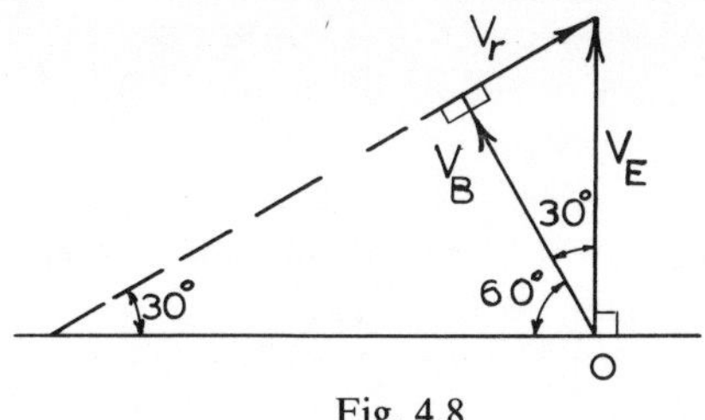

Fig. 4.8

Example 4.4. Determine the velocity of block D in Example 4.3 analytically, by using the basic vector equation (4.2).

Solution. Introducing the position vector $\mathbf{R}_1$ (Fig. 4.7), we find $\mathbf{R}_1 = 0{\cdot}915\mathbf{i} + 0{\cdot}529\mathbf{j}$ m. The angular velocity is $\boldsymbol{\omega} = \omega\mathbf{k} = 3\mathbf{k}$. Equation (4.2) states that $\mathbf{V}_E = \mathbf{V}_B + \mathbf{V}_r$, and we have $\mathbf{V}_B = \boldsymbol{\omega} \times \mathbf{R}_1 = -1{\cdot}587\mathbf{i} + 2{\cdot}745\mathbf{j}$ m/s. The relative velocity $\mathbf{V}_r$ is the velocity of sliding, which can only be along OC, so that $\mathbf{V}_r$ is inclined at 30° to the horizontal at the instant considered. We may then take

$$\mathbf{V}_r = V_r \cos 30\,\mathbf{i} + V_r \cos 60\,\mathbf{j} = 0{\cdot}866 V_r \mathbf{i} + 0{\cdot}50 V_r \mathbf{j}$$

Substituting in (4.2) gives

$$\mathbf{V}_E = \mathbf{V}_B + \mathbf{V}_r = (0{\cdot}866 V_r - 1{\cdot}587)\mathbf{i} + (0{\cdot}50 V_r + 2{\cdot}745)\mathbf{j}$$

However, $\mathbf{V}_E$ is vertical, so that $\mathbf{V}_E = 0\,\mathbf{i} + V_E\,\mathbf{j}$; equating the $\mathbf{i}$ terms gives $0{\cdot}866 V_r - 1{\cdot}587 = 0$, or $V_r = 1{\cdot}83$ m/s at 30°, so that $\mathbf{V}_r = 1{\cdot}587\mathbf{i} + 0{\cdot}915\mathbf{j}$ m/s. Equating the $\mathbf{j}$ terms for V_E, we find $V_E = 0{\cdot}50 V_r + 2{\cdot}745 = 3{\cdot}66$ m/s, and $V_E = 3{\cdot}66\mathbf{j}$ m/s.

4.4 Kinematics of relative motion. Acceleration equation

Consider again the situation shown in Fig. 4.4. The relative velocity of sliding of point A along its path on the base is $\dot{\mathbf{r}}_r$, which is found as the time derivative of the position vector $\mathbf{r}$ of A from a point P fixed on the base. The differentiation of $\mathbf{r}$ is performed in the system (x, y) fixed in the base. The direction of $\dot{\mathbf{r}}_r$ is as shown, along the tangent to the path of A across the base.

A second time derivative of $\mathbf{r}$ in the rotating system (x, y) gives the *relative acceleration* of sliding of A along its path $\ddot{\mathbf{r}}_r$. Since the relative motion of A is curvilinear motion, the relative acceleration $\ddot{\mathbf{r}}_r$ is directed as shown, to give a tangential component, and a normal component directed towards the centre of curvature of the relative path on the base.

From the velocity equation developed in Section 4.3, we have the equation $\dot{\mathbf{R}}_A = \dot{\mathbf{R}}_P + \boldsymbol{\omega} \times \mathbf{r} + \dot{\mathbf{r}}_r$; taking the time derivative of this equation in the fixed inertial system (X, Y), we find

$$\ddot{\mathbf{R}}_A = \ddot{\mathbf{R}}_P + \boldsymbol{\omega} \times \dot{\mathbf{r}} + \dot{\boldsymbol{\omega}} \times \mathbf{r} + d\dot{\mathbf{r}}_r/dt$$

Applying eq. (4.1) to $\dot{\mathbf{r}}$ and $d\dot{\mathbf{r}}_r/dt$, we have

$$\dot{\mathbf{r}} = \dot{\mathbf{r}}_r + \boldsymbol{\omega} \times \mathbf{r} \qquad \text{and} \qquad d\dot{\mathbf{r}}_r/dt = \ddot{\mathbf{r}}_r + \boldsymbol{\omega} \times \dot{\mathbf{r}}_r$$

Substituting this gives

$$\begin{aligned} \ddot{\mathbf{R}}_A &= \ddot{\mathbf{R}}_P + \boldsymbol{\omega} \times (\dot{\mathbf{r}}_r + \boldsymbol{\omega} \times \mathbf{r}) + \dot{\boldsymbol{\omega}} \times \mathbf{r} + (\ddot{\mathbf{r}}_r + \boldsymbol{\omega} \times \dot{\mathbf{r}}_r) \\ &= [\ddot{\mathbf{R}}_P + \boldsymbol{\omega} \times (\boldsymbol{\omega} \times \mathbf{r}) + \dot{\boldsymbol{\omega}} \times \mathbf{r}] + \ddot{\mathbf{r}}_r + 2\boldsymbol{\omega} \times \dot{\mathbf{r}}_r \end{aligned} \tag{4.3}$$

Applying this general formula to a base point B for point A, as explained under the velocity equation, the relative position vector $\mathbf{r}$ for point B is fixed in the base and therefore of constant magnitude and in a fixed direction in the system (x, y), so that $\dot{\mathbf{r}}_r = \mathbf{0}$ and $\ddot{\mathbf{r}}_r = \mathbf{0}$ for point B. We find then that

$$\ddot{\mathbf{R}}_B = [\ddot{\mathbf{R}}_P + \boldsymbol{\omega} \times (\boldsymbol{\omega} \times \mathbf{r}) + \dot{\boldsymbol{\omega}} \times \mathbf{r}]$$

If we introduce the notations

$$\ddot{\mathbf{R}}_B = \mathbf{A}_B,\ \ddot{\mathbf{R}}_P = \mathbf{A}_P$$

$$\mathbf{A}_{BP}^{\mathrm{N}} = \boldsymbol{\omega} \times (\boldsymbol{\omega} \times \mathbf{r}) \qquad \text{and} \qquad \mathbf{A}_{BP}^{\mathrm{T}} = \dot{\boldsymbol{\omega}} \times \mathbf{r}$$

so that $\mathbf{A}_{BP} = \mathbf{A}_{BP}^{\mathrm{N}} + \mathbf{A}_{BP}^{\mathrm{T}}$, we have $\mathbf{A}_B = \mathbf{A}_P + \mathbf{A}_{BP}$, which is the result previously obtained in formula (3.3) for the acceleration of a point fixed in a rigid body.

Substituting $\mathbf{A}_B$ for the square bracket in eq. (4.3), introducing the notation $\mathbf{A}_r$ for the relative acceleration $\ddot{\mathbf{r}}_r$, and the notation $\mathbf{C} = 2\boldsymbol{\omega} \times \dot{\mathbf{r}}_r = 2\boldsymbol{\omega} \times \mathbf{V}_r$, the formula (4.3) simplifies to the following:

$$\mathbf{A}_A = \mathbf{A}_B + \mathbf{A}_r + \mathbf{C} \tag{4.4}$$

Equation (4.4) indicates that the *absolute* acceleration $\mathbf{A}_A$ of a point A moving on a translating and rotating rigid body, may be found as the sum of three acceleration vectors $\mathbf{A}_B$, $\mathbf{A}_r$ and $\mathbf{C}$.

The acceleration $\mathbf{A}_B$ of the base point is determined by *fixing the moving point* on the base, at the instant and position considered, and determining its acceleration as a point fixed in a rigid body by the formula $\mathbf{A}_B=\mathbf{A}_P+\mathbf{A}_{BP}$ in the usual way. The relative acceleration $\mathbf{A}_r$ is determined by *fixing the body*, and considering the point in its motion across the body.

The acceleration vector $\mathbf{C}=2\boldsymbol{\omega}\times\mathbf{V}_r$ has been given the name *Coriolis acceleration* after the French mathematician who first published a discussion of this *component* of the absolute acceleration. It is convenient to have a name for this component, just as we use the names 'normal' and 'tangential' components of acceleration. The direction and magnitude of the Coriolis acceleration is completely determined by the vector cross product $2\boldsymbol{\omega}\times\mathbf{V}_r$, where $\boldsymbol{\omega}$ is the instantaneous angular velocity of the base, and $\mathbf{V}_r$ is the instantaneous relative velocity of the point in question.

From the expression for Coriolis acceleration, it may be seen that this component of the absolute acceleration vanishes if the body is translating, so that $\boldsymbol{\omega}=\mathbf{0}$, or if the point is fixed in the body, so that $\mathbf{V}_r=\mathbf{0}$. We also have $\mathbf{C}=\mathbf{0}$ if $\boldsymbol{\omega}$ and $\mathbf{V}_r$ are parallel; this happens, for instance, in the case of a point moving along a generator of a right circular cylinder which is rotating about its fixed geometrical axis. In this case $\mathbf{V}_r$ and $\boldsymbol{\omega}$ are always parallel, so that $2\boldsymbol{\omega}\times\mathbf{V}_r=\mathbf{0}$.

Since the formula $\dot{\mathbf{r}}=\dot{\mathbf{r}}_r+\boldsymbol{\omega}\times\mathbf{r}$ is valid in three-dimensional motion, the formulae (4.3) and (4.4) may be extended at once to *three-dimensional motion*.

In general it will be found that the formula (4.4) is more convenient to use than the formula (4.3), since the sum of the first three terms of (4.3) may usually be determined directly as the base-point acceleration.

Example 4.5. Determine the *absolute* acceleration of point A in Example 4.1, Fig. 4.5.

Solution. Equations (4.4) and (3.3) states that $\mathbf{A}_A=\mathbf{A}_B+\mathbf{A}_r+\mathbf{C}$ and $\mathbf{A}_B=\mathbf{A}_O+\mathbf{A}_{BO}$, where the centre O of the disc has been taken as a pole for the motion of the base point B. We have here $\mathbf{A}_O=\mathbf{0}$, and $\mathbf{A}_B=\mathbf{A}_{BO}=r\omega^2\leftarrow$.

The relative acceleration $\mathbf{A}_r$ is found by fixing the base and considering the relative motion of A. This is circular motion with

radius r and centre O, and with constant speed V_r, so that $\mathbf{A}_r = V_r^2/r \leftarrow$. The Coriolis acceleration is $\mathbf{C} = 2\boldsymbol{\omega} \times \mathbf{V}_r$, so that $\mathbf{C} = 2\omega V_r \rightarrow$.
We have now

$$A_A = r\omega^2 + \frac{V_r^2}{r} - 2\omega V_r = r\left(\frac{V_r}{r} - \omega\right)^2 > 0$$

so that $\mathbf{A}_A$ is always directed towards O.

The absolute motion of A is circular motion, and the centripetal acceleration is directed towards the centre, independently of the direction of rotation.

If $V_r = r\omega$, we find $\mathbf{A}_A = \mathbf{0}$; the point A is then stationary in space.

Example 4.6. Determine the absolute acceleration of point A in Example 4.2, Fig. 4.6, if the disc has an angular acceleration $\dot{\omega}$ in the same direction as ω.

Solution. We have from eqs. (4.2) and (3.2),

$$\mathbf{A}_A = \mathbf{A}_B + \mathbf{A}_r + \mathbf{C} \qquad \text{and} \qquad \mathbf{A}_B = \mathbf{A}_P + \mathbf{A}_{BP}$$

The chosen pole P is in circular motion with radius a, so that its acceleration is $\mathbf{A}_P = a\omega^2 \downarrow + a\dot{\omega} \leftarrow$.

The relative motion of the base point B about P is circular motion with radius x, so that $\mathbf{A}_{BP} = x\omega^2 \leftarrow + x\dot{\omega} \uparrow$. The relative acceleration $\mathbf{A}_r = \ddot{x} \rightarrow$. The Coriolis acceleration $\mathbf{C} = 2\boldsymbol{\omega} \times \mathbf{V}_r$, so that $\mathbf{C} = 2\omega\dot{x} \uparrow$.

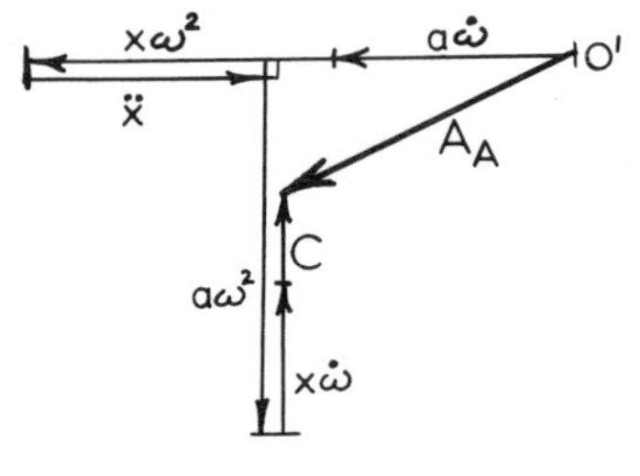

Fig. 4.9

A *sketch* of the acceleration diagram is shown in Fig. 4.9. To determine the magnitude and direction of $\mathbf{A}_A$, the numerical values of a, x, ω, $\dot{x}$, $\dot{\omega}$ and $\ddot{x}$ must, of course, be known at the instant considered.

Example 4.7. Determine the acceleration of the block D at the instant shown in Fig. 4.7, Example 4.3, by constructing an acceleration diagram.

Solution. Equations (4.2) and (3.2) are $\mathbf{A}_E = \mathbf{A}_B + \mathbf{A}_r + \mathbf{C}$ and $\mathbf{A}_B = \mathbf{A}_O + \mathbf{A}_{BO}$. We know that $\mathbf{A}_E$ is vertical$\updownarrow$ and $\mathbf{A}_O = \mathbf{0}$. Point B is in a circular motion about O, so that $\mathbf{A}_B = (OB)\omega^2 = 1{\cdot}058 \times 3^2 = 9{\cdot}52$ m/s^2 at 210°. The relative acceleration $\mathbf{A}_r$ is the acceleration of sliding along the base OC, so that the direction of $\mathbf{A}_r$ is at 30° or 210°. The Coriolis acceleration is $\mathbf{C} = 2\boldsymbol{\omega} \times \mathbf{V}_r$, with magnitude $C = 2\omega V_r$, where V_r is 1·83 m/s at 30°, so that $\mathbf{C} = 2 \times 3 \times 1{\cdot}83 = 10{\cdot}98$ m/s^2 at 120°.

The acceleration diagram may now be constructed as shown in Fig. 4.10. From the diagram we find

$$\mathbf{A}_E = 12{\cdot}7 \text{ m/s}^2 \uparrow$$

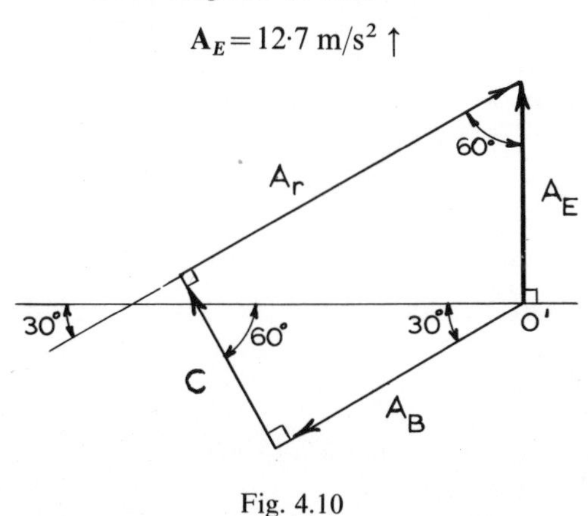

Fig. 4.10

The acceleration of point E was determined analytically in Example 2.1, as a case of rectilinear motion. The result was $\mathbf{A}_E = 2l\omega^2 \sin\theta/\cos^3\theta \uparrow$, where $l = OF = 0{\cdot}915$ m and $\theta = 30°$. The result is

$$\mathbf{A}_E = 2 \times 0{\cdot}915 \times 3^2 \sin 30/\cos^3 30 = 12{\cdot}66 \text{ m/s}^2 \uparrow$$

as found before.

Example 4.8. Determine the acceleration of the block D in Example 4.7 analytically by using the basic vector equation (4.2).

Solution. From Example 4.4, we have $\mathbf{R}_1 = 0{\cdot}915\mathbf{i} + 0{\cdot}529\mathbf{j}$, $\boldsymbol{\omega} = 3\mathbf{k}$,

$\mathbf{V}_B = -1{\cdot}587\mathbf{i} + 2{\cdot}745\mathbf{j}$ and $\mathbf{V}_r = 1{\cdot}587\mathbf{i} + 0{\cdot}915\mathbf{j}$. Equation (4.2) is $\mathbf{A}_E = \mathbf{A}_B + \mathbf{A}_r + \mathbf{C}$. The acceleration

$$\mathbf{A}_B = \omega \times (\omega \times \mathbf{R}_1) = \omega \times \mathbf{V}_B = -8{\cdot}235\mathbf{i} - 4{\cdot}761\mathbf{j}$$

The relative acceleration $\mathbf{A}_r$ is along OC, so that

$$\mathbf{A}_r = A_r \cos 30\mathbf{i} + A_r \cos 60\,\mathbf{j} = 0{\cdot}866A_r\mathbf{i} + 0{\cdot}50A_r\mathbf{j}$$

The Coriolis acceleration is

$$\mathbf{C} = 2\omega \times \mathbf{V}_r = -5{\cdot}49\mathbf{i} + 9{\cdot}522\mathbf{j}$$

Combining the components gives

$$\mathbf{A}_E = \mathbf{A}_B + \mathbf{A}_r + \mathbf{C} = (0{\cdot}866A_r - 13{\cdot}725)\mathbf{i} + (0{\cdot}50A_r + 4{\cdot}761)\mathbf{j}$$

$\mathbf{A}_E$ is in the vertical direction: $\mathbf{A}_E = 0\mathbf{i} + A_E\mathbf{j}$, and equating the items in the two expressions for $\mathbf{A}_E$ gives $0{\cdot}866A_r - 13{\cdot}725 = 0$, or $A_r = 15{\cdot}86$ m/s^2. Equating the $\mathbf{j}$ terms for $\mathbf{A}_E$ results in $\mathbf{A}_E = 0{\cdot}50A_r + 4{\cdot}761 = 12{\cdot}69$ m/s^2 $\uparrow$.

4.5 Equation of relative motion. Relative equilibrium. D'Alembert's principle in relative motion

Newton's second law for the motion of a particle, or the centre of mass of a rigid body is $\mathbf{F} = M\ddot{\mathbf{r}}_C = M\mathbf{A}$, where $\mathbf{F}$ is the resultant acting force, M the total mass, and $\mathbf{A}$ the *absolute* acceleration of the particle or the centre of mass of a rigid body. The law is valid only in an inertial coordinate system.

If the motion is referred to a rotating coordinate system fixed in a rigid body, we have from eq. (4.4), that $\mathbf{A} = \mathbf{A}_B + \mathbf{A}_r + \mathbf{C}$, where $\mathbf{A}_B$ is the base-point acceleration, $\mathbf{A}_r$ the relative acceleration and $\mathbf{C} = 2\omega \times \mathbf{V}_r$ is the Coriolis acceleration. Newton's second law may now be stated as

$$\mathbf{F} = M\mathbf{A} = M(\mathbf{A}_B + \mathbf{A}_r + \mathbf{C})$$

If we want to write the law in a rotating coordinate system in the usual form where mass times acceleration is equal to the resultant force, we find, from the above equation, that

$$M\mathbf{A}_r = \mathbf{F} + (-M\mathbf{A}_B) + (-M\mathbf{C}) \qquad (4.5)$$

Equation (4.5) is a *vector equation of relative motion*. The terms $(-M\mathbf{A}_B)$ and $(-M\mathbf{C})$ are 'inertia forces', as defined before under D'Alembert's principle, due to the base-point acceleration $\mathbf{A}_B$ and the Coriolis acceleration $\mathbf{C}$; these inertia forces must be added vectorially to the resultant external force, $\mathbf{F}$, in order to write Newton's second law for a particle in a rotating, that is, *non-inertial*, coordinate system.

The vector equation (4.5) may be projected on any axis to give the scalar equation of relative motion along that axis.

If the base is translating with constant velocity, we have $\mathbf{A}_B=\mathbf{0}$ and $\boldsymbol{\omega}=\mathbf{0}$, so that $\mathbf{C}=\mathbf{0}$; the vector equation of motion (4.5) then takes the form $M\mathbf{A}_r=\mathbf{F}$, which is the usual form for Newton's second law, since in this case $\mathbf{A}_r=\mathbf{A}$. A coordinate system which is translating with constant velocity is therefore *also an inertial system*, in which Newton's law holds without corrections. Dynamics experiments performed in a laboratory that is moving in translation with constant velocity then gives the same results as the experiments performed in a stationary laboratory.

For *relative equilibrium* of a particle, we must have the particle in a fixed position on the base. This is the case if $\mathbf{V}_r=\mathbf{0}$ and $\mathbf{A}_r=\mathbf{0}$, in which case we also have $\mathbf{C}=2\boldsymbol{\omega}\times\mathbf{V}_r=\mathbf{0}$, and from (4.5) we find

$$\mathbf{F}+(-M\mathbf{A}_B) = \mathbf{0} \tag{4.6}$$

Equation (4.6) is the *equation for relative equilibrium*, and it states that the *resultant external force* $\mathbf{F}$ must balance the inertia force $(-M\mathbf{A}_B)$ due to the base-point acceleration, in order to have relative equilibrium.

Equation (4.5) may be stated in the following form:

$$\mathbf{F}+(-M\mathbf{A}_B)+(-M\mathbf{A}_r)+(-M\mathbf{C}) = \mathbf{0} \tag{4.7}$$

Equation (4.7) is the *D'Alembert equation* for *dynamic equilibrium* in relative motion. Besides the *resultant external force* $\mathbf{F}$, the inertia forces due to the base-point acceleration, relative acceleration and Coriolis acceleration must be added as acting forces to the system of forces, to create dynamic equilibrium.

Example 4.9. Figure 4.11 shows a horizontal disc rotating at a constant angular velocity ω about a fixed vertical axis through its centre O. A particle of mass m is moving without friction in a groove cut along a diameter of the disc.

Determine the pressure from the side of the groove on the particle, the equation of relative motion and the solution of this equation. Assume that the particle is initially placed in the groove a distance x_0 from the centre O of the disc and with zero relative velocity.

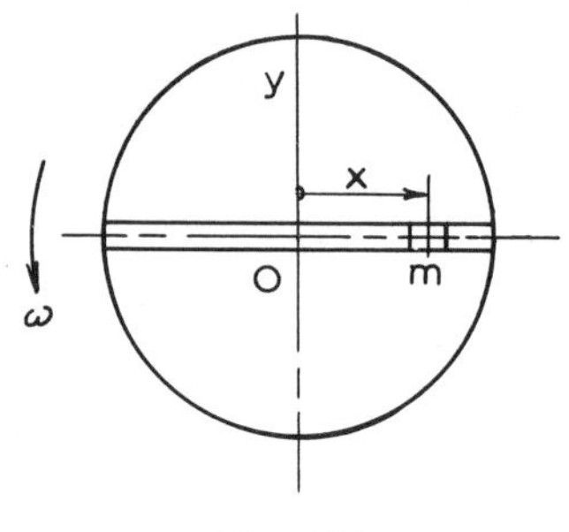

Fig. 4.11

Solution. The vector equation of relative motion (4.5) is

$$m\mathbf{A}_r = \mathbf{F} + (-m\mathbf{A}_B) + (-m\mathbf{C})$$

Fixing a coordinate system (x, y) in the disc as shown, we have $\mathbf{A}_r = \ddot{x} \rightarrow$, $\mathbf{A}_B = x\omega^2 \leftarrow$, $\mathbf{V}_r = \dot{x} \rightarrow$ and $\mathbf{C} = 2\boldsymbol{\omega} \times \mathbf{V}_r$, so that $\mathbf{C} = 2\omega\dot{x} \uparrow$.

The gravity force is balanced by a vertical pressure from the bottom of the groove, and with no friction these forces have no effect on the motion. The resultant external force $\mathbf{F}$ is the normal pressure from the side of the groove, so that $\mathbf{F}$ is in the y-direction.

Projecting all vectors on the y-axis, we find $0 = F + (-2m\omega\dot{x})$, so that $F = 2m\omega\dot{x}$. This is the force *on* the particle, and it is in the positive y-direction. The pressure *on* the side of the groove is equal and opposite to the force on the particle.

The motion of the particle then results in a retarding torque $T = Fx = 2mx\omega\dot{x}$; this torque must be supplied by the external driving torque to keep ω constant. Projecting on the x-axis gives the *equation of relative motion*:

$$m\ddot{x} = mx\omega^2 \qquad \text{or} \qquad \ddot{x} - x\omega^2 = 0$$

The characteristic equation is $R^2 - \omega^2 = 0$, with roots $R = \pm\omega$, the general solution to the equation of motion is then $x = C_1 \cosh \omega t + C_2 \sinh \omega t$, where C_1 and C_2 are arbitrary constants; this may also be shown by direct substitution in the differential equation.

The starting conditions at $t=0$ are $x=x_0$ and $\dot{x}=0$; substituting $t=0$ and $x=x_0$ in the general solution gives $x_0=C_1$, so that

$$x = x_0 \cosh \omega t + C_2 \sinh \omega t$$

and

$$\dot{x} = x_0\omega \sinh \omega t + C_2\omega \cosh \omega t$$

Substitution of $t=0$ and $\dot{x}=0$ gives $0=C_2\omega$, or $C_2=0$; the solution is then $x=x_0 \cosh \omega t$, $\dot{x}=x_0\omega \sinh \omega t$, and $\ddot{x}=x_0\omega^2 \cosh \omega t=x\omega^2$.

The force on the side of the groove is now $F=2m\omega\dot{x}$, as a function of the relative velocity; $F=2mx_0\omega^2 \sinh \omega t$ as a function of time, and $F=2m\omega^2\sqrt{(x^2-x_0^2)}$ as a function of the relative position.

Example 4.10. The straight pipe in Fig. 4.12 discharges 265 litres of water per minute from each end, while rotating with constant angular velocity of 600 rev/min. Determine the magnitude of the driving torque to maintain this uniform rotation.

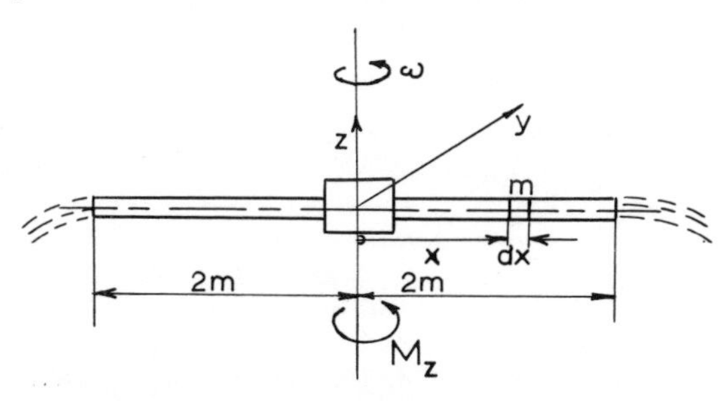

Fig. 4.12

The weight of water may be taken as 9·78 N/l, and 1 litre $=10^{-3}$ m^3. Wind resistance and friction may be neglected.

Solution. The vector equation of relative motion (4.5) is $m\mathbf{A}_r=\mathbf{F}+(-m\mathbf{A}_B)+(-m\mathbf{C})$, for a water particle m as shown in the figure.

Introducing a coordinate system (x, y, z) fixed in the pipe, we have $A_r=\ddot{x}=0$, since the relative velocity $\dot{x}$ is constant. The acceleration of the base point is $\mathbf{A}_B=x\omega^2 \leftarrow$, and the Coriolis acceleration is $\mathbf{C}=2\boldsymbol{\omega}\times\mathbf{V}_r$, so that $\mathbf{C}=2\omega\dot{x} \nearrow$ in the positive y-direction.

Taking the resultant force in the y-direction as Ry, we find by projecting on the y-axis that $0=Ry-2m\omega\dot{x}$, so that $Ry=2m\omega\dot{x}$.

This is in the positive y-direction on the water, and so in the negative y-direction on the pipe.

For a length dx of the pipe, the retarding torque is $dM_z = 2m\omega\dot{x}\,x$. Taking the cross-sectional area of the pipe as A, and the mass per unit volume of water as ρ, we have $m = A\rho\;dx$, and $dM_z = 2A\rho\omega\dot{x}x\;dx$; for the two pipes the result is $M_z = 2\int_0^l 2A\rho\omega\dot{x}x\;dx$, or since $\dot{x}$ and ω are constant,

$$M_z = 4A\rho\omega\dot{x}\int_0^l x\;dx = 2A\rho\omega\dot{x}l^2 \text{ Nm}$$

We have now

$$A\dot{x} = 0{\cdot}265 \text{ m}^3/\text{min} \qquad \dot{x} = \frac{0{\cdot}265}{A}\text{ m/min} = \frac{0{\cdot}004\,41}{A}\text{ m/s}$$

$$M_z = 2A\rho\omega\frac{0{\cdot}004\,41}{A}l^2 = 0{\cdot}008\,82\rho\omega l^2 \text{ Nm}$$

$$l = 2\text{ m} \qquad \omega = 2\pi\frac{600}{60} = 62{\cdot}8 \text{ rad/s} \qquad \rho = \frac{9{\cdot}78\times 10^3}{9{\cdot}81}$$

$$= 997 \text{ kg/m}^3$$

Introducing this gives $M_z = 0{\cdot}008\,82\times 997\times 62{\cdot}8\times 4 = 2206$ Nm.

The force distribution on each pipe is a uniformly distributed load, since $\dot{x}$ and ω are constants.

Example 4.11. Figure 4.13 shows a horizontal circular disc of radius r which rotates with constant angular velocity ω about a fixed vertical axis through its centre O. A shallow circular well of radius $r/2$ is cut in the top face of the disc as shown. A particle of mass m slides without friction in the well.

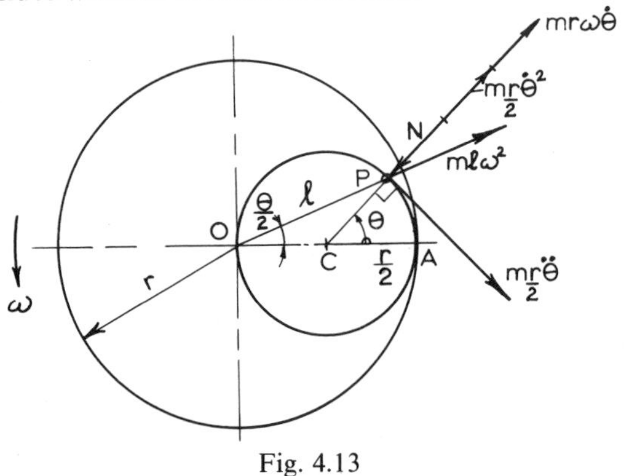

Fig. 4.13

Show that the position at A is stable relative equilibrium for the particle. Apply D'Alembert's principle to create dynamic equilibrium, and develop the equation of relative motion of the particle. Determine the frequency for small relative displacements θ.

Solution. Without friction the particle can only stay along the rim of the well, since the necessary force for circular motion cannot be established otherwise. The gravity force is balanced by a normal pressure from the bottom of the well. Without friction these forces have no effect on the motion in the horizontal plane, and the resultant external force is the normal pressure N from the rim of the well as shown in the figure.

The position of the particle in the well is determined by the *relative* coordinate θ.

For relative equilibrium we must have $\mathbf{N}+(-m\mathbf{A}_B)=\mathbf{0}$, according to (4.6). The base acceleration has magnitude $A_B=l\omega^2$, directed from P towards O, the inertia force is then $ml\omega^2$ directed as shown; clearly $\mathbf{N}$ and the inertia force have a resultant $\mathbf{R}$, which moves the particle back towards position A; this position is then a position of *stable* relative equilibrium with $\mathbf{R}=\mathbf{0}$.

The position at O is a position of *unstable* relative equilibrium, since a small motion away from O results in a force $\mathbf{R}$ which drives the particle towards A.

For dynamic equilibrium, we have (4.7):

$$\mathbf{N}+(-m\mathbf{A}_B)+(-m\mathbf{A}_r)+(-m\mathbf{C}) = \mathbf{0}$$

The inertia force due to base acceleration has already been introduced. The relative acceleration $\mathbf{A}_r$ has the two components $\frac{1}{2}r\dot{\theta}^2$ and $\frac{1}{2}r\ddot{\theta}$; the inertia forces due to these components are shown in the figure. The Coriolis acceleration is $\mathbf{C}=2\boldsymbol{\omega}\times\mathbf{V}_r$; $V_r=\frac{1}{2}r\dot{\theta}$ and $C=2\omega(r/2)\dot{\theta}=r\omega\dot{\theta}$; the inertia force due to the Coriolis acceleration is then $mr\omega\dot{\theta}$ directed as shown.

We have now a system of forces in a plane in equilibrium, and we may take a moment equation about any point in the plane. Taking moments about point C, we find

$$\left(m\frac{r}{2}\ddot{\theta}\right)\frac{r}{2}+\left(ml\omega^2\sin\frac{\theta}{2}\right)\frac{r}{2} = 0 \qquad \text{or} \qquad \ddot{\theta}+\frac{2l}{r}\omega^2\sin\frac{\theta}{2} = 0$$

This is the equation of relative motion of the particle. From the geometry of the figure we find that $l/2=(r/2)\cos(\theta/2)$. Introducing this gives the equation of motion the form $\ddot{\theta}+\omega^2 \sin\theta=0$, which is a non-linear equation of vibratory motion.

If we take θ as a small angle, we have $\sin\theta \simeq \theta$, and $\ddot{\theta}+\omega^2\theta=0$, which indicates simple harmonic motion with $f=\omega/2\pi$ cycles/s. For one rotation of the disc, the particle performs one complete vibration about point A.

4.6 Effect of the Earth's motion on the relative motion on the Earth's surface

Newton's laws are claimed to be valid only in a *primary inertial system*, that is a coordinate system with axes in fixed directions in space, stationary or moving in a parallel translation with constant speed.

A coordinate system *fixed on the earth's surface* partakes in the earth's revolution about its axis, and in its motion about the sun, so it is *not* an inertial system.

To investigate the effect of the earth's motion, we shall use the formula for acceleration (4.3):

$$\mathbf{A}_A = \ddot{\mathbf{R}}_A = [\ddot{\mathbf{R}}_P+\boldsymbol{\omega}\times(\boldsymbol{\omega}\times\mathbf{r})+\dot{\boldsymbol{\omega}}\times\mathbf{r}]+\ddot{\mathbf{r}}_r+2\boldsymbol{\omega}\times\dot{\mathbf{r}}_r=\mathbf{A}_B+\mathbf{A}_r+\mathbf{C}$$

where $\mathbf{A}_A$ is the *absolute* acceleration of a point A, that is the acceleration in an inertial system.

Consider the situation shown in Fig. 4.14. The figure shows the earth in its orbit about the sun. A coordinate system (X, Y, Z) with origin at the centre of the sun and fixed directions of axes may be taken as a primary inertial system for all practical purposes, so that we can consider the centre S of the sun as fixed in space.

The centre C of the earth moves in an elliptical orbit about the sun, with the sun at one focal point. The orbit is, however, very nearly circular with the sun in the centre. The mean distance from the sun to the earth is $R_C=150\times10^6$ km.

The orbit of the earth is in a plane called the ecliptic, and the earth's axis of rotation CN in inclined at about 23·5° to the normal to the ecliptic.

A coordinate system (x, y, z) is introduced *fixed* on the earth's surface with origin P as shown. Motion on or near the earth's surface is measured in this system; for convenience the z-axis has been taken parallel to the axis of rotation CN.

The earth's orbital speed may be taken as practically constant, and may be determined by the fact that the earth moves around its orbit once in about 365 days; the length of the orbit is $2\pi R_C = 2\pi \times 150 \times 10^6$ km, and the orbital speed is

$$V = \frac{2\pi \times 150 \times 10^6}{365 \times 24 \times 3600} = 29{\cdot}8 \text{ km/s} = 29{\cdot}8 \times 10^3 \text{ m/s}$$

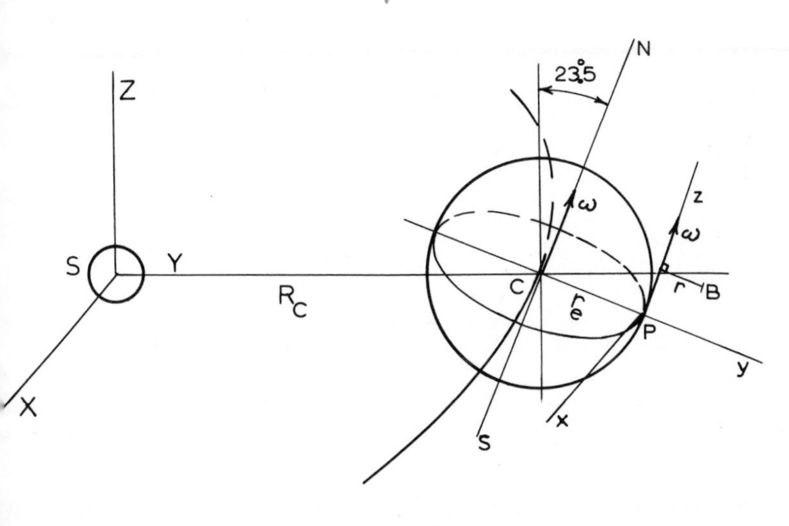

Fig. 4.14

Since V may be taken as constant, the acceleration of C is thus the normal component of acceleration only, and is

$$A_C = V^2/R_C = \frac{29{\cdot}8^2 \times 10^6}{150 \times 10^9} = 0{\cdot}005\ 92 \text{ m/s}^2$$

directed from C towards S.

The earth rotates about its axis once in about 24 hours, which gives the angular velocity $\omega = 7{\cdot}27 \times 10^{-5}$ rad/s; a more accurate value is $\omega = 7{\cdot}29 \times 10^{-5}$ rad/s. For all practical purposes this angular velocity may be taken as constant in magnitude and in a fixed direction, so that $\dot{\boldsymbol{\omega}} = \mathbf{0}$.

The radius of the earth at the equator is $r_e = 6378$ km $= 6{\cdot}378 \times 10^6$ m.

The acceleration of the origin P of the system (x, y, z) is $\mathbf{A}_P = \mathbf{A}_C + \mathbf{A}_{PC}$; since $\dot{\boldsymbol{\omega}} = \mathbf{0}$, the magnitude of $\mathbf{A}_{PC}$ is $A_{PC} = r_e \omega^2$, directed from P towards C, and $A_{PC} = 6{\cdot}378 \times 10^6 \times 7{\cdot}29^2 \times 10^{-10} = 0{\cdot}0338$ m/s^2. The *magnitude* of $\mathbf{A}_P$ is found from this to be $A_P = 0{\cdot}0392$ m/s^2.

For a base point B fixed in the system (x, y, z), the acceleration is $\mathbf{A}_B = \mathbf{A}_P + \mathbf{A}_{BP}$; since $\dot{\boldsymbol{\omega}} = \mathbf{0}$, we have the magnitude $A_{BP} = r\omega^2$, where r is the distance from B to the z-axis; even for a distance of $r = 100$ m, this magnitude is only $A_{BP} = 100 \times 7{\cdot}29^2 \times 10^{-10} = 5{\cdot}31 \times 10^{-7}$, which is very small compared to A_P. For any base point near to P, we find then that $A_B \simeq A_P = 0{\cdot}0392$ m/s^2.

The Coriolis acceleration is $\mathbf{C} = 2\boldsymbol{\omega} \times \dot{\mathbf{r}}_r$; suppose we have a relative velocity as high as $\dot{\mathbf{r}}_r = 1000$ km/h $= 278$ m/s in the positive x-direction; the Coriolis acceleration is then in the positive y-direction and of magnitude $C = 2\omega V_r = 2 \times 7{\cdot}29 \times 10^{-5} \times 278 = 0{\cdot}041$ m/s^2.

If the relative acceleration $\mathbf{A}_r$ is of the order of g, we have $A_r = g = 9{\cdot}81$ m/s^2; in that case the base-point acceleration $\mathbf{A}_B$ and Coriolis acceleration $\mathbf{C}$ are both of magnitude about 0·4% of A_r. We have now $\mathbf{A}_A = \mathbf{A}_B + \mathbf{A}_r + \mathbf{C} \simeq \mathbf{A}_r$, so that the acceleration $\mathbf{A}_r$ measured in the system (x, y, z) is practically identical to the absolute acceleration $\mathbf{A}_A$ measured in an inertial system. A coordinate system fixed on the surface of the earth may now be considered as a *secondary inertial system* in which Newton's law may be applied without correction in the great majority of cases. As mentioned in Chapter 2, the main exceptions to this are problems with large time intervals of interest, orbital motion, space travel, long-range rocket flight and certain large-scale problems of fluid and air flows across the earth's surface.

In the case of orbital motion, a coordinate system with origin at the earth centre and translating with the earth, but with the earth rotating in the system, was applied. From the figures used in this discussion, about 85% of the base-point acceleration was due to the earth's rotation, while about 15% was due to the earth's motion in its orbit; a coordinate system translating with the earth is, therefore, closer to an inertial system than a system fixed on the earth's surface.

Problems

4.1 In Fig. 4.15 a horizontal turntable rotates about O with constant angular velocity ω. A man starting from O at $t=0$ walks with constant relative speed $V=3$ m/s outwards along the rotating radial line OA. Find the velocity and acceleration of the man at the instant when $t=2\pi/\omega$. Assume $\omega=1$ rad/s.

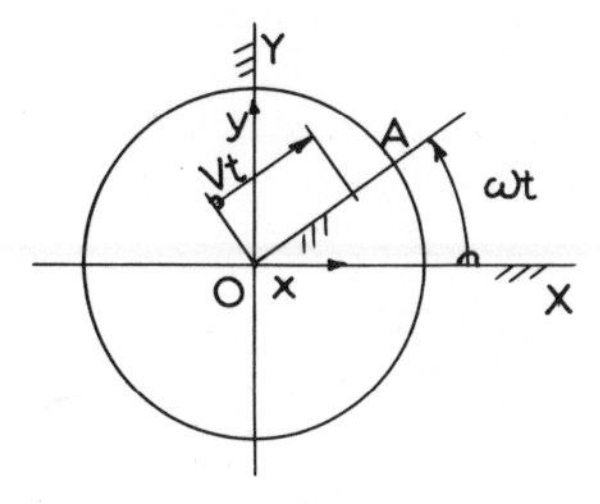

Fig. 4.15

4.2 A point P_1 (Fig. 4.16) moves with constant $V_r=3{\cdot}66$ m/s along AB while the disc rotates with constant $\omega=62{\cdot}9$ rad/s. Find V and A of P_1 for the position shown, if $r=0{\cdot}305$ m and $h=0{\cdot}153$ m. Assume that P_1 at this instant is moving towards B.

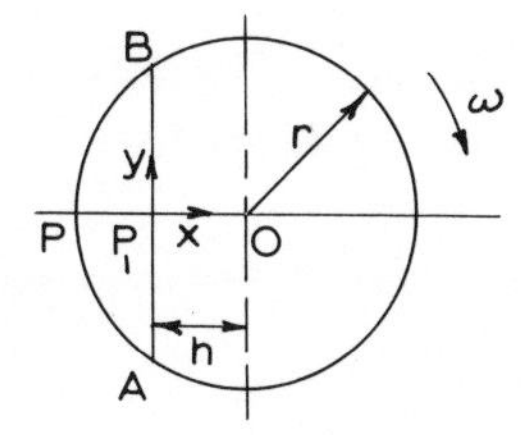

Fig. 4.16

4.3 A gun fires a shell with muzzle velocity $V_r=3000$ m/s ($A_r\simeq 0$), while rotating in a horizontal plane about a vertical axis (Fig. 4.17) at $\omega=0{\cdot}1$ rev/s. If the gun barrel is of length $l=15$ m, find the velocity and acceleration of the shell just as it leaves the barrel.

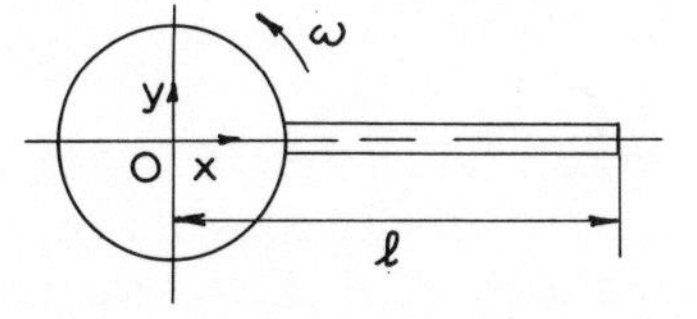

Fig. 4.17

4.4 The crank 2 in the plane-motion mechanism (Fig. 4.18) has a constant speed of 60 rev/min counter-clockwise. Find the velocity of point B, given the dimensions $O_2O_4=30{\cdot}5$ cm, $O_2A=17{\cdot}8$ cm, $O_4B=71{\cdot}2$ cm, $O_4C=14{\cdot}1$ cm.

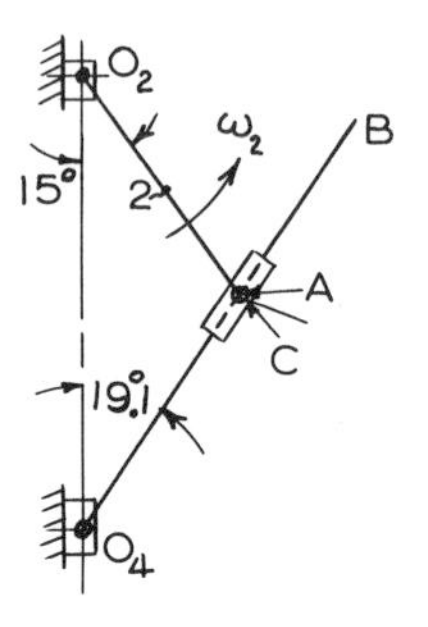

Fig. 4.18

4.5 In the plane mechanism shown in Fig. 4.19, the crank arm AB of length 5·08 cm rotates about A in the clockwise direction with a constant angular velocity $\omega=12$ rad/s.

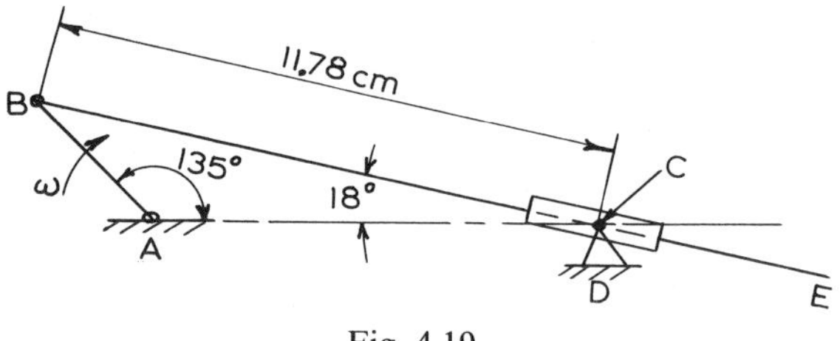

Fig. 4.19

The link BE is pinned to AB at B and slides freely in the guide at D; BE is of length 17·8 cm. The guide at D rotates freely in the plane about D, and AD is of length 7·62 cm.

At the position where angle $BAD=135°$, find the velocity and acceleration of point C on BE at the centre of the guide D, and the velocity and acceleration of point E.

Find also the angular velocity and angular acceleration of BE.

4.6 In Fig. 4.20, a small ball of weight W rides in a groove cut in the top face of a disc rotating with constant angular velocity about its vertical geometric axis. Find the magnitude of the absolute velocity with which it leaves the disc, if $h=r/2$ and friction is neglected. Given $W=1$ N, $r=2$ m and $n=300$ rev/min, find the reaction of the ball on the side of the groove when it reaches the rim. When $t=0$, $\dot{x}_1=0$ and $x_1=r/100$.

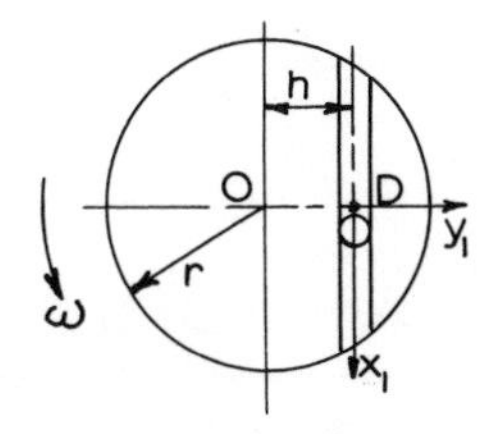

Fig. 4.20

4.7 A mass m is free to vibrate in the direction of two restraining springs K and is guided in frictionless supports. The assembly is attached to a horizontal disc (Fig. 4.21) which rotates with uniform angular velocity ω. Find the frequency of vibrations of m relative to the disc.

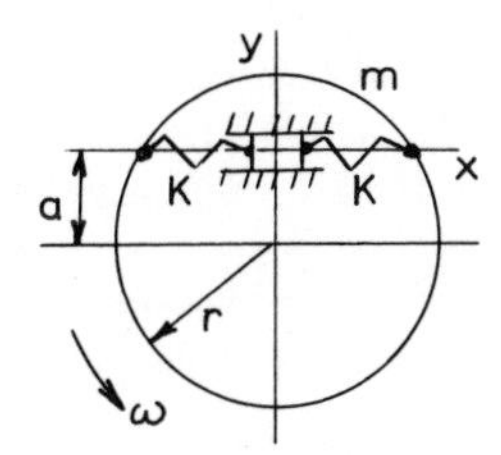

Fig. 4.21

CHAPTER FIVE

Moment of Momentum of a Rigid Body

5.1 Motion of the centre of mass of a rigid body

Newton's second law for a rigid body is of the form $\mathbf{F} = M\ddot{\mathbf{r}}_C$, with the corresponding three scalar equations of motion. The law deals only with the motion of the centre of mass, and nothing is stated about the possible rotation of the body.

Consider, as an example on the motion of the centre of mass, a thin disc rolling in a vertical plane down an incline *without slipping* as shown in Fig. 5.1. The coefficient of friction between the disc and the plane is μ; air resistance may be neglected for speeds encountered in this problem. The forces acting are the gravity force Mg through C vertically downwards, and a reaction at the point of contact, which has been resolved into components F and N for convenience; N is the *normal pressure* and F is the *friction force*.

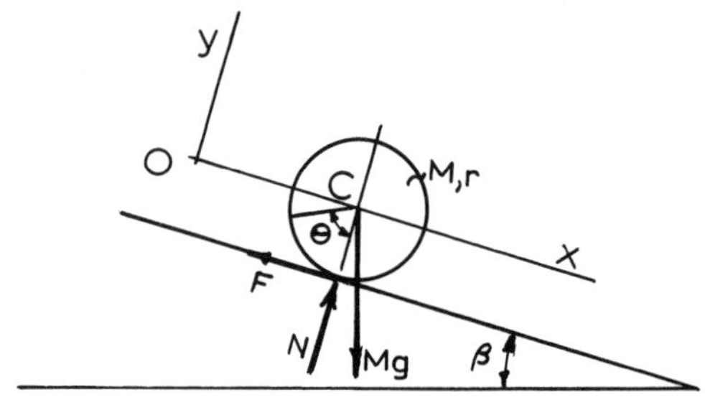

Fig. 5.1

The *fixed* coordinate system (x, y) (assumed inertial) is introduced in *the simplest possible manner*. The motion of the centre of mass is now determined by Newton's second law, by considering a particle of mass M accumulated at C as shown in Fig. 5.2.

$$F_x = M\ddot{x}_C \quad \text{gives} \quad Mg \sin\beta - F = M\ddot{x}$$

$$F_y = M\ddot{y}_C \quad \text{gives} \quad N - Mg\cos\beta = M\ddot{y} = 0 \quad \text{or}$$

$$N = Mg\cos\beta$$

$$F_z = M\ddot{z}_C = 0$$

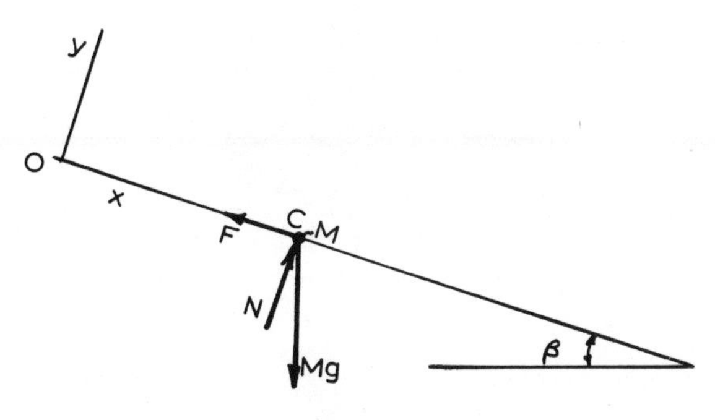

Fig. 5.2

Since the friction force F is unknown (it is only equal to μN when there is slipping, otherwise $F<\mu N$), the problem cannot be solved since we have two unknowns F and $\ddot{x}$; introducing the *equation of constraint* $x=r\theta+x_0$(for no slip), or $\ddot{x}=r\ddot{\theta}=r\dot{\omega}$, gives an additional equation, but also introduces an extra unknown $\dot{\omega}$. It is fairly obvious that to solve the problem completely, the rotation $\boldsymbol{\omega}$ of the disc must be taken into account. We shall use the concept of *moment of momentum*, to be discussed in the next paragraph, to derive equations of motion for a rigid body that involve $\boldsymbol{\omega}$ and $\dot{\boldsymbol{\omega}}$.

The moment of momentum was introduced for a single particle in Chapter 2.

5.2 Moment of momentum of a rigid body

The concept of the *moment of a force or a vector* was widely used in statics and found to be of great importance, since the moment determined the 'turning action' of a force.

The moment of any vector $\mathbf{F}$ about a *point* P is defined as the vector cross product $\mathbf{M}_P=\mathbf{r}\times\mathbf{F}$, where $\mathbf{r}$ is a vector from P to *any point* on the line of action of **F**; the scalar magnitude of the component of $\mathbf{M}_P$ along *any line* through P is defined as the *moment of* **F** *about that line.*

To investigate the moments of forces acting on a rigid body in a dynamics case, consider the particle m in a rigid body (Fig. 5.3) rotating and translating under the action of forces not shown.

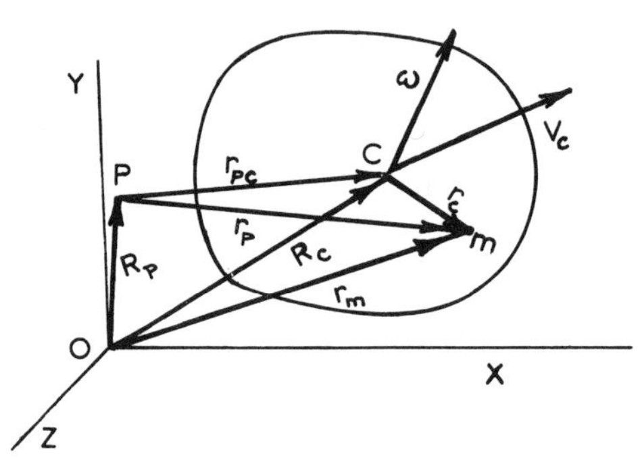

Fig. 5.3

Newton's second law gives $\mathbf{F}_m = \mathbf{F}_{em} + \mathbf{F}_{im} = m\ddot{\mathbf{r}}_m$, where $\mathbf{F}_m$ is the resultant force on m, and $\mathbf{F}_{em}$ is the resultant *external* force, while $\mathbf{F}_{im}$ is the resultant *internal* force. The moment of $\mathbf{F}_m$ about point P is defined by

$$\mathbf{r}_P \times \mathbf{F}_m = \mathbf{r}_P \times (\mathbf{F}_{em} + \mathbf{F}_{im}) = \mathbf{r}_P \times \mathbf{F}_{em} + \mathbf{r}_P \times \mathbf{F}_{im} = \mathbf{r}_P \times m\ddot{\mathbf{r}}_m$$

summing this type of equations for all particles gives

$$\sum \mathbf{r}_P \times \mathbf{F}_{em} + \sum \mathbf{r}_P \times \mathbf{F}_{im} = \sum \mathbf{r}_P \times m\ddot{\mathbf{r}}_m$$

the first term in this expression is the moment $\mathbf{M}_P$ in P of all *external* forces acting on the body. Therefore $\mathbf{M}_P = \sum \mathbf{r}_P \times \mathbf{F}_{em}$; the internal forces occur in equal and opposite pairs and therefore cancel in the moment calculation. Therefore

$$\sum \mathbf{r}_P \times \mathbf{F}_{im} = \mathbf{0}$$

Therefore

$$\mathbf{M}_P = \sum \mathbf{r}_P \times m\ddot{\mathbf{r}}_m = \frac{d}{dt}\left(\sum \mathbf{r}_P \times m\dot{\mathbf{r}}_m\right) - \sum \dot{\mathbf{r}}_P \times m\dot{\mathbf{r}}_m$$

Introducing $\mathbf{r}_m = \mathbf{R}_P + \mathbf{r}_P$, hence $\dot{\mathbf{r}}_m = \dot{\mathbf{R}}_P + \dot{\mathbf{r}}_P$, in the second term gives

$$\sum \dot{\mathbf{r}}_P \times m\dot{\mathbf{r}}_m = \sum \dot{\mathbf{r}}_P \times m(\dot{\mathbf{R}}_P + \dot{\mathbf{r}}_P) = \sum \dot{\mathbf{r}}_P \times m\dot{\mathbf{R}}_P = -\dot{\mathbf{R}}_P \times (\sum m\dot{\mathbf{r}}_P)$$

so that

$$\mathbf{M}_P = \frac{d}{dt}(\sum \mathbf{r}_P \times m\dot{\mathbf{r}}_m) + \dot{\mathbf{R}}_P \times (\sum m\dot{\mathbf{r}}_P)$$

The vector $m\dot{\mathbf{r}}_m$ is the absolute linear *momentum vector* of the particle m. The vector $\mathbf{r}_P \times m\dot{\mathbf{r}}_m$ is the moment of the momentum vector about P as introduced in Chapter 2.

We define now the *moment of momentum* $\mathbf{H}_P$ of a rigid body about a point P as the sum of the moment of momentum vectors about P of all particles in the body, so that

$$\mathbf{H}_P = \sum \mathbf{r}_P \times m\dot{\mathbf{r}}_m \tag{5.1}$$

$\mathbf{H}_P$ is sometimes called the *angular momentum*, but this name should be used only when the body is rotating.

Introducing (5.1) in the expression for $\mathbf{M}_P$ gives

$$\mathbf{M}_P = \dot{\mathbf{H}}_P + \dot{\mathbf{R}}_P \times (\sum m\dot{\mathbf{r}}_P)$$

From Fig. 5.3 it may be seen that $\mathbf{r}_P = \mathbf{r}_m - \mathbf{R}_P = \mathbf{R}_C + \mathbf{r}_C - \mathbf{R}_P$, so that

$$\dot{\mathbf{r}}_P = \dot{\mathbf{R}}_C + \dot{\mathbf{r}}_C - \dot{\mathbf{R}}_P$$

Substituting this in the second term in the expression for $\mathbf{M}_P$ gives

$$\begin{aligned}\dot{\mathbf{R}}_P \times (\sum m\dot{\mathbf{r}}_P) &= \dot{\mathbf{R}}_P \times (\sum m\dot{\mathbf{R}}_C + \sum m\dot{\mathbf{r}}_C - \sum m\dot{\mathbf{R}}_P)\\ &= \dot{\mathbf{R}}_P \times \dot{\mathbf{R}}_C \sum m + \dot{\mathbf{R}}_P \times \sum m\dot{\mathbf{r}}_C - \dot{\mathbf{R}}_P \times \dot{\mathbf{R}}_P \sum m\end{aligned}$$

Now $\sum m\dot{\mathbf{r}}_C = \mathbf{0}$ according to (2.9), and $\dot{\mathbf{R}}_P \times \dot{\mathbf{R}}_P = \mathbf{0}$; introducing $\sum m = M =$ total mass of the body, we get $\mathbf{M}_P = \dot{\mathbf{H}}_P + \dot{\mathbf{R}}_P \times \dot{\mathbf{R}}_C M$, which simplifies to $\mathbf{M}_P = \dot{\mathbf{H}}_P$ in the following cases:

1. $\dot{\mathbf{R}}_P = \mathbf{0}$ (*P is a fixed point*).
2. $\mathbf{R}_P = \mathbf{R}_C$ (P is *the centre of mass*).
3. $\dot{\mathbf{R}}_C = \mathbf{0}$ (The centre of mass is fixed; P may then be any point).
4. $\dot{\mathbf{R}}_P$ is parallel to $\dot{\mathbf{R}}_C$ (The velocity of P is parallel to the velocity of the centre of mass).

Since differentiation is involved in $\mathbf{M}_P = \dot{\mathbf{H}}_P$, these conditions must be valid during a finite element of time, it is not sufficient that a condition is valid instantaneously.

The expression $\mathbf{M}_P = \sum \mathbf{r}_P \times m\ddot{\mathbf{r}}_m$ may also be transformed in a different manner. From Fig. 5.3 we have $\mathbf{r}_m = \mathbf{R}_P + \mathbf{r}_P$, or $\ddot{\mathbf{r}}_m = \ddot{\mathbf{R}}_P + \ddot{\mathbf{r}}_P$, so that

$$\mathbf{M}_P = \sum \mathbf{r}_P \times m\ddot{\mathbf{R}}_P + \sum \mathbf{r}_P \times m\ddot{\mathbf{r}}_P = (\sum m\mathbf{r}_P) \times \ddot{\mathbf{R}}_P + \sum \mathbf{r}_P \times m\ddot{\mathbf{r}}_P$$

Introducing $\mathbf{r}_P = \mathbf{r}_{PC} + \mathbf{r}_C$ in the first term gives $\sum m\mathbf{r}_P = \sum m\mathbf{r}_{PC} + \sum m\mathbf{r}_C = \sum m\mathbf{r}_{PC}$, since $\sum m\mathbf{r}_C = \mathbf{0}$ according to (2.8). Now

$$\sum m\mathbf{r}_P = \sum m\mathbf{r}_{PC} = (\sum m)\mathbf{r}_{PC} = M\mathbf{r}_{PC}$$

The second term in $\mathbf{M}_P$ is

$$\sum \mathbf{r}_P \times m\ddot{\mathbf{r}}_P = \frac{d}{dt}(\sum \mathbf{r}_P \times m\dot{\mathbf{r}}_P) - \sum \dot{\mathbf{r}}_P \times m\dot{\mathbf{r}}_P$$

$$= \frac{d}{dt}(\sum \mathbf{r}_P \times m\dot{\mathbf{r}}_P) = \dot{\mathbf{H}}'_P$$

where $\mathbf{H}'_P = \sum \mathbf{r}_P \times m\dot{\mathbf{r}}_P$ is the *relative moment of momentum* of the body about P, that is the moment of momentum about P determined by using the relative velocities of the mass particles with respect to point P.

We have now $\mathbf{M}_P = M\mathbf{r}_{PC} \times \ddot{\mathbf{R}}_P + \dot{\mathbf{H}}'_P$, which reduces to the simple form $\mathbf{M}_P = \dot{\mathbf{H}}'_P$ in the following cases:

1. $\mathbf{r}_{PC} = \mathbf{0}$ (P is the *centre of mass*).
2. $\ddot{\mathbf{R}}_P = \mathbf{0}$, so that $\dot{\mathbf{R}}_P$ is of constant magnitude and in a fixed direction (P is moving at a constant speed in a fixed direction).
3. $\mathbf{r}_{PC}$ is parallel to $\ddot{\mathbf{R}}_P$ (P is accelerating towards or away from the centre of mass).
4. $\dot{\mathbf{R}}_P = \mathbf{0}$ and $\ddot{\mathbf{R}}_P = \mathbf{0}$ (P is a *fixed point*).

It follows from this that $\dot{\mathbf{H}}_P = \dot{\mathbf{H}}'_P$ for a fixed point, which is obvious in any case, since the velocities relative to a fixed point are also the absolute velocities.

It also follows that $\dot{\mathbf{H}}_C = \dot{\mathbf{H}}'_C$; this may also be seen from the expression $\mathbf{H}_C = \sum \mathbf{r}_C \times m\dot{\mathbf{r}}_m$. Introducing $\mathbf{r}_m = \mathbf{R}_C + \mathbf{r}_C$, or $\dot{\mathbf{r}}_m = \dot{\mathbf{R}}_C + \dot{\mathbf{r}}_C$, we have

$$\mathbf{H}_C = \sum \mathbf{r}_C \times m\dot{\mathbf{R}}_C + \sum \mathbf{r}_C \times m\dot{\mathbf{r}}_C = (\sum m\mathbf{r}_C) \times \dot{\mathbf{R}}_C + \sum \mathbf{r}_C \times m\dot{\mathbf{r}}_C$$

$$= \sum \mathbf{r}_C \times m\dot{\mathbf{r}}_C = \mathbf{H}'_C$$

In calculating the moment of momentum in the centre of mass, the absolute *or* relative velocities may be used.

Of the various points for which we have $\mathbf{M}=\dot{\mathbf{H}}$, we shall essentially use only a *fixed point* or the *centre of mass*; the other special points are usually not available, or too difficult to locate in any particular problem.

In the case of translation, any point of the body has a velocity parallel to the velocity of the centre of mass, and $\mathbf{M}=\dot{\mathbf{H}}$ may then be taken about any point in the body; in general we shall have to use Newton's law for the centre of mass and usually we shall also use the centre of mass as moment centre.

In a few cases *the instantaneous centre of rotation* may be used as a moment centre with $\mathbf{M}=\dot{\mathbf{H}}'$; these are the cases where the instantaneous centre is *accelerating* towards the centre of mass. Two examples of this are shown in Fig. 5.4(a) and (b). Fig. 5.4(a) shows a wheel or cylinder *rolling without slipping*. The instantaneous centre is the contact point P, and it may be shown that P is always accelerating towards the geometrical centre C. If this point is *also* the *centre of mass*, we may write $\mathbf{M}=\dot{\mathbf{H}}_P'$; since the velocities with respect to P are also absolute velocities, we have $\dot{\mathbf{H}}'=\dot{\mathbf{H}}$ in this case.

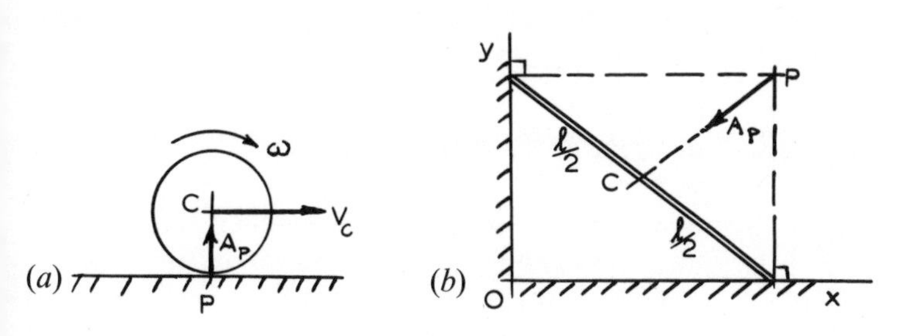

Fig. 5.4

A second example is shown in Fig. 5.4(b), which shows a bar with ends sliding along the x- and y-axes. Again it may be shown that the instantaneous centre P is accelerating towards the middle point C of the bar; if C is also the centre of mass, we may take $\mathbf{M}_P=\dot{\mathbf{H}}_P$, where $\dot{\mathbf{H}}_P=\dot{\mathbf{H}}_P'$ in this case.

In these examples, the equation $\mathbf{M}=\dot{\mathbf{H}}$ fails if the centre of mass is *not* in the geometrical centre shown.

5.3 Transfer formula for the moment of momentum

In certain impact problems in plane motion, it is convenient to have a transfer formula relating the moment of momentum about any point P to the moment of momentum about the centre of mass; such a formula may be developed as follows.

The moment of momentum about point P (Fig. 5.3) is $\mathbf{H}_P = \sum \mathbf{r}_P \times m\dot{\mathbf{r}}_m$. From the figure we have $\mathbf{r}_m = \mathbf{R}_C + \mathbf{r}_C$, or $\dot{\mathbf{r}}_m = \dot{\mathbf{R}}_C + \dot{\mathbf{r}}_C$, and also $\mathbf{r}_P = \mathbf{r}_{PC} + \mathbf{r}_C$; substituting this gives

$$\begin{aligned} \mathbf{H}_P &= \sum (\mathbf{r}_{PC} + \mathbf{r}_C) \times m(\dot{\mathbf{R}}_C + \dot{\mathbf{r}}_C) \\ &= \sum \mathbf{r}_{PC} \times m\dot{\mathbf{R}}_C + \sum \mathbf{r}_{PC} \times m\dot{\mathbf{r}}_C + \sum \mathbf{r}_C \times m\dot{\mathbf{R}}_C + \sum \mathbf{r}_C \times m\dot{\mathbf{r}}_C \end{aligned}$$

Taking vectors that are the same for all particles at any instant outside the summation signs, we find

$$\mathbf{H}_P = \mathbf{r}_{PC} \times \dot{\mathbf{R}}_C \sum m + \mathbf{r}_{PC} \times (\sum m\dot{\mathbf{r}}_C) + (\sum m\mathbf{r}_C) \times \dot{\mathbf{R}}_C + \sum \mathbf{r}_C \times m\dot{\mathbf{r}}_C$$

We have $\sum m\mathbf{r}_C = \mathbf{0}$ and $\sum m\dot{\mathbf{r}}_C = \mathbf{0}$ from (2.8) and (2.9); writing $\dot{\mathbf{R}}_C = \mathbf{V}_C$, $\sum m = M$ and $\mathbf{H}'_C = \sum \mathbf{r}_C \times m\dot{\mathbf{r}}_C$ $(= \mathbf{H}_C)$, we find

$$\mathbf{H}_P = \mathbf{r}_{PC} \times M\mathbf{V}_C + \mathbf{H}_C \tag{5.2}$$

The first term in (5.2) is the moment of momentum about P of a particle of mass M, where M is the total mass of the body, located in the centre of mass and moving with the velocity of the centre of mass. The second term is the absolute *or* relative moment of momentum of the body about the centre of mass.

Example 5.1. Figure 5.5(a) shows a massless circular rim of radius r which rolls *without sliding* on a horizontal plane under the action of a force P. At the instant considered the distance $OS = x_C$, and the angular velocity is ω. Four point masses, each of mass m, are attached to the rim as shown.

Determine the moment of momentum vector about the following points: (1) the centre of mass C, (2) the fixed point O, (3) the point of contact S. Establish the equations of motion of the rim, and determine the acceleration $\ddot{x}_C$ of the centre of mass and the angular acceleration $\dot{\omega}$, as functions of the external force P.

Solution. Since S is the instantaneous centre of rotation, we may establish the velocity of each point mass and its momentum vector as shown in Fig. 5.5(b). Taking moments about point C we find

$$\mathbf{H}_C = -2mr\dot{x}_C\mathbf{k} - \sqrt{2}\, m\dot{x}_C r\frac{\sqrt{2}}{2}\mathbf{k} - \sqrt{2}\, m\dot{x}_C r\frac{\sqrt{2}}{2}\mathbf{k} = -4rm\dot{x}_C\mathbf{k}$$

Since the total mass $M = 4m$, we have

$$\mathbf{H}_C = -Mr\dot{x}_C\mathbf{k} \qquad \text{and} \qquad \dot{\mathbf{H}}_C = -Mr\ddot{x}_C\mathbf{k}$$

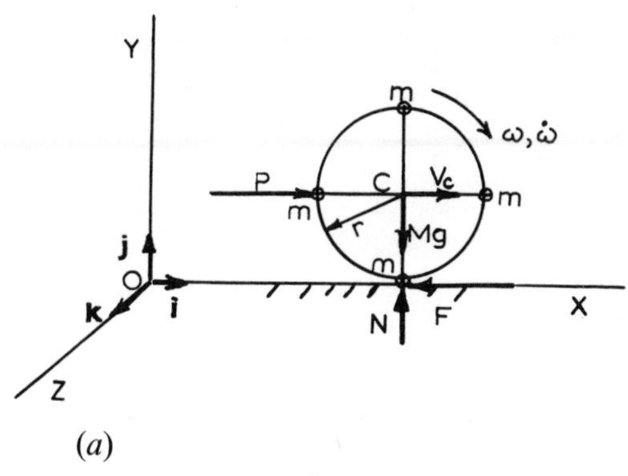

(*a*)

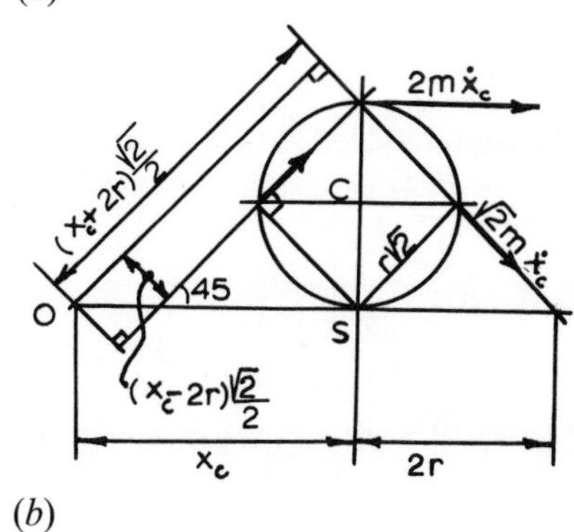

(*b*)

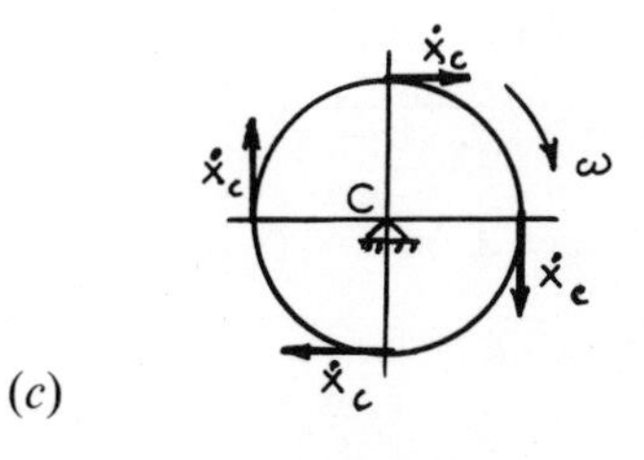

(*c*)

Fig. 5.5

The relative velocities with respect to C are as shown in Fig. 5.5(c), and the magnitude of the relative velocities are $r\omega = \dot{x}_C$. Taking moments about C of the relative momentum vectors, we find

$$\mathbf{H}'_C = -4rm\dot{x}_C\mathbf{k} = -Mr\dot{x}_C\mathbf{k} = \mathbf{H}_C \qquad \dot{\mathbf{H}}'_C = -Mr\ddot{x}_C\mathbf{k}$$

Taking moments about point O gives

$$\mathbf{H}_O = -2r(2\dot{x}_C m)\mathbf{k} - (x_C + 2r)\frac{\sqrt{2}}{2}(\sqrt{2}\dot{x}_C m)\mathbf{k} + (x_C - 2r)\frac{\sqrt{2}}{2}(\sqrt{2}\dot{x}_C m)\mathbf{k}$$

$$= -8rm\dot{x}_C\mathbf{k} = -2Mr\dot{x}_C\mathbf{k}$$

and

$$\dot{\mathbf{H}}_O = -2Mr\ddot{x}_C\mathbf{k}$$

$\mathbf{H}_O$ may also be determined from (5.2) which states that $\mathbf{H}_O = \mathbf{r}_{OC} \times M\mathbf{V}_C + \mathbf{H}_C$. We have $\mathbf{r}_{OC} = x_C\mathbf{i} + r\mathbf{j}$, so that

$$\mathbf{r}_{OC} \times M\mathbf{V}_C = (x_C\mathbf{i} + r\mathbf{j}) \times M\dot{x}_C\mathbf{i} = Mr\dot{x}\mathbf{j} \times \mathbf{i} = -Mr\dot{x}_C\mathbf{k}$$

and

$$\mathbf{H}_O = -Mr\dot{x}_C\mathbf{k} - Mr\dot{x}_C\mathbf{k} = -2Mr\dot{x}_C\mathbf{k}$$

as before.

Taking moments of the momentum vectors about the contact point S leads to

$$\mathbf{H}_S = -2r(2m\dot{x}_C)\mathbf{k} - r\sqrt{2}(\sqrt{2}\,m\dot{x}_C)\mathbf{k} - r\sqrt{2}(\sqrt{2}\,m\dot{x}_C)\mathbf{k}$$
$$= -8rm\dot{x}_C\mathbf{k} = -2Mr\dot{x}_C\mathbf{k}$$

from which $\dot{\mathbf{H}}_S = -2Mr\ddot{x}_C\mathbf{k}$.

$\mathbf{H}_S$ may also be found from (5.2): $\mathbf{H}_S = \mathbf{r}_{SC} \times M\mathbf{V}_C + \mathbf{H}_C$. We have

$$\mathbf{r}_{SC} = r\mathbf{j} \qquad \mathbf{r}_{SC} \times M\mathbf{V}_C = r\mathbf{j} \times M\dot{x}_C\mathbf{i} = -Mr\dot{x}_C\mathbf{k}$$

so that

$$\mathbf{H}_S = -Mr\dot{x}_C\mathbf{k} - Mr\dot{x}_C\mathbf{k} = -2Mr\dot{x}_C\mathbf{k}$$

as before.

Using Newton's second law for the motion of the centre of mass, we find by summing the vertical forces:

$$N - Mg = M\ddot{y}_C = 0 \tag{a}$$

or $N = Mg$. Summation of horizontal forces gives the equation

$$P - F = M\ddot{x}_C \tag{b}$$

Since there is no slip, the equation of constraint is $x_C = r\theta + x_O$, so that $\dot{x}_C = r\dot{\theta} = r\omega$, and

$$\ddot{x}_C = r\dot{\omega} \tag{c}$$

Since we have only the two equations (b) and (c) with three unknown quantities $\ddot{x}_C$, F and $\dot{\omega}$, we need one additional equation; taking this as $\mathbf{M}_C = \dot{\mathbf{H}}_C$, we have $\mathbf{M}_C = -Fr\mathbf{k}$, so that

$$Fr = Mr\ddot{x}_C \tag{d}$$

The solution of equations (b), (c) and (d) is

$$F = \frac{P}{2} \qquad \ddot{x}_C = \frac{P}{2M} \qquad \dot{\omega} = \frac{P}{2rM}$$

Equation (d) may also be taken as $\mathbf{M}_S = \dot{\mathbf{H}}_S$ or $\mathbf{M}_O = \dot{\mathbf{H}}_O$.

Since there is no slip in this problem $F < \mu N = \mu Mg$, where μ is the coefficient of friction; since $F = P/2$, we find the maximum value of P to be $P = 2\mu Mg$; for $P > 2\mu Mg$ the rim will start to slip.

In Example 5.1 the moment of momentum vector $\mathbf{H}$ was easy to determine, since we only had to sum the contributions from four particles. In the case of a rigid body we have to sum the contributions from all the particles of the body; to do this it is necessary to develop a formula for $\mathbf{H}$ in which the summation is done by integration over the volume of the body.

Figure 5.6(a) shows a body rotating about a *fixed point* O. A special, and much more important, case of this is a body *rotating about a fixed axis* (Fig. 5.6(b)); in this case any point on the axis of rotation is a fixed point, and the body may be considered to be rotating about any point on this axis.

We have, by definition, in these two cases $\mathbf{H}_O = \int \mathbf{r}_O \times \mathbf{V}_m \, dm$. Using the fixed point O as a pole, we have $\mathbf{V}_m = \mathbf{V}_O + \mathbf{V}_{mO} = \mathbf{V}_{mO} = \dot{\mathbf{r}}_O = \boldsymbol{\omega} \times \mathbf{r}_O$, so that

$$\mathbf{H}_O = \int \mathbf{r}_O \times (\boldsymbol{\omega} \times \mathbf{r}_O) \, dm \tag{5.3}$$

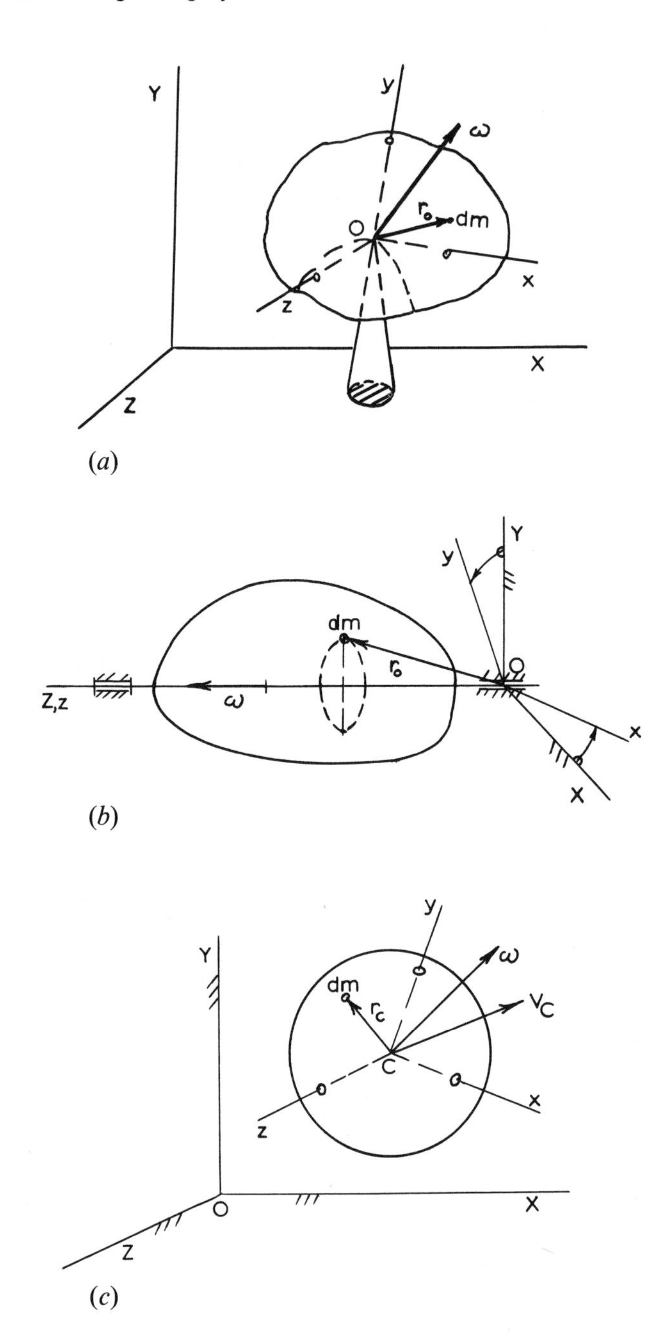
Y
y
ω
r_o
dm
O
x
z
X
Z
(a)
Y
y
dm
r_o
O
Z,z
ω
x
X
(b)
y
ω
Y
dm
r_c
V_C
C
x
z
O
X
Z
(c)

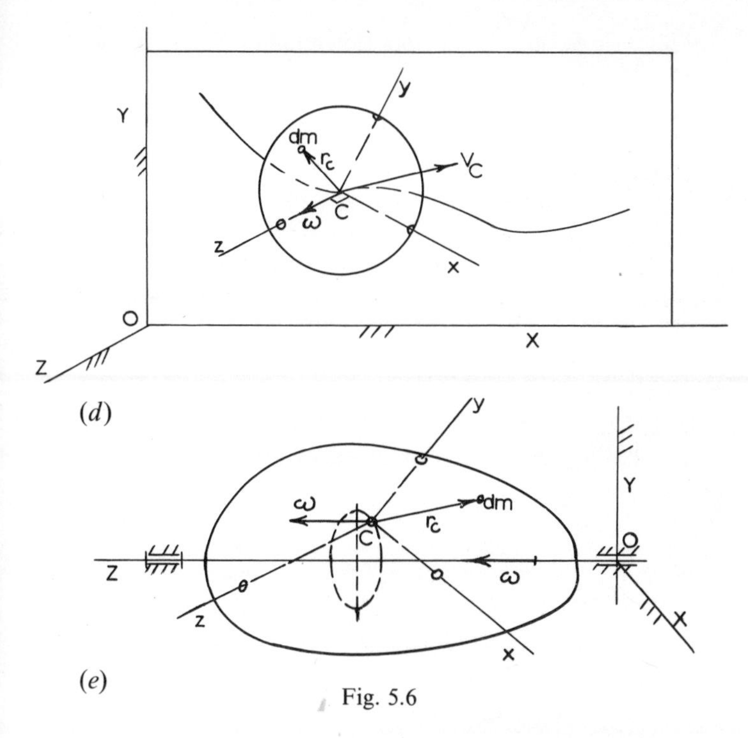

Fig. 5.6

For a body in *general motion* (Fig. 5.6(c)), we shall use the centre of mass C as a moment centre. A special case of general motion is *plane motion*, shown in Fig. 5.6(d). In Fig. 5.6(e) we shall use the centre of mass C as a moment centre, instead of a point O on the axis of rotation. The case shown in Fig. 5.6(e) is *rotation about a fixed axis*, which is a special case of plane motion.

For the three cases of motion shown in Figs. 5.6(c)–(e), we have by definition

$$\mathbf{H}_C = \mathbf{H}'_C = \int \mathbf{r}_C \times \mathbf{V}m \, dm$$

Taking C as a pole for the motion, we have

$$\mathbf{V}_m = \mathbf{V}_C + \mathbf{V}_{mC} = \mathbf{V}_C + \dot{\mathbf{r}}_C = \mathbf{V}_C + \boldsymbol{\omega} \times \mathbf{r}_C$$

so that

$$\mathbf{H}_C = \int \mathbf{r}_C \times (\mathbf{V}_C + \boldsymbol{\omega} \times \mathbf{r}_C) \, dm = \int \mathbf{r}_C \times \mathbf{V}_C \, dm + \int \mathbf{r}_C \times (\boldsymbol{\omega} \times \mathbf{r}_C) \, dm$$

The first term is $\int \mathbf{r}_C \times \mathbf{V}_C \, dm = [\int \mathbf{r}_C \, dm] \times \mathbf{V}_C = \mathbf{0}$, so that

$$\mathbf{H}_C = \mathbf{H}'_C = \int \mathbf{r}_C \times (\boldsymbol{\omega} \times \mathbf{r}_C) \, dm \tag{5.4}$$

For all the five cases of motion shown in Fig. 5.6(a)–(e), we have now $\mathbf{M}=\dot{\mathbf{H}}$, and $\mathbf{H}=\int \mathbf{r}\times(\boldsymbol{\omega}\times\mathbf{r})\,dm$, where the moment centre is the fixed point O in the first two cases, and the centre of mass C in the last three cases.

Introducing a coordinate system (x, y, z) fixed in the body and therefore rotating with it, and with origin O or C as shown in the various cases, we may express the position vector $\mathbf{r}$ to a mass particle dm and the angular velocity vector $\boldsymbol{\omega}$, in this rotating coordinate system, by the expressions

$$\mathbf{r} = x\mathbf{i}+y\mathbf{j}+z\mathbf{k} \qquad \text{and} \qquad \boldsymbol{\omega} = \omega_x\mathbf{i}+\omega_y\mathbf{j}+\omega_z\mathbf{k}$$

To determine the components of the moment of momentum vector $\mathbf{H}$, we have the formula $\mathbf{H}=\int \mathbf{r}\times(\boldsymbol{\omega}\times\mathbf{r})\,dm$.

Using the formula for the triple vector product,

$$\mathbf{A}\times(\mathbf{B}\times\mathbf{C}) = (\mathbf{A}\cdot\mathbf{C})\mathbf{B}-(\mathbf{A}\cdot\mathbf{B})\mathbf{C}$$

we find

$$\begin{aligned}\mathbf{r}\times(\boldsymbol{\omega}\times\mathbf{r}) &= (\mathbf{r}\cdot\mathbf{r})\boldsymbol{\omega}-(\mathbf{r}\cdot\boldsymbol{\omega})\mathbf{r}\\ &= (x^2+y^2+z^2)\,(\omega_x\mathbf{i}+\omega_y\mathbf{j}+\omega_z\mathbf{k})-(x\omega_x+y\omega_y+z\omega_z)\\ &\qquad\qquad (x\mathbf{i}+y\mathbf{j}+z\mathbf{k})\\ &= [(y^2+z^2)\omega_x-xy\omega_y-xz\omega_z]\mathbf{i}+[-xy\omega_x+(x^2+z^2)\omega_y\\ &\qquad -yz\omega_z]\mathbf{j}+[-xz\omega_x-yz\omega_y+(x^2+y^2)\omega_z]\mathbf{k}\end{aligned}$$

Multiplying by dm and integrating, and using the notations

$$\begin{aligned}I_x &= \int (y^2+z^2)\,dm \qquad I_y = \int (x^2+z^2)\,dm \qquad I_z = \int (x^2+y^2)\,dm\\ I_{xy} &= \int xy\,dm \qquad I_{xz} = \int xz\,dm \qquad I_{yz} = \int yz\,dm\end{aligned}$$

we find the components of $\mathbf{H}$:

$$\left.\begin{aligned}H_x &= I_x\omega_x-I_{xy}\omega_y-I_{xz}\omega_z\\ H_y &= -I_{xy}\omega_x+I_y\omega_y-I_{yz}\omega_z\\ H_z &= -I_{xz}\omega_x-I_{yz}\omega_y+I_z\omega_z\end{aligned}\right\} \tag{5.5}$$

The six integrals I_x, I_y, I_z, I_{xy}, I_{xz} and I_{yz} in these expressions are *constants* for the body, since the system (x, y, z) is fixed in the body and moving with it. The terms I_x, I_y and I_z are called *moments of inertia* of the body with respect to the x-, y- and z-axes respectively. The terms I_{xy}, I_{xz} and I_{yz} are called *products of inertia* of the body with respect to the xy-, xz- and yz-planes respectively.

These six integrals determine the mass distribution of the body. It is necessary to discuss methods of calculation of the moments and products of inertia of a rigid body before any further work in dynamics of rigid bodies can be attempted. A full discussion of the moments of inertia is given in Chapter 6.

The equations of rotational motion will be developed from the formula $\mathbf{M}=\dot{\mathbf{H}}$. Once these equations are available, we shall have no particular use for the moment of momentum vector $\mathbf{H}$, and from then on it rather fades out of the picture.

CHAPTER SIX

Moments of Area and Moments of Inertia

6.1 Moments of area of a plane figure

6.1.1 First moment of area. Centre of area

The *first moment of area* of the plane figure in Fig. 6.1 with respect to the x-axis is defined by the integral $S_x = \int_A y\, dA$, and for the y-axis: $S_y = \int_A x\, dA$; the units are m^3 and S_x and S_y may be positive, negative or zero.

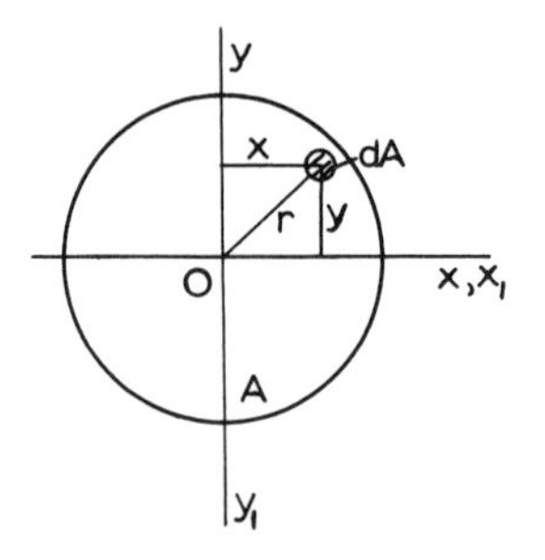

Fig. 6.1

By writing $S_x = y_C A$, $S_y = x_C A$, where A is the total area of the figure, we define a point C (x_C, y_C) called the *centre of area* or *centroid*; the centre of area is then determined by the expressions (both in metres)

$$x_C = \frac{\int x\, dA}{A} \qquad y_C = \frac{\int y\, dA}{A} \tag{6.1}$$

For an axis through C, $x_C = y_C = 0$, so for centroidal axes

$$\int x\,dA = \int y\,dA = 0 \tag{6.2}$$

If the y-axis is an *axis of symmetry*, we get contributions $x\,dA$ and $-x\,dA$ from symmetrical elements; hence $\int x\,dA = 0$ or $x_C = 0$, which means that the centre of area is on the axis of symmetry. For two axes of symmetry the centre of area is at the intersection of the axes.

For the rectangular area shown in Fig. 6.2, $dA = b\,dy$. Hence $S_x = \int_A y\,dA = \int_0^a by\,dy = \frac{1}{2}ba^2 = Ay_C = Aa/2$.

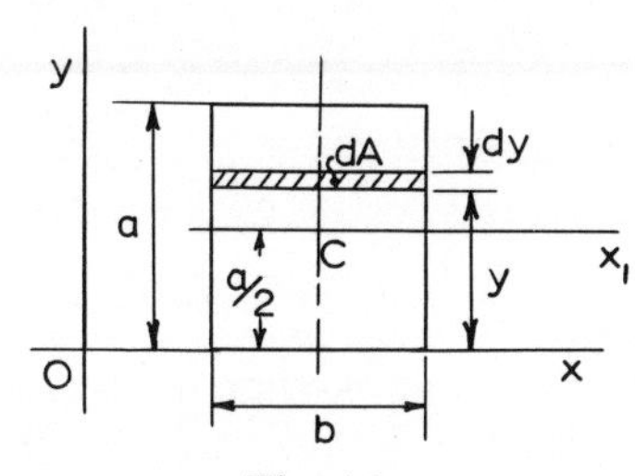

Fig. 6.2

For an area under a given curve $y = f(x)$ (Fig. 6.3)

$$dS_x = (y\,dx)\,y/2 \qquad S_x = \tfrac{1}{2}\int_{x1}^{x2} y^2\,dx = y_C A$$

$$dS_y = (y\,dx)x \qquad S_y = \int_{x1}^{x2} xy\,dx = x_C A$$

these two formulae determine $C(x_C, y_C)$ for the area.

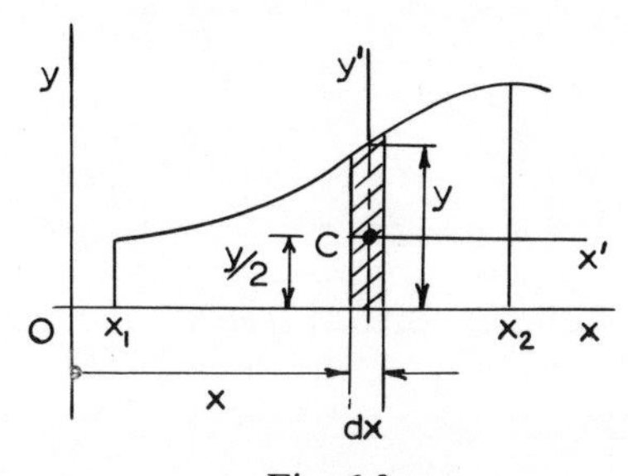

Fig. 6.3

6.1.2 Second moments of area of a plane area

The *second moments of area* are defined by the expressions

$$I_x = \int_A y^2\,dA \qquad I_y = \int_A x^2\,dA$$

these are always positive. The units for second moments of area are m^4. Analytical calculation may be done by dividing the area (Fig. 6.2) into rectangles:

$$I_{x1} = 2\int_0^{a/2} y^2 b\,dy = \tfrac{2}{3}b\,(a/2)^3 = \tfrac{1}{12}ba^3.$$
$$I_x = \int_0^a (b\,dy)y^2 = \tfrac{1}{3}ba^3$$

Writing $I_x = Ar_x^2$, we call r_x the *radius of gyration* with respect to the x-axis, and r_x is defined and calculated from this expression; similarly $I_y = Ar_y^2$.

Second moment of area with respect to an axis perpendicular to the area

This second moment is called the *polar moment of area* J_0, and is defined by the expression $J_0 = \int_A r^2\,dA$ (Fig. 6.1), so

$$J_0 = \int_A (x^2+y^2)\,dA = \int_A x^2\,dA + \int_A y^2\,dA = I_y + I_x \quad (6.3)$$

The polar moment of area J_0 is equal to the sum of I_x and I_y for *any* set of perpendicular axes in the area through O. Hence $I_x + I_y =$ constant for any set of perpendicular axes in the plane of the area through the same point; this is called the *perpendicular-axis theorem.*

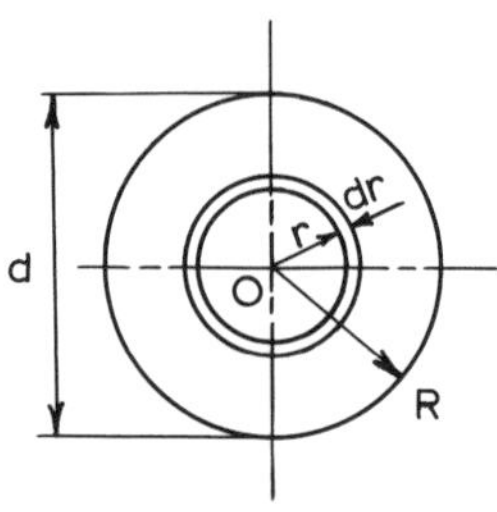

Fig. 6.4

Example 6.1. For a circular area (Fig. 6.4), J_0 for an axis through the centre O is determined by dividing the area in elements as shown; the contribution of one element $dA = 2\pi r\,dr$ is $dJ_0 = dA(r^2) = (2\pi r\,dr)r^2$, so

$$J_0 = \int_0^R (2\pi r\,dr)r^2 = \int_0^R 2\pi r^3\,dr = \frac{\pi}{2}[r^4]_0^R = \frac{\pi}{2}R^4$$

Since $I_x + I_y = J_0$, and $I_y = I_x$, we find $I_x = J_0/2 = (\pi/4)R^4 = (\pi/64)d^4$; this is much simpler than integrating directly for I_x by rectangular division of the area.

Parallel-axis theorem

Figure 6.5 shows an area with centre of area C and a set of co-ordinate axes (x, y) through C; suppose that $I_x = \int_A y^2\, dA$ is known for the x-axis; for a parallel axis x_1 we find

$$I_{x_1} = \int_A y_1^2\, dA = \int_A (y+b)^2\, dA = \int_A y^2\, dA + 2b \int_A y\, dA + b^2 \int dA$$
$$= I_x + b^2 A \quad (6.4)$$

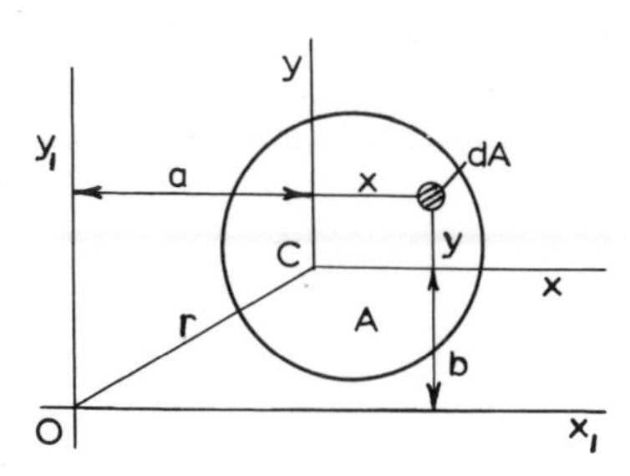

Fig. 6.5

In a similar manner

$$I_{y_1} = I_y + a^2 A \qquad (6.5)$$

These formulae constitute the *parallel-axis theorem*; the x- and y-axes *must* be through the centre of area.

The formulae are often used in the form

$$I_x = I_{x_1} - b^2 A \qquad \text{and} \qquad I_y = I_{y_1} - a^2 A$$

Adding the expressions (6.4) and (6.5) gives

$$I_{x_1} + I_{y_1} = I_x + I_y + (a^2 + b^2)A = I_x + I_y + r^2 A$$

Hence

$$J_O = J_C + r^2 A \qquad (6.6)$$

The parallel-axis theorem holds also for *polar moments* of area. The theorem is particularly useful for calculations of second moments of area of composite areas.

Products of area

The *product* of an area A (Fig. 6.1) with respect to the x and y axes is *defined* by the summation: $I_{xy} = \int_A xy\, dA$; the units are m^4. It follows at once that $I_{xy} = I_{yx}$; I_{xy} may be positive, negative or vanish; that $I_{xy}=0$ for a certain position of xy may be seen by changing the y axis to y_1; we find $x_1=x$, $y_1=-y$, so

$$I_{x_1y_1} = \int_A x_1y_1\, dA = \int_A x(-y)\, dA = -I_{xy}$$

Since I_{xy} changes as we rotate the axes about O, this shows that at some position $I_{xy}=0$; these particular axes are called *principal axes* for the area at point O.

If a figure has an axis of symmetry (Fig. 6.6) the contribution from dA to I_{xy} is $dI_{xy}=xy\, dA+(-x)y\, dA=0$, so that $I_{xy}=0$; the x-axis may be positioned anywhere perpendicular to the axis of symmetry; such a set of axes are *principal axes.*

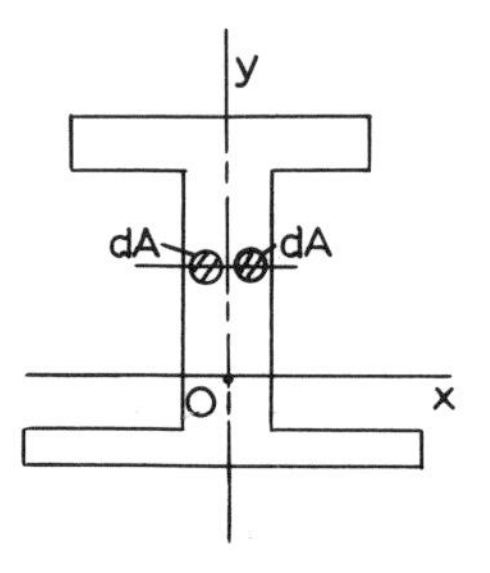

Fig. 6.6

Parallel-axis theorem for products of area
Suppose $I_{xy}=\int_A xy\, dA$ is known for a set of axes xy *through the centre of area C* (Fig. 6.5); for a parallel set of axes (x_1, y_1), we find

$$I_{x_1y_1} = \int_A x_1y_1\, dA = \int_A (x+a)\,(y+b)\, dA$$
$$= \int_A xy\, dA+b\int_A x\, dA+a\int_A y\, dA+ab\int_A dA$$

Therefore

$$I_{x_1y_1}= I_{xy}+abA \tag{6.7}$$

a and b are coordinates of C in the system (x_1, y_1) (positive or negative), *or* coordinates of O in the system (x, y).

For an area under a curve $y=f(x)$ (Fig. 6.3), we find for the rectangular strip with centre C

$$I_{xys} = I_{x'y's}+A_s(x)\,(y/2)$$
$$I_{x'y's} = 0 \qquad \text{from symmetry}$$

Hence $I_{xys}=A_sxy/2=(y\, dx)\, xy/2$, and therefore $I_{xy}=\frac{1}{2}\int_{x1}^{x2} xy^2\, dx$.

6.1.3 Determination of principal axes for a point of a plane area

Assume that for the given area A in Fig. 6.7, we know $I_x = \int_A y^2 \, dA$, $I_y = \int_A x^2 \, dA$ and $I_{xy} = \int_A xy \, dA$. The formulae connecting the two coordinate systems are

$$\begin{aligned} x_1 &= x \cos \theta + y \sin \theta \\ y_1 &= y \cos \theta - x \sin \theta \end{aligned} \tag{6.8}$$

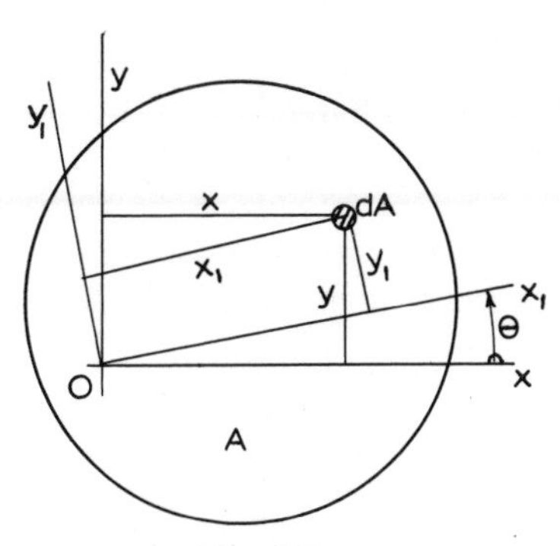

Fig. 6.7

Substituting (6.8) in $I_{x_1} = \int_A y_1^2 \, dA$ and $I_{y_1} = \int_A x_1^2 \, dA$ leads to

$$\begin{aligned} I_{x_1} &= I_x \cos^2 \theta + I_y \sin^2 \theta - I_{xy} \sin 2\theta \\ I_{y_1} &= I_x \sin^2 \theta + I_y \cos^2 \theta + I_{xy} \sin 2\theta \end{aligned}$$

The sum and difference of these expressions give

$$\begin{aligned} I_{x_1} + I_{y_1} &= I_x + I_y \\ I_{x_1} - I_{y_1} &= (I_x - I_y) \cos 2\theta - 2I_{xy} \sin 2\theta \end{aligned} \tag{6.9}$$

Equations (6.9) give the simplest numerical calculation of I_{x_1} and I_{y_1}. In the same manner we find

$$I_{x_1 y_1} = \int_A x_1 y_1 \, dA = \tfrac{1}{2}(I_x - I_y) \sin 2\theta + I_{xy} \cos 2\theta \tag{6.10}$$

The principal axes are found for $I_{x_1 y_1} = 0$:

$$\tfrac{1}{2}(I_x - I_y) \sin 2\theta + I_{xy} \cos 2\theta = 0$$

that is at the position determined by

$$\tan 2\theta = \frac{2I_{xy}}{I_y - I_x} \tag{6.11}$$

If we sum the expressions (6.9), the result is

$$I_{x_1} = \frac{I_x+I_y}{2}+\frac{(I_x-I_y)}{2}\cos 2\theta - I_{xy}\sin 2\theta$$

Taking $dI_{x_1}/d\theta=0$, gives $\tan 2\theta = 2I_{xy}/(I_y-I_x)$, which shows that I_{x_1} is a maximum or minimum where $I_{x_1y_1}=0$ (the same result is found for I_{y_1}). Therefore the *principal axes* also determine the maximum and minimum second moments of area for any set of orthogonal axes through O. The actual maximum or minimum axis is determined from the numerical result for I_{x_1} and I_{y_1} for this value of θ.

The formulae relate to the axes as shown in Fig. 6.7. If one of the axes is in the opposite direction, the sign of I_{xy} and $I_{x_1y_1}$ must be changed in the formulae; if the rotation θ is clockwise, $-\theta$ must be substituted for θ in the formulae.

6.2 Moments of inertia of a lamina

Consider a lamina, as shown in Fig. 6.8, of homogeneous material of density ρ and uniform thickness t; a coordinate system (x, y, z) is fixed at O in the middle plane of the lamina. We assume that t is small compared to the other dimensions.

By definition the *moment of inertia* of the lamina with respect to the z-axis is

$$I_z = \int_V (x^2+y^2)\,dm = \int_V r^2\,dm$$

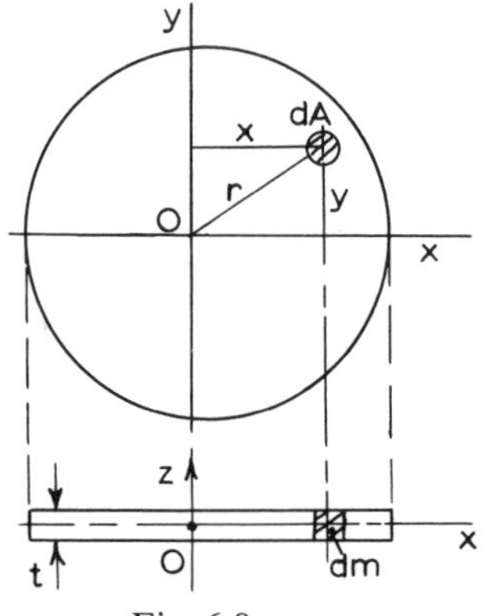

Fig. 6.8

where $dm = (dA)t\rho$ and V denotes the volume. Hence

$$I_z = \int_A r^2 \rho t \, dA = (\rho t) \int_A r^2 \, dA = \rho t J_0 \qquad (6.12)$$

where J_O is the *polar moment of area* with respect to the z-axis. Now

$$I_x = \int_V (y^2 + z^2) \, dm$$

and with $|z| \leqslant t/2$ we have $|z| \ll |y|$. Neglecting z^2, we find

$$I_x \simeq \int_V y^2 \, dm = (\rho t) \int_A y^2 \, dA = (\rho t) I_{xA} \qquad (6.13)$$

where I_{xA} is the second *moment of area* with respect to the x-axis; similarly

$$I_y = \int_V (x^2 + z^2) \, dm \simeq \int_V x^2 \, dm = \rho t \int_A x^2 \, dA = (\rho t) I_{yA}$$

which means that second *moments of area* may be used directly to find *moments of inertia* of a lamina by multiplication by the constant factor ρt.

Perpendicular-axis theorem
For second moments of area, Fig. 6.8, eq. (6.3), we have $I_{xA} + I_{yA} = J_O = I_{zA}$. Multiplying by ρt gives

$$I_x + I_y = I_z \qquad (6.14)$$

Hence the perpendicular-axis theorem holds also for *moments of inertia* of a lamina with xy in the middle plane. (The theorem does *not* hold for three-dimensional bodies in general.)

Parallel-axis theorem
For the area in Fig. 6.5 we have formula (6.4), $I_{x_1 A} = I_{xA} + Ab^2$. Multiplying by ρt gives

$$I_{x_1} = I_x + Mb^2 \qquad (6.15)$$

In (6.15) M is the total mass of the lamina.

It is sometimes convenient to use expressions of the form $I_x = Mr_x^2$, where r_x is called the *radius of gyration* with respect to the x-axis; r_x is defined and calculated from this expression, which is particularly useful when it is possible to cancel the mass M in an equation.

6.3 Moments of inertia of three-dimensional bodies

6.3.1 Definition of moments and products of inertia

The moment of inertia, with respect to the x-axis, of the body in Fig. 6.9 was *defined* as follows (see Chapter 5, p. 144):

$$I_x = \int_V (y^2 + z^2)\, dm$$

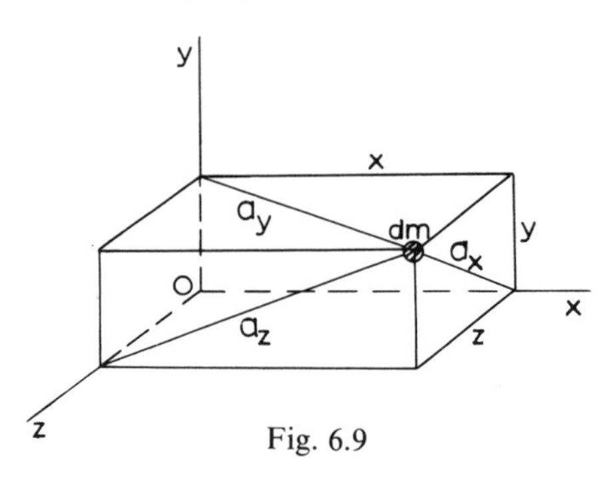

Fig. 6.9

Hence the sum of each mass element multiplied by the square of its distance from the axis

$$I_x = \int_V a_x^2\, dm$$

In the same way

$$I_y = \int_V (x^2 + z^2)\, dm = \int_V a_y^2\, dm$$
$$I_z = \int_V (x^2 + y^2)\, dm = \int_V a_z^2\, dm$$

These expressions are always positive.

It is sometimes convenient to write $I_x = Mr_x^2$, in the same way as for a lamina, where M is the total mass of the body and r_x the radius of gyration; similarly $I_y = Mr_y^2$ and $I_z = Mr_z^2$.

The *product of inertia* with respect to the xy-plane was *defined* by the expression (see Chapter 5, p. 144)

$$I_{xy} = \int_V xy\, dm = \int_V yx\, dm = I_{yx}$$

The product of inertia is the sum of each mass element multiplied by the product of its x- and y-coordinates, that is by the product of its distances from the yz- and xz-planes; similarly

$$I_{xz} = I_{zx} = \int_V xz\, dm \qquad I_{yz} = I_{zy} = \int_V yz\, dm$$

Products of inertia may be positive, negative or vanish, as will be shown in the following.

6.3.2 Parallel-axis theorem for a rigid body

Figure 6.10 shows the centre of mass C of a rigid body of total mass M; a coordinate system (x, y, z) has been introduced with origin at C and a particular mass element dm of the body is shown at position (x, y, z); a coordinate system (x_1, y_1, z_1) with origin O has been introduced with axes *parallel* to (x, y, z).

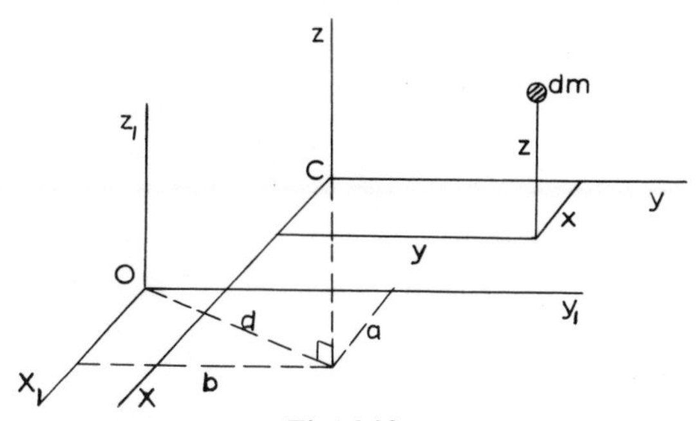

Fig. 6.10

By definition

$$I_{z_1} = \int_V (x_1^2+y_1^2)\, dm = \int_V [(a+x)^2+(b+y)^2]\, dm$$
$$= \int_V (x^2+y^2)\, dm + 2a \int_V x\, dm + 2b \int_V y\, dm + (a^2+b^2) \int_V dm$$
$$= I_z + d^2 M \quad (6.16)$$

The formula (6.16) states the parallel-axis theorem for a rigid body; it is essential to remember that the z-axis, in this case, *must be through the centre of mass*; the formula is often used in the form $I_z = I_{z_1} - d^2 M$.

The product of inertia $I_{x_1 y_1}$ is by definition

$$I_{x_1 y_1} = \int_V x_1 y_1\, dm = \int_V (x+a)(y+b)\, dm$$
$$= \int_V xy\, dm + b \int_V x\, dm + a \int_V y\, dm + ab \int_V dm = I_{xy} + abM \quad (6.17)$$

In (6.17) a and b are coordinates, with proper signs, of C in the system (x_1, y_1), or the coordinates of O in the system (x, y); the formula is often used in the form $I_{xy} = I_{x_1 y_1} - abM$; the x- and y- axes *must be through the centre of mass C*.

The formulae (6.16) and (6.17) constitute the parallel-axis theorem for a rigid body; the *perpendicular-axis theorem* developed for areas and laminae does *not* hold in general for a three-dimensional body.

Example 6.2. Figure 6.11 shows a right circular homogeneous cylinder of density ρ and total mass $M=\pi R^2 H\rho$. The determination of the *moment of inertia* I_x with respect to a diameter in the base is

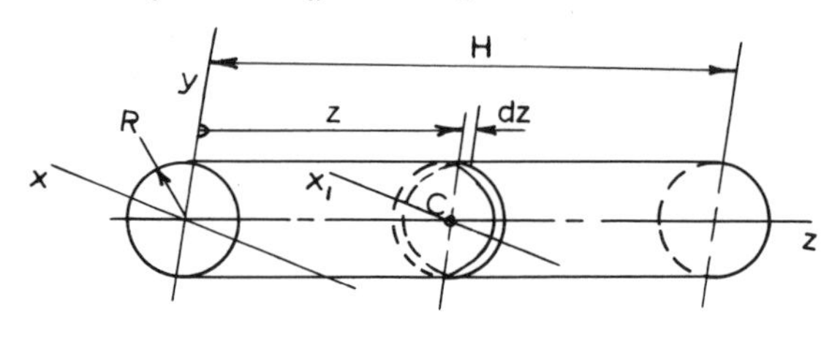

Fig. 6.11

done most simply in the following manner: for the lamina shown parallel to the base the second moment of area of the circular surface with respect to the x_1-axis parallel to the x-axis is $I_{x_1A}=(\pi/4)R^4$; the moment of inertia is then $I_{x_1}=I_{x_1A}\ (\rho t)=(\pi/4)R^4\rho\ dz$; the mass of the lamina is $m=\pi R^2\ dz\ \rho$; the contribution of the lamina to I_x is from (6.16):

$$dI_x = I_{x_1}+mz^2 = (\pi/4)\rho R^4\ dz+\pi R^2\ dz\ \rho z^2$$

Therefore

$$I_x = (\pi/4)\rho R^4 \int_0^H dz+\pi R^2\rho \int_0^H z^2\ dz$$

$$= (\pi R^2 H)\rho(\tfrac{1}{4}R^2+\tfrac{1}{3}H^2) = M\left[\frac{3R^2+4H^2}{12}\right] = Mr_x^2$$

6.3.3 Principal axes for a rigid body

Figure 6.12 shows a coordinate system (x, y, z) and a mass element dm of a rigid body; dm is located at position (x, y, z); suppose that the body is *homogeneous* and has a *plane of symmetry yz*; corresponding to the element dm at (x, y, z), there will always be an element dm at $(-x, y, z)$; the contribution to I_{xy} from these elements is then $dm\ (x)(y)+dm\ (-x)(y)=0$. Therefore

$$I_{xy} = \int_V xy\ dm = 0$$

for the body; similarly $I_{xz}=\int_V xz\ dm=0$.

If the xy-plane is also a plane of symmetry, we find that $I_{yz}=\int_V yz\ dm=0$, so for a *homogeneous* body with two perpendicular planes of symmetry, the line of intersection between the planes (here the y-axis), together with two perpendicular axes, one in each plane of symmetry (here the x- and z- axes), will determine

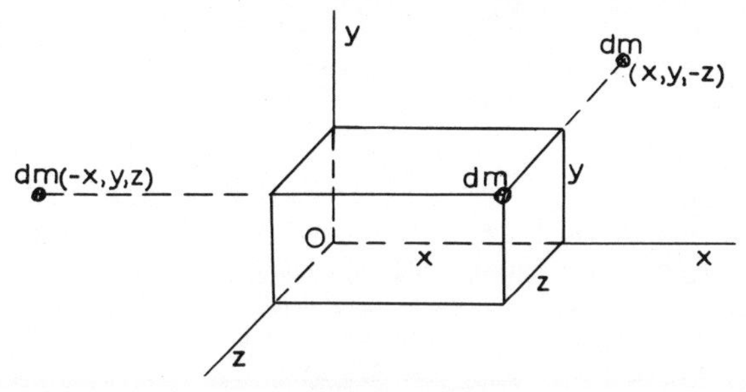

Fig. 6.12

a set of axes for which $I_{xy}=I_{xz}=I_{yz}=0$; such axes are called *principal axes* for the body at O and the moments of inertia I_x, I_y and I_z are called *principal moments of inertia* for the body at point O.

Principal axes may be determined by inspection for regular homogeneous bodies with two planes of symmetry, for instance the geometrical centre line of a right circular cylinder or cone, any diameter in a sphere, and the centre line of a right rectangular prism are all principal axes.

Some useful principal moments of inertia

These are shown in Figs. 6.13–6.18 and their captions. All bodies are assumed to be homogeneous and of total mass M.

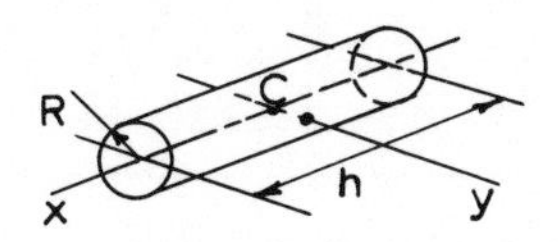

Fig. 6.13. Solid right circular cylinder or disc: $I_x=MR^2/2$, $I_y=M(h^2+3R^2)/12$

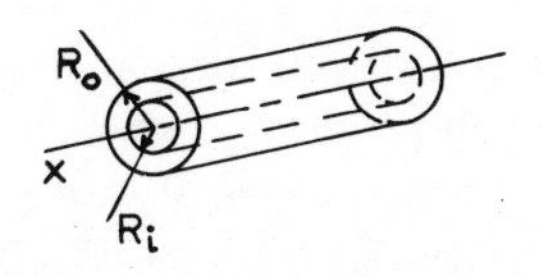

Fig. 6.14. Hollow right circular cylinder or disc: $I_x=M(R_0^2+R_i^2)/2$; thin cylindrical shell or hoop of mean radius R: $I_x=MR^2$

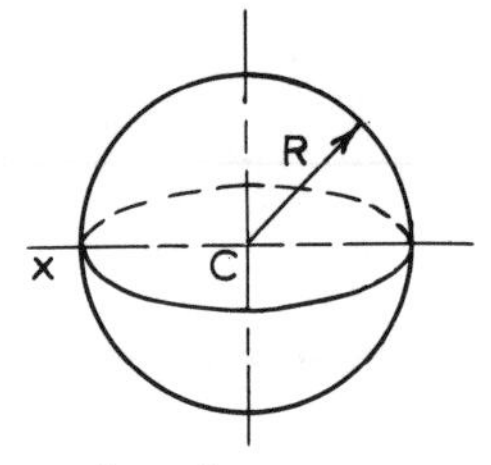

Fig. 6.15. Solid sphere: $I_x = \frac{2}{5}MR^2$; hollow sphere with inner and outer radii R_i and R_o: $I_x = \frac{2}{5}M(R_o^5 - R_i^5)/(R_o^3 - R_i^3)$; thin spherical shell with mean radius R: $I_x = \frac{2}{3}MR^2$

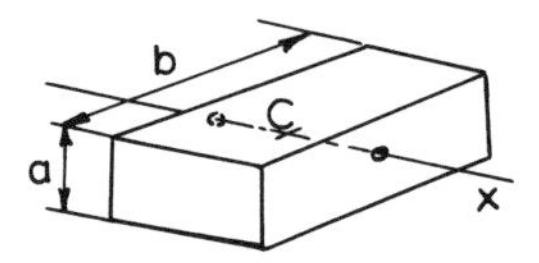

Fig. 6.16. Right regular rectangular prism: $I_x = M(a^2 + b^2)/12$

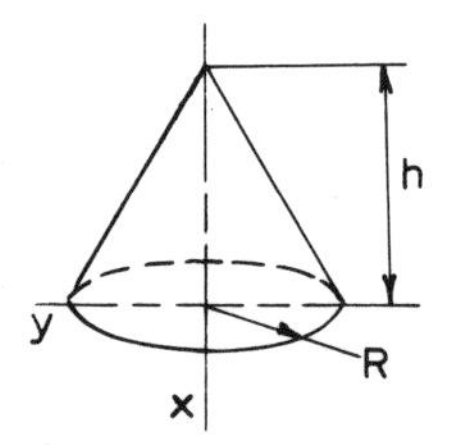

Fig. 6.17. Right circular cone: $I_x = \frac{3}{10}MR^2$, $I_y = M(3R^2 + 2h^2)/20$

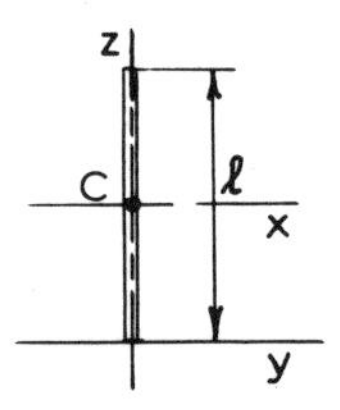

Fig. 6.18. Slender uniform prismatic bar: $I_x = Ml^2/12$, $I_y = Ml^2/3$, $I_z = 0$

6.3.4 Moments of inertia about any axis through O

Figure 6.19 shows a coordinate system (x_1, y_1, z_1) and a mass element dm of a rigid body; dm is located at position (x_1, y_1, z_1) with position vector $\mathbf{r}$; a second set of axes (x, y) is shown rotated to a different position from (x_1, y_1) about O.

Suppose that I_{x_1}, I_{y_1}, I_{z_1} and $I_{x_1y_1}$, $I_{x_1z_1}$, $I_{y_1z_1}$ are known for the body, and that the moment of inertia I_x is wanted. The x-axis is determined by its direction cosines: $\cos\angle(x, x_1)=l$, $\cos\angle(x, y_1)=m$ and $\cos\angle(x, z_1)=n$; for the y-axis the direction cosines are l', m' and n'.

The unit vectors $\mathbf{i}$ and $\mathbf{j}$ are determined by

$$\mathbf{i} = l\mathbf{i}_1+m\mathbf{j}_1+n\mathbf{k}_1 \qquad \mathbf{j} = l'\mathbf{i}_1+m'\mathbf{j}_1+n'\mathbf{k}_1$$

The position vector $\mathbf{r}$ is $\mathbf{r}=x_1\mathbf{i}_1+y_1\mathbf{j}_1+z_1\mathbf{k}_1$, and by definition $I_x=\int_V h^2\,dm$, where

$$h = |\mathbf{r}|\sin\alpha = |\mathbf{r}\times\mathbf{i}|$$

$$\mathbf{r}\times\mathbf{i} = (y_1n-z_1l)\mathbf{i}_1+(z_1l-x_1n)\mathbf{j}_1+(x_1m-y_1l)\mathbf{k}_1$$

$$h^2 = |\mathbf{r}\times\mathbf{i}|^2 = (y_1n-z_1l)^2+(z_1l-x_1n)^2+(x_1m-y_1l)^2$$

$$= (x_1^2+y_1^2)n^2+(x_1^2+z_1^2)m^2+(y_1^2+z_1^2)l^2 - 2(x_1y_1lm+x_1z_1ln+y_1z_1mn)$$

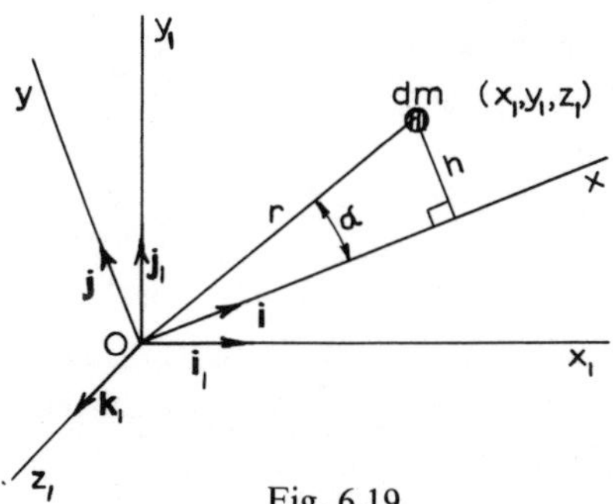

Fig. 6.19

Therefore

$$I_x = l^2I_{x_1}+m^2I_{y_1}+n^2I_{z_1}-2(lmI_{x_1y_1}+lnI_{x_1z_1}+mnI_{y_1z_1}) \qquad (6.18)$$

Similar expressions may be developed for I_y and I_z. The use of (6.18) to determine I_x is much simpler than a complicated integration for an inclined axis.

If the x_1-, y_1-, z_1-axes are *principal axes*, $I_{x_1y_1}=I_{x_1z_1}=I_{y_1z_1}=0$, and (6.18) takes the simpler form:

$$I_x = l^2I_{x_1}+m^2I_{y_1}+n^2I_{z_1} \qquad (6.19)$$

For the *product* of inertia I_{xy} we have by definition $I_{xy}=\int_V xy\,dm$; introducing

$$x = \mathbf{r}\cdot\mathbf{i} = x_1l+y_1m+z_1n \qquad \text{and} \qquad y = \mathbf{r}\cdot\mathbf{j} = x_1l'+y_1m'+z_1n'$$

we find

$$xy = x_1^2 ll' + y_1^2 mm' + z_1^2 nn' + x_1 y_1 (lm' + ml') \\ + x_1 z_1 (ln' + nl') + y_1 z_1 (mn' + nm')$$

subtracting from xy the expression

$$(\mathbf{r} \cdot \mathbf{r})(\mathbf{i} \cdot \mathbf{j}) = (x_1^2 + y_1^2 + z_1^2)(ll' + mm' + nn') = 0$$

the first three terms in xy change to

$$-(y_1^2 + z_1^2) ll' - (x_1^2 + z_1^2) mm' - (x_1^2 + y_1^2) nn'$$

therefore

$$I_{xy} = -(I_{x_1} ll' + I_{y_1} mm' + I_{z_1} nn') \\ + [I_{x_1 y_1}(lm' + ml') + I_{x_1 z_1}(ln' + nl') + I_{y_1 z_1}(mn' + nm')] \tag{6.20}$$

Similar expressions may be developed for I_{xz} and I_{yz}. If the x_1- y_1- and z_1-axes are principal axes, (6.20) reduces to the simpler form

$$I_{xy} = -(I_{x_1} ll' + I_{y_1} mm' + I_{z_1} nn') \tag{6.21}$$

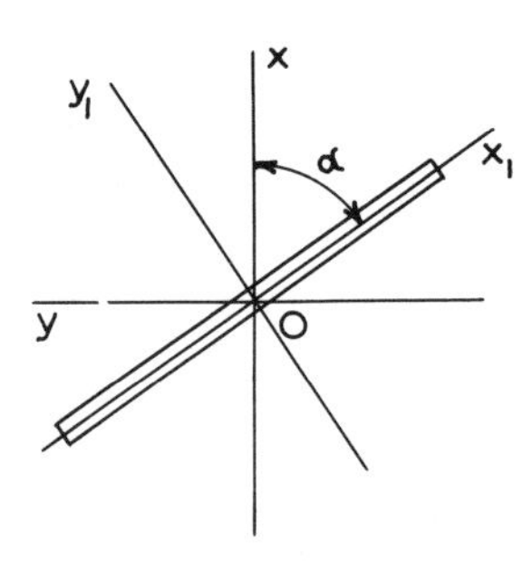

Fig. 6.20

Example 6.3. Figure 6.20 shows a thin circular disc of radius r and mass M; determine the moment of inertia about the inclined axis x and the product of inertia with respect to the plane xy. The z- and z_1-axes are perpendicular to the paper.

Solution. From symmetry,

$$I_{x_1 y_1} = I_{x_1 z_1} = I_{y_1 z_1} = 0$$

so x_1, y_1 and z_1 are principal axes.

$$I_{x_1} = I_{z_1} = Mr^2/4 \qquad I_{y_1} = Mr^2/2$$

$$l = \cos \angle(x, x_1) = \cos \alpha$$

$$m = \cos \angle(x, y_1) = \cos (90-\alpha) = \sin \alpha$$

$$n = \cos \angle(x, z_1) = \cos 90° = 0$$

$$l' = \cos \angle(y, x_1) = \cos (270-\alpha) = -\sin \alpha$$

$$m' = \cos \angle(y, y_1) = \cos \alpha$$

$$n' = \cos \angle(y, z_1) = \cos 90° = 0$$

Substituting in (6.18) gives

$$I_x = (\cos^2 \alpha)\frac{Mr^2}{4} + (\sin^2 \alpha)\frac{Mr^2}{2} = \frac{Mr^2(1+\sin^2 \alpha)}{4}$$

Equation (6.21) gives

$$I_{xy} = -\left[\frac{Mr^2}{4}\cos \alpha(-\sin \alpha) + \frac{Mr^2}{2}\sin \alpha \cos \alpha\right]$$

$$= -\frac{Mr^2}{4}\sin \alpha \cos \alpha = -\frac{Mr^2}{8}\sin 2\alpha$$

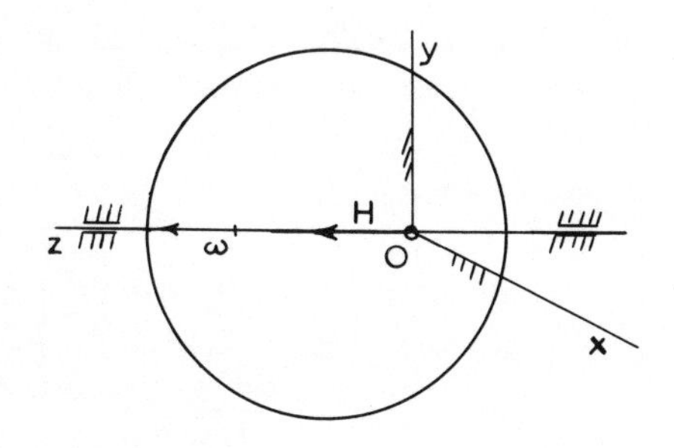

Fig. 6.21

6.3.5 Determination of principal axes

Figure 6.21 shows a rigid body rotating about a fixed *principal* axis, which has been taken as the z-axis; this means that $I_{xz} = I_{yz} = 0$ and the $\boldsymbol{\omega}$ vector is on the axis of rotation, so that in the rotating system (x, y, z) which is fixed in the body we have $\omega_x = \omega_y = 0$, and $\omega_z = \omega$. The formula (5.5) for the moment of momentum vector **H** now gives $H_x=0$, $H_y=0$ and $H_z=I_z\omega_z$, so for rotation about a principal axis, the moment of momentum vector **H** is along the axis and equal to $\mathbf{H}=I\boldsymbol{\omega}$; this may in fact be taken as a definition of a principal axis.

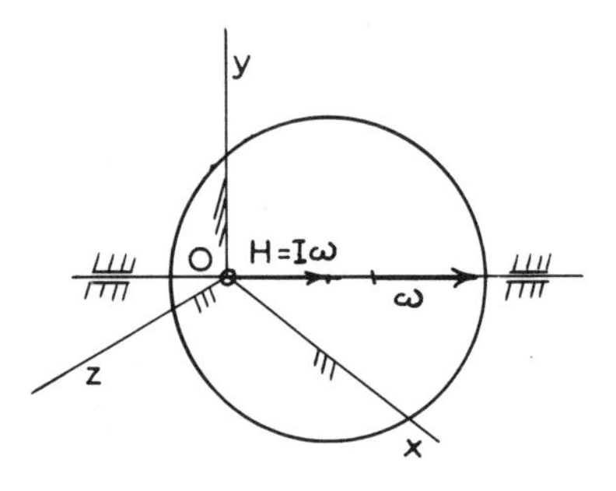

Fig. 6.22

Suppose the body in Fig. 6.22 is rotating with angular velocity $\boldsymbol{\omega}$ about a *fixed principal axis*, so that $\mathbf{H}=I\boldsymbol{\omega}$, where I is the principal moment of inertia for the axis of rotation; for the system (x, y, z) fixed in the body with origin O on the axis of rotation, we find from formula (5.5)

$$\begin{aligned} H_x &= I_x\omega_x - I_{xy}\omega_y - I_{xz}\omega_z &&= I\omega_x \\ H_y &= -I_{xy}\omega_x + I_y\omega_y - I_{yz}\omega_z &&= I\omega_y \\ H_z &= -I_{xz}\omega_x - I_{yz}\omega_y + I_z\omega_z &&= I\omega_z \end{aligned}$$

or

$$\left.\begin{aligned} (I_x - I)\omega_x - I_{xy}\omega_y - I_{xz}\omega_z &= 0 \\ -I_{xy}\omega_x + (I_y - I)\omega_y - I_{yz}\omega_z &= 0 \\ -I_{xz}\omega_x - I_{yz}\omega_y + (I_z - I)\omega_z &= 0 \end{aligned}\right\} \qquad (6.22)$$

Since we have real solutions for ω_x, ω_y, ω_z, the determinant of the coefficients of (6.22) must vanish:

$$\begin{vmatrix} I_x - I & -I_{xy} & -I_{xz} \\ -I_{xy} & I_y - I & -I_{yz} \\ -I_{xz} & -I_{yz} & I_z - I \end{vmatrix} = 0$$

Expanding the determinant gives a cubic equation of the form $I^3 + bI^2 + cI + d = 0$; it may be shown that this equation always has three real roots, which also follows from the physical significance of the principal moments of inertia. The coefficients of the equation are

$$\left.\begin{aligned} b &= -(I_x+I_y+I_z) \\ c &= I_xI_y+I_xI_z+I_yI_z-I_{xy}^2-I_{xz}^2-I_{yz}^2 \\ d &= I_xI_{yz}^2+I_yI_{xz}^2+I_zI_{xy}^2-I_xI_yI_z+2I_{xy}I_{xz}I_{yz} \end{aligned}\right\} \qquad (6.23)$$

By substituting numerical values for the inertia terms, we may find the three roots of the equation as follows: by calculating

$$P = \frac{c}{3}-\left(\frac{b}{3}\right)^2 \qquad Q = \frac{d}{2}-\frac{3}{2}\left(\frac{b}{3}\right)\left(\frac{c}{3}\right)+\left(\frac{b}{3}\right)^3 \qquad R = P^3+Q^2$$

it follows from the theory of the cubic equation that there are two possible cases:

1. If $R=0$, the roots are

$$I_1 = (2\sqrt[3]{\ }-Q)-\frac{b}{3} \qquad I_2 = I_3 = (-\sqrt[3]{\ }-Q)-\frac{b}{3}$$

2. If $R<0$, $I_1=(2\sqrt{(-P)})\cos\theta/3-b/3$, where $\cos\theta=Q/P\sqrt{(-P)}$, and $0<\theta<180°$. The other two roots may be found by dividing out I_1 from the cubic equation and solving the remaining quadratic equation; alternatively they may be calculated from

$$I_2 = (-2\sqrt{(-P)})\sin\left(\frac{\theta}{3}+30°\right)-\frac{b}{3}$$

$$I_3 = (-2\sqrt{(-P)})\cos\left(\frac{\theta}{3}+60°\right)-\frac{b}{3}$$

If the x-axis is a principal axis, $I_{xy}=I_{xz}=0$, the determinant shows at once that $I_1=I_x$, and the other two roots may be found from the remaining second-order determinant, which gives the equation $(I_y-I)(I_z-I)-I_{yz}^2=0$; in a similar manner if the y-axis is principal, $I_1=I_y$ and I_2 and I_3 may be found from $(I_x-I)(I_z-I)-I_{xz}^2=0$; for the z-axis principal $I_1=I_z$ and $(I_x-I)(I_y-I)-I_{xy}^2=0$ for the remaining roots.

By substituting the value for I_1 in eq. (6.22), the *ratios* of ω_x, ω_y and ω_z may be calculated. The direction of the principal axis for I_1 is then given by the vector $\boldsymbol{\omega}=\omega_x\mathbf{i}+\omega_y\mathbf{j}+\omega_z\mathbf{k}$, and the direction of the other two principal axes may be determined in a similar manner. The procedure is best described by an example.

Example 6.4. Figure 6.23 shows a homogeneous right rectangular prism. A coordinate system (x, y, z) has been introduced with origin O at one corner of the prism; the dimensions are as shown and the density $\rho = 7760\ \mathrm{kg/m^3}$. Determine the principal moments of inertia at O and the direction of the principal axes.

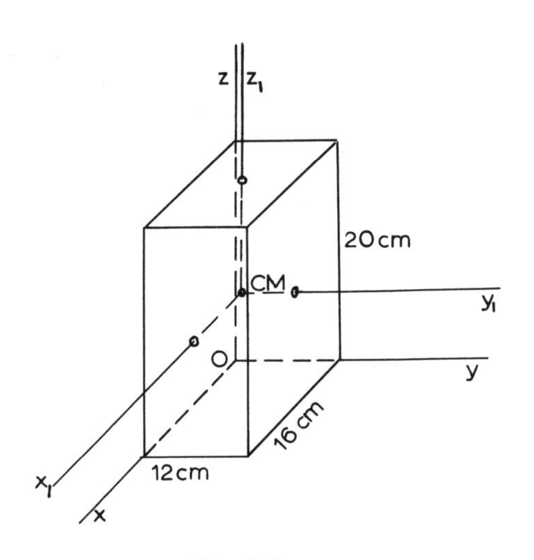

Fig. 6.23

Solution. Slide-rule accuracy is generally not sufficient for this type of calculation: depending on the figures the calculation must be carried to five or more figures after the point.

The mass of the prism is $M = 0{\cdot}12 \times 0{\cdot}16 \times 0{\cdot}20 \times 7760 = 29{\cdot}798\ 40$ kg.

The centre of mass C is located at (0·08, 0·06, 0·10) m. Introducing parallel axes (x_1, y_1, z_1) at C, we find from the formula (Fig. 6.16) $I = M(a^2 + b^2)/12$, that

$$I_{x_1} = 2{\cdot}483\ 20\ (0{\cdot}12^2 + 0{\cdot}20^2) = 0{\cdot}135\ 09\ \mathrm{kg\ m^2}$$
$$I_{y_1} = 2{\cdot}483\ 20\ (0{\cdot}16^2 + 0{\cdot}20^2) = 0{\cdot}162\ 90\ \mathrm{kg\ m^2}$$
$$I_{z_1} = 2{\cdot}483\ 20\ (0{\cdot}16^2 + 0{\cdot}12^2) = 0{\cdot}099\ 33\ \mathrm{kg\ m^2}$$

From symmetry $I_{x_1y_1}=I_{x_1z_1}=I_{y_1z_1}=0$. Using the parallel-axis theorem (6.16 and 6.17), we find

$$I_x = 0{\cdot}135\,09+29{\cdot}798\,40\times0{\cdot}0136 = 0{\cdot}540\,34 \text{ kg m}^2$$
$$I_y = 0{\cdot}162\,90+29{\cdot}798\,40\times0{\cdot}0164 = 0{\cdot}651\,59 \text{ kg m}^2$$
$$I_z = 0{\cdot}099\,33+29{\cdot}798\,40\times0{\cdot}0100 = 0{\cdot}397\,31 \text{ kg m}^2$$

$$\begin{cases} I_{xy} = 29{\cdot}798\,40\times0{\cdot}08\times0{\cdot}06 = 0{\cdot}143\,03 \text{ kg m}^2 \\ I_{xz} = 29{\cdot}798\,40\times0{\cdot}08\times0{\cdot}10 = 0{\cdot}238\,39 \text{ kg m}^2 \\ I_{yz} = 29{\cdot}798\,40\times0{\cdot}06\times0{\cdot}10 = 0{\cdot}178\,79 \text{ kg m}^2 \end{cases}$$

Using these values in (6.23) we find

$$b = -1{\cdot}589\,25 \qquad c = 0{\cdot}716\,40 \qquad d = -0{\cdot}065\,27$$
$$P = -0{\cdot}041\,83 \qquad Q = 0{\cdot}008\,46 \qquad R = -0{\cdot}000\,001\,7<0$$

$$\cos\theta = \frac{Q}{P\sqrt{(-P)}} = -0{\cdot}988\,55 \qquad \theta = 171{\cdot}32^\circ$$

Hence

$$I_1 = 0{\cdot}751\,88 \text{ kg m}^2 \quad I_2 = 0{\cdot}121\,21 \text{ kg m}^2 \quad I_3 = 0{\cdot}716\,16 \text{ kg m}^2$$

these are the three principal moments of inertia at O.

Equation (6.22) may now be used to find the direction of the principal axes. The determinant of the coefficients is known to vanish for I equal to any one of the three principal moments of inertia; this means that one of ω_x, ω_y or ω_z may be taken arbitrarily, so that only two of the equations need to be used.

Substituting $I_1=0{\cdot}75188$ in the first two equations of (6.22) and solving for ω_x and ω_y gives

$$\omega_x = 2{\cdot}199\,33\omega_z \qquad \omega_y = -4{\cdot}919\,38\omega_z$$

the direction of the principal axis for the maximum principal moment of inertia I_1 is now determined by the vector

$$\boldsymbol{\omega} = \omega_x\mathbf{i}+\omega_y\mathbf{j}+\omega_z\mathbf{k} = \omega_z(2{\cdot}199\,33\mathbf{i}-4{\cdot}919\,38\mathbf{j}+\mathbf{k})$$

Therefore the direction is given by the vector $2{\cdot}199\,33\mathbf{i}-4{\cdot}919\,38\mathbf{j}+\mathbf{k}$, or by the unit vector $0{\cdot}401\,29\mathbf{i}-0{\cdot}897\,59\mathbf{j}+0{\cdot}182\,46\mathbf{k}$, so that the direction cosines are $l=0{\cdot}401\,29$, $m=-0{\cdot}897\,59$, $n=0{\cdot}182\,46$.

In a similar manner the direction of the axis for the minimum principal moment I_2 is found to be $\boldsymbol{\omega}=\omega_z(0{\cdot}753\,10\mathbf{i}+0{\cdot}540\,19\mathbf{j}+\mathbf{k})$, which is in the direction of the unit vector $0{\cdot}552\,35\mathbf{i}+0{\cdot}396\,20\mathbf{j}+0{\cdot}733\,44\mathbf{k}$, which directly gives the direction cosines.

The direction of the intermediate axis I_3 may be found from the fact that it is perpendicular to the other two axes, or by a similar calculation, which gives $\boldsymbol{\omega}=\omega_z(-1{\cdot}117\,99\mathbf{i}-0{\cdot}292\,44\mathbf{j}+\mathbf{k})$; the corresponding unit vector is $-0{\cdot}731\,57\mathbf{i}-0{\cdot}191\,36\mathbf{j}+0{\cdot}654\,36\mathbf{k}$, which directly gives the direction cosines.

The three principal axes are shown in Fig. 6.24; the I_1- and I_3 axes only touch the body at O; the minimum axis I_2 goes through the body from O to a point on the top surface of the body as shown.

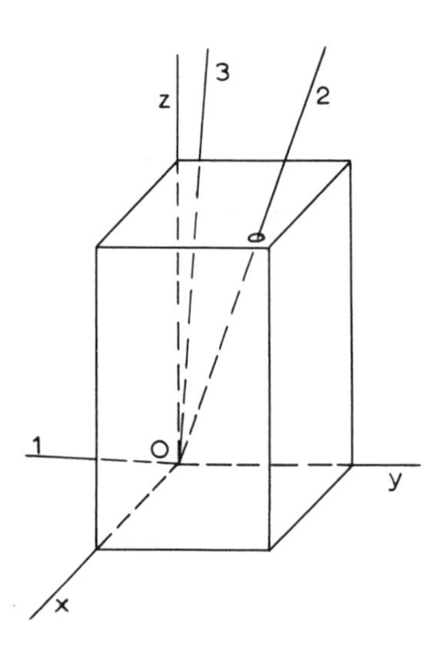

Fig. 6.24

6.3.6 Maximum and minimum moments of inertia

Let us assume that we have determined the principal axes x_1, y_1, z_1 at a point O, and the principal moments of inertia I_{x_1}, I_{y_1} and I_{z_1} for a particular rigid body. We shall also assume that the notation of the axes has been arranged so that $I_{x_1}>I_{y_1}>I_{z_1}$.

The moment of inertia of the body about some other axis $-x$ through O is now given by the formula (6.19):

$$I_x = l^2 I_{x_1} + m^2 I_{y_1} + n^2 I_{z_1}$$

Introducing $l^2+m^2+n^2=1$, or $m^2=1-l^2-n^2$ leads to

$$I_x = l^2 I_{x_1}+(1-l^2-n^2)I_{y_1}+n^2 I_{z_1}$$
$$= [(I_{x_1}-I_{y_1})l^2]-[(I_{y_1}-I_{z_1})n^2]+[I_{y_1}]$$

Each of the brackets inside the square brackets is positive, so that I_x must have its maximum value when l is a maximum, $l=1$, and n is a minimum, $n=0$, so that

$$I_{x\ \mathrm{max}} = (I_{x_1}-I_{y_1})+I_{y_1} = I_{x_1}$$

This shows that the largest principal moment of inertia is also the largest possible moment of inertia that can be obtained for any axis through O.

In a similar way, I_x is a minimum if $l=0$ and $n=1$, or

$$I_{x\ \mathrm{min}} = -(I_{y_1}-I_{z_1})+I_{y_1} = I_{z_1}$$

The smallest possible moment of inertia for any axis through O is the smallest principal moment of inertia at O. These results correspond to the results previously obtained for principal axes of an area or a lamina.

It follows from the parallel-axis theorem, which states that $I_z=I_{z_1}+Md^2$, where I_{z_1} is the minimum principal moment of inertia at the centre of mass, that I_{z_1} is also the absolute minimum moment of inertia possible for any axis in space for the body in question.

6.3.7 Invariance of the sum of the moments of inertia at a point

Consider the rigid body shown in Fig. 6.25. The sum of the moments of inertia with respect to the x-, y- and z-axes at point O is:

$$I_x+I_y+I_z = \int (y^2+z^2)\,dm+\int (x^2+z^2)\,dm+\int (x^2+y^2)\,dm$$
$$= 2\int (x^2+y^2+z^2)\,dm = 2\int r^2\,dm = \text{constant}$$

The sum $(I_x+I_y+I_z)$ depends only on the position of point O relative to the body, independently of the *orientation* of the x-, y- and z-axes. It is therefore the same for any set of perpendicular axes at point O.

Example 6.5. Determine the moment of inertia about a diameter for a homogeneous solid sphere of radius R and mass M.

Solution. Introducing a coordinate system (x, y, z) at the centre of the sphere, we have $I_x+I_y+I_z=3I_x=2\int r^2\,dm$. Dividing the sphere into thin spherical shells of thickness dr and surface $4\pi r^2$, and taking the density as ρ, we have $dm=(4\pi r^2)dr\,\rho$, so that

$$I_x = \tfrac{2}{3}\int_0^R 4\pi r^4\rho\,dr = \tfrac{8}{3}\,\pi\rho\tfrac{1}{5}R^5$$

The volume of a sphere is $\frac{4}{3}\pi R^3$, and the total mass is $M=\frac{4}{3}\pi R^3\rho$; introducing this gives $I_x=\frac{2}{5}MR^2$.

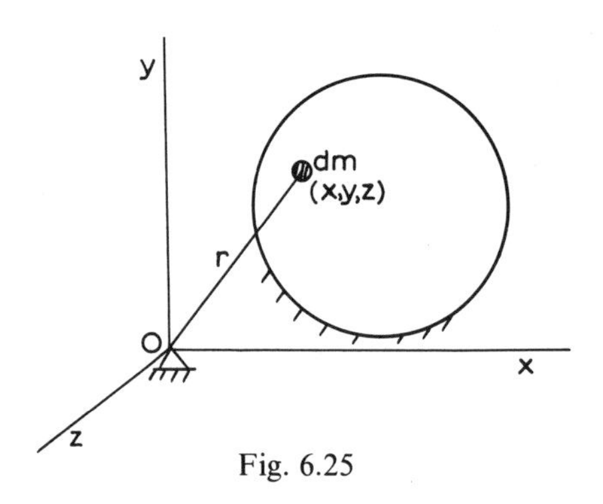

Fig. 6.25

Example 6.6. Determine the principal moment of inertia I_3 in Example 6.4 from the fact that $I_x+I_y+I_z=$constant at a point.

Solution. From Example 6.4, we have

$$I_x+I_y+I_z = 1{\cdot}589\,24 = I_1+I_2+I_3$$

so that

$$I_3 = 1{\cdot}589\,24-(I_1+I_2) = 1{\cdot}589\,24-0{\cdot}873\,09 = 0{\cdot}716\,15\text{ kg m}^2$$

in agreement with the previous result in Example 6.4. The direction of this principal axis may be determined by the fact that it is perpendicular to the other two principal axes.

Consider now a set of perpendicular axes (x, y, z) in a rigid body at a point O. If the *z-axis* is *fixed* in the body, we have $I_z=$constant, and since $I_x+I_y+I_z=$constant, we find also $I_x+I_y=$constant, even if the x- and y-axes are *rotated about the z-axis.*

This corresponds to the perpendicular-axis theorem for plane areas or laminae, the important difference being that for bodies in general the sum $I_x + I_y$ is *not* equal to I_z. If the point O is moved along the z-axis the sum $I_{x_1} + I_{y_1}$, for axes x_1 and y_1 at the new position of point O, will be different from $I_x + I_y$..

Example 6.7. Determine the moment of inertia I_y in Example 6.3, Fig. 6.20.

Solution. Since the z-axis is fixed in the body, we have

$$I_x + I_y = \text{constant} = I_{x_1} + I_{y_1} = \frac{Mr^2}{4} + \frac{Mr^2}{2} = \tfrac{3}{4}Mr^2$$

I_x was determined in Example 6.3 as

$$I_x = \frac{Mr^2}{4}(1 + \sin^2 \alpha)$$

We have now

$$I_y = \tfrac{3}{4}Mr^2 - I_x = \tfrac{3}{4}Mr^2 - \frac{Mr^2}{4}(1 + \sin^2 \alpha) = \frac{Mr^2}{4}(2 - \sin^2 \alpha)$$

$$= \frac{Mr^2}{4}(1 + \cos^2 \alpha)$$

In this particular example, where I_x is determined as a function of the angle α, we may also find I_y by substituting $\alpha + 90°$ for α in the expression for I_x.

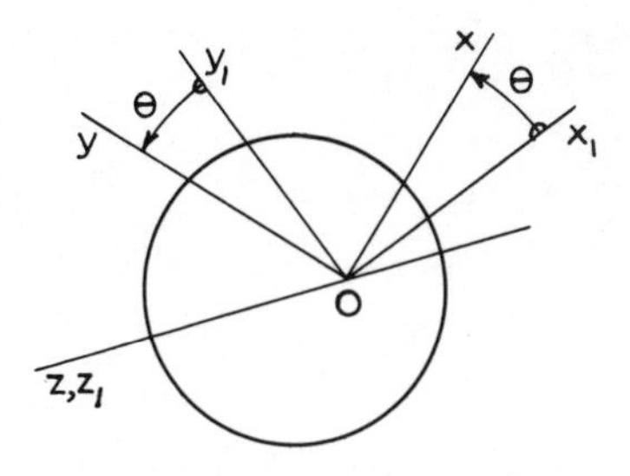

Fig. 6.26

6.3.8 A principal axis through the centre of mass

Figure 6.26 shows a rigid body with a set of *principal axes* (x_1, y_1, z_1) at a point O. A second coordinate system (x, y, z) is also shown rotated through an angle θ about the z_1-axis.

We have now, from (6.21)

$$I_{xz} = -(I_{x_1}ll'' + I_{y_1}mm'' + I_{z_1}nn'')$$

and

$$I_{yz} = -(I_{x_1}l'l'' + I_{y_1}m'm'' + I_{z_1}n'n'')$$

The direction cosines are (l, m, n) for the x-axis, (l', m', n') for the y-axis, and (l'', m'', n'') for the z-axis.

It may be seen from the figure that $l''=\cos 90=0$; $m''=\cos 90=0$ and $n''=\cos 90=1$, and also $n=\cos 90=0$ and $n'=\cos 90=0$; substituting this gives $I_{xz}=0$ and $I_{yz}=0$. This means that if the *z-axis is a principal axis at O*, the product moments I_{xz} and I_{yz} are zero *for all positions of* the x- and y-axes.

If it is known that $I_{xz}=I_{yz}=0$ for a set of axes at O, we find from the equation for principal moments of inertia in Section 6.3.5 that

$$\begin{vmatrix} I_x - I & -I_{xy} & 0 \\ -I_{xy} & I_y - I & 0 \\ 0 & 0 & I_z - I \end{vmatrix} = 0$$

Expanding the determinant gives $(I_z-I)[(I_x-I)(I_y-I)-I_{xy}^2]=0$, which is known to have three real positive roots, one of which is $I=I_z$. Hence the z-axis is a principal axis.

Figure 6.27 shows a rigid body of mass M with a coordinate system (x, y, z) with origin O and with axes taken along the principal axes at O. A parallel coordinate system (x_1, y_1, z_1) is shown with origin at a point O on the z-axis, a distance b from O.

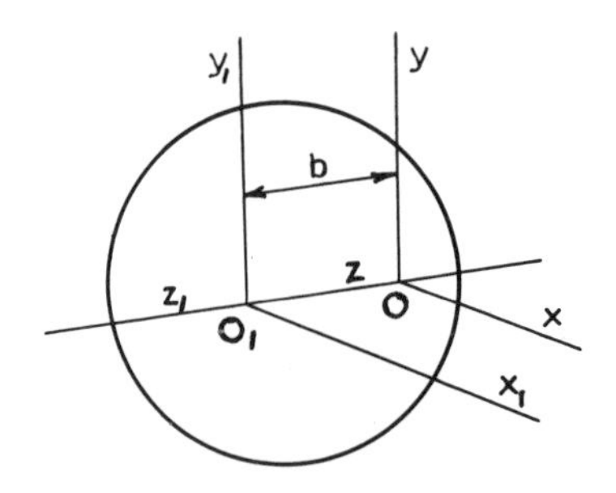

Fig. 6.27

We have $I_{x_1z_1} = \int x_1 z_1 \, dm$ and $I_{y_1z_1} = \int y_1 z_1 \, dm$. Introducing $x_1 = x$, $y_1 = y$ and $z_1 = z - b$, we find

$$I_{x_1z_1} = \int x(z-b) \, dm = \int xz \, dm - b \int x \, dm = I_{xz} - b \int x \, dm$$
$$I_{y_1z_1} = \int y(z-b) \, dm = \int yz \, dm - b \int y \, dm = I_{yz} - b \int y \, dm$$

Since the z-axis is principal at O, we have $I_{xz} = I_{yz} = 0$. The centre of mass of the body is determined by the expressions

$$x_C M = \int x \, dm \qquad y_C M = \int y \, dm \qquad z_C M = \int z \, dm$$

Assuming now that the centre of mass is *on the z-axis*, we have $x_C = y_C = 0$, or $\int x \, dm = \int y \, dm = 0$; under these conditions then $I_{x_1z_1} = I_{y_1z_1} = 0$, and the z_1-axis is a principal axis at O_1. Since O_1 was taken an arbitrary distance b from O, this holds for any point O_1 on the z_1-axis. We have hereby shown that if the z-axis is a principal axis for a point on the axis, it will be a principal axis for any point on the axis, provided that the *centre of mass* is on the axis.

Problems

6.1 The section of a 40 mm by 5 mm angle iron may be regarded as a square of side 40 mm from one corner of which a square of side 35 mm has been removed (Fig. 6.28). Determine the position of the centroid of the figure and find the second moment of area about (a) the common diagonal of the two squares, (b) an axis in the plane of the section passing through the centroid and arranged at right angles to the common diagonal.

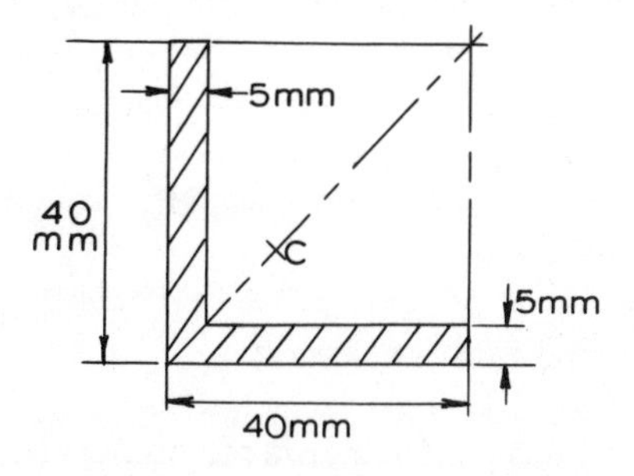

Fig. 6.28

6.2 For the triangular area in Fig. 6.29 with centre of area C, find the products of area I_{xy} and $I_{x_1y_1}$.

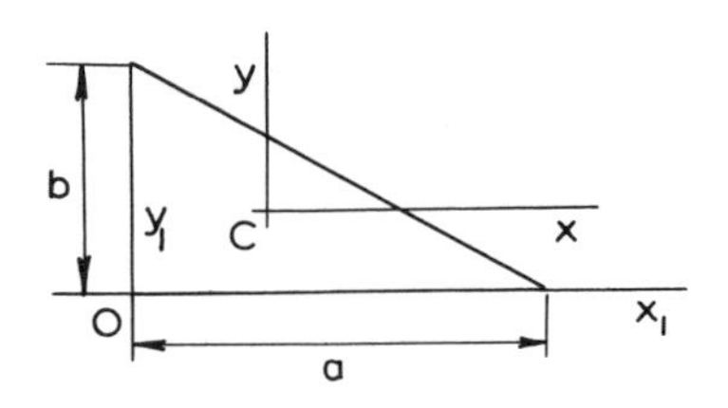

Fig. 6.29

6.3 For the Z-section shown in Fig. 6.30, the dimensions are: $a=160$ mm, $a_1=138$ mm, $b=70$ mm, $b_1=59$ mm. Find the location of the centroidal principal axis and the principal second moments of area.

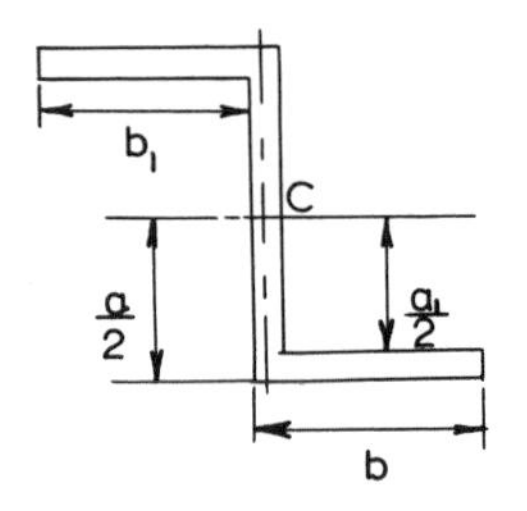

Fig. 6.30

6.4 Calculate the moment of inertia of a thin square plate of side length $2b$ and mass M with respect to an axis perpendicular to the plate through the mid-point of one side. Calculate the moment of inertia with respect to an axis perpendicular to the plane of a semi-circular thin plate of radius r and mass M, through a corner.

6.5 Locate the centre of gravity of a homogeneous right circular cone of height h and radius of base r.

6.6 Calculate the moment of inertia of a homogeneous rectangular body with respect to an axis through one edge (Fig. 6.31).

6.7 Show that the moments of inertia of a solid cone of mass M and semi-vertical angle α at its vertex are

$$I_x = \frac{3Mh^2 \tan^2 \alpha}{10} \qquad I_y = I_z = \frac{3Mh^2}{20}(4+\tan^2 \alpha)$$

where the axis of the cone is chosen as the axis OX with O at the vertex.

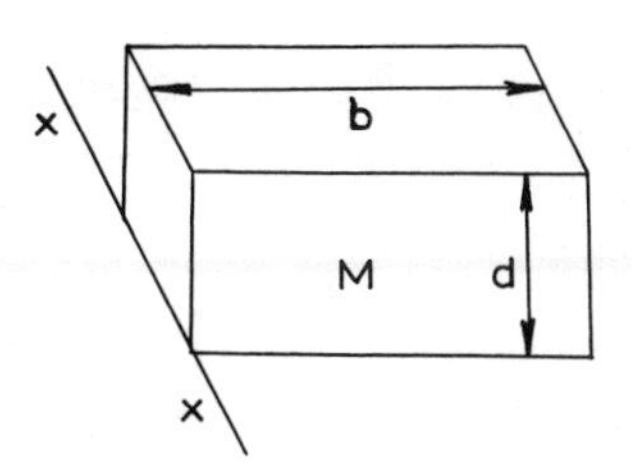

Fig. 6.31

6.8 The moment of inertia of a solid sphere of mass M about a diameter is $2Ma^2/5$ and the mass centre of a solid hemisphere is at a distance $3a/8$ from its base, a being the radius in each case. Show from these facts that the moment of inertia of a solid hemisphere about a tangent at its pole is $(13/20)\ ma^2$, where m is the mass of the hemisphere.

CHAPTER SEVEN

Rotation of a Rigid Body about a Fixed Axis

Rotation about a fixed axis is one of the most common and important types of motion encountered in engineering.

7.1 Equations of motion

The general situation is shown in Fig. 7.1, where the axis of rotation has been taken as the Z-axis for the fixed inertial coordinate system (X, Y, Z). A rotating system (x, y, z) is fixed in the body with origin O, which is a fixed point on the axis of rotation. The system has one degree of freedom, and the position may be given by the angle θ as shown.

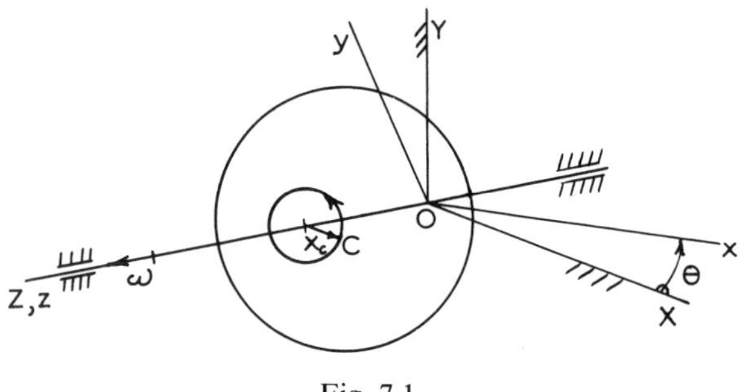

Fig. 7.1

The centre of mass C of the body is in circular motion with radius x_C, since the x- and y-axes have been arranged so that the

(x, y)-coordinates of point C are $(x_C, 0)$. The absolute acceleration components of point C in the system (x, y, z) are

$$\ddot{x}_C = -x_C\omega^2, \ddot{y}_C = x_C\dot{\omega} \text{ and } \ddot{z}_C = 0$$

Newton's second law for the motion of the centre of mass gives the force equations

$$\left.\begin{aligned} F_x &= M\ddot{x}_C = -Mx_C\omega^2 \\ F_y &= M\ddot{y}_C = Mx_C\dot{\omega} \\ F_z &= M\ddot{z}_C = 0 \end{aligned}\right\} \quad (7.1)$$

where M is the total mass of the body. If the centre of mass is on the axis of rotation, we find $F_x = F_y = F_z = 0$. The angular velocity vector $\boldsymbol{\omega}$ is placed on the axis of rotation so that $\omega_x = \omega_y = 0$ and $\boldsymbol{\omega} = \omega_z\mathbf{k}$.

Taking moments in the fixed point O, eqs. (5.5) for the moment of momentum vector components are

$$H_x = -I_{xz}\omega_z \qquad H_y = -I_{yz}\omega_z \qquad H_z = I_z\omega_z$$

Using the equation $\mathbf{M} = \dot{\mathbf{H}}$, we have $\mathbf{M} = \dot{\mathbf{H}}_r + \boldsymbol{\omega} \times \mathbf{H}$ from eq. (4.1). Since $\mathbf{i}$, $\mathbf{j}$ and $\mathbf{k}$ are of fixed directions in the system (x, y, z), we find $\dot{\mathbf{H}}_r = \dot{H}_x\mathbf{i} + \dot{H}_y\mathbf{j} + \dot{H}_z\mathbf{k}$.
The cross product

$$\begin{aligned} \boldsymbol{\omega} \times \mathbf{H} &= \omega_z\mathbf{k} \times (H_x\mathbf{i} + H_y\mathbf{j} + H_z\mathbf{k}) \\ &= \omega_z H_x\mathbf{k} \times \mathbf{i} + \omega_z H_y\mathbf{k} \times \mathbf{j} + \omega_z H_z\mathbf{k} \times \mathbf{k} = \omega_z H_x\mathbf{j} - \omega_z H_y\mathbf{i} \end{aligned}$$

We have now $\mathbf{M} = (\dot{H}_x - \omega_z H_y)\mathbf{i} + (\dot{H}_y + \omega_z H_x)\mathbf{j} + \dot{H}_z\mathbf{k}$. Introducing $\dot{H}_x = -I_{xz}\dot{\omega}_z$, $\dot{H}_y = -I_{yz}\dot{\omega}_z$ and $\dot{H}_z = I_z\dot{\omega}_z$, we find the scalar moment equations

$$\left.\begin{aligned} M_x &= -I_{xz}\dot{\omega}_z + I_{yz}\omega_z^2 \\ M_y &= -I_{yz}\dot{\omega}_z - I_{xz}\omega_z^2 \\ M_z &= I_z\dot{\omega}_z \end{aligned}\right\} \quad (7.2)$$

These equations are called *Euler's equations* after the Swiss mathematician who first developed these equations in about 1750. The rotating system (x, y, z) is often called an Euler system. If the axis of rotation is a *principal axis*, we have $I_{xz} = I_{yz} = 0$, and the equations take the simple form $M_x = 0$, $M_y = 0$, $M_z = I_z\dot{\omega}_z$.

The third Euler equation determines the rotation of the body.

The first two moment equations and the force eqs. (7.1) determine the force necessary to keep the axis of rotation stationary.

7.2 Determination of the rotational motion from the third Euler equation

In the same way as for linear motion of a particle, some simple cases of rotational motion may be determined from the equation of motion $M_z = I_z\dot{\omega}_z$; these are cases where the angle θ or the moment M_z are given functions of time, or where M_z is constant or proportional to θ. We shall use the notations $M_z = M_0$, $I_z = I$ and $\dot{\omega}_z = \dot{\omega}$, so that the equation of motion is

$$M_0 = I\dot{\omega} \tag{7.3}$$

Example 7.1. A rotor of moment of inertia I is rotated by a moment or torque M_0 so that the angular position is given by the function $\theta = At^2 + Bt$, where A and B are constants. Determine the necessary moment M_0.

Solution. From $\theta = At^2 + Bt$, we find by differentiation that $\dot{\theta} = \omega = 2At + B$, and $\ddot{\theta} = \dot{\omega} = 2A$. Equation (7.3) then gives the result $M_0 = 2AI$.

If the acting moment is given as a function of time $M_0 = f(t)$, we may again determine the motion $\theta = F(t)$ from eq. (7.3) as long as it is possible to integrate the function $f(t)$, otherwise numerical or graphical methods must be applied.

Example 7.2. A rotor of inertia I is rotating under the action of a torque $M_0 = \sin kt$, where k is a constant. Determine the resultant motion if $\omega = \omega_0$ and $\theta = 0$ when $t = 0$.

Solution. The equation of motion, from eq. (7.3) is $M_0 = I\dot{\omega} = \sin kt$, so that the angular acceleration is $\dot{\omega} = (1/I) \sin kt$. Integrating gives $\omega = -(1/kI) \cos kt + C_1$; when $t = 0$, $\omega = \omega_0$ and $C_1 = \omega_0 + 1/kI$. A second integration gives

$$\theta = \left(\frac{1}{kI} + \omega_0\right)t - \frac{1}{k^2 I} \sin kt + C_2$$

When $t = 0$, $\theta = 0$ and we find $C_2 = 0$.

In some important cases of rotation, the *applied moment* M_0 is *constant*, in which case the equation of motion is

$$\dot{\omega} = M_0/I = \alpha \text{ (constant)}$$

Comparing this equation in Section 2.4.2, $\ddot{x} = E/m = a = \text{constant}$, we see that the two equations are mathematically equivalent, so that the solution may be taken directly from (2.10):

$$\left.\begin{aligned} \theta &= \tfrac{1}{2}\alpha t^2 + \omega_0 t + \theta_0 \\ \dot{\theta} &= \omega = \alpha t + \omega_0 \\ \ddot{\theta} &= \dot{\omega} = M_0/I = \alpha \end{aligned}\right\} \qquad (7.4)$$

In (7.4) the angular velocity is ω_0 and the angular displacement θ_0, when $t = 0$.

Example 7.3. A flywheel of moment of inertia $I = 4{\cdot}7$ kg m^2 is initially rotating at 200 rev/min, when the driving torque is removed. The wheel comes to rest under the action of a constant friction torque of 2 Nm. Determine the time taken for the wheel to stop and the number of revolutions performed.

Solution. The constant angular deceleration is

$$\alpha = -\frac{M_0}{I} = -\frac{2}{4{\cdot}7} = -0{\cdot}43 \text{ rad/s}^2$$

The initial angular velocity is $\omega_0 = 200 \times \pi/30 = 20{\cdot}9$ rad/s. Equations (7.4) now give the result:

$$\begin{aligned} \omega &= -0{\cdot}43t + 20{\cdot}9 \text{ rad/s} \\ \theta &= -0{\cdot}215t^2 + 20{\cdot}9t = t(20{\cdot}9 - 0{\cdot}215t) \end{aligned}$$

where θ_0 has been taken equal to zero when $t = 0$.

The wheel stops when $\omega = 0$, so that the time elapsed is $t = 20{\cdot}9/0{\cdot}43 = 48{\cdot}6$ s. The angle rotated through is

$$\theta = 48{\cdot}6(20{\cdot}9 - 0{\cdot}215 \times 48{\cdot}6) = 508 \text{ rad}$$

The number of revolutions is $508/2\pi = 81$.

Since ω is a *linear* function of time, we may also find the angle of revolution: $\theta = \omega_{av} \times t = \tfrac{1}{2}(20{\cdot}9 + 0) \times 48{\cdot}6 = 508$.

We sometimes meet cases of rotational motion, where the applied moment is a *resisting torque proportional to the angle of rotation* $M_0 = -k\theta$, where k is a constant. The equation of motion is then, from (7.3),

$$M_0 = I\ddot{\theta} = -k\theta \qquad \text{or} \qquad \ddot{\theta} + (k/I)\theta = 0$$

Substituting $\omega_0^2 = k/I$, we have $\ddot{\theta} + \omega_0^2\theta = 0$. This equation is of the same form as eq. (2.10), and the general solution is

$$\theta = A \cos \omega_0 t + B \sin \omega_0 t$$

From the result of the solution of (2.10), we may give the general solution in the form

$$\theta = A_0 \cos (\omega_0 t - \varphi)$$

where

$$A_0 = \sqrt{\left(\theta_0^2 + \frac{\dot{\theta}_0^2}{\omega_0^2}\right)} \qquad \text{and} \qquad \tan \varphi = \frac{\dot{\theta}_0}{\theta_0 \omega_0}$$

θ_0 and $\dot{\theta}_0$ being the angular displacement and velocity when $t = 0$.

If $\dot{\theta}_0 = 0$, we find $A_0 = \theta_0$ and $\varphi = 0$ so that $\theta = \theta_0 \cos \omega_0 t$. This is a *vibratory rotational motion* with *frequency* $f = \omega_0/2\pi = (1/2\pi)\sqrt{(k/I)}$ cycles/s, and maximum amplitude θ_0.

Example 7.4. Figure 7.2 shows a so-called *torsional pendulum* consisting of a uniform circular bar or wire, which is fixed at the upper end and attached to a mass of moment of inertia I at the lower end.

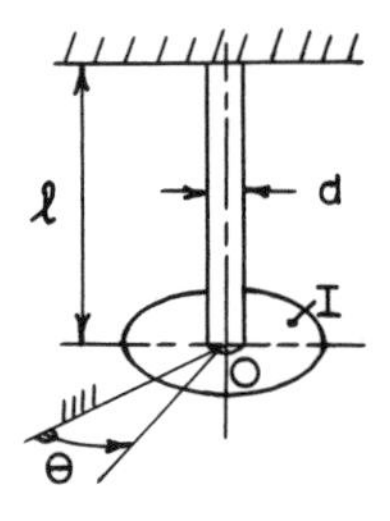

Fig. 7.2

Determine the equation of motion of the inertia and its solution, if the bar is of length l, diameter d and the modulus of shear of the bar material is G.

Solution. The torque is proportional to the angle of twist θ and may be expressed as $M_0 = -k\theta$, where k is the torque necessary to twist the shaft 1 rad. We call k the *torsional spring constant* of the shaft. It is known from the theory of strength of materials, that $k = GI_p/l$ Nm/rad, where I_p is the polar moment of the cross-sectional area. For a circular shaft $I_p = (\pi/32)d^4$ m^4. If we start the disc vibrating torsionally by releasing it at an angle of twist θ_0, the motion is given by $\theta = \theta_0 \cos \omega_0 t$; this is simple harmonic motion with frequency

$$f = \frac{\omega_0}{2\pi} = \frac{1}{2\pi}\sqrt{\frac{k}{I}} = \frac{1}{2\pi}\sqrt{\frac{GI_p}{lI}} \text{ cycles/s}$$

If G, I_p and l are known, we may determine I from this formula by measuring the frequency with a stop watch. If the system constants are not known, we may still determine I by first replacing the body with a body with a known moment of inertia I_1, usually a cylinder, and measuring the frequency f_1 for this system $f_1 = (1/2\pi)\sqrt{(k/I_1)}$. Repeating the process with both bodies we measure a frequency $f_2 = (1/2\pi)\sqrt{[k/(I_1 + I)]}$; eliminating k between these two expressions and solving for I gives

$$I = I_1\left[\left(\frac{f_1}{f_2}\right)^2 - 1\right]$$

The torsional pendulum in Example 7.4 is useful for the determination of moments of inertia where the calculation creates difficulties, for instance in the case of a rotor of an electric motor. Another useful method is the so-called *trifilar pendulum method* shown in the next example.

Example 7.5. Figure 7.3 shows a trifilar suspension of a symmetrical body. The body of moment of inertia I and mass M is suspended on three evenly spaced vertical wires at a distance r_p = pitch circle radius from C. The body is given an angular displacement θ_0 in the horizontal plane. Determine the equation of motion if the length l of the wires is large, so that the angle φ may be assumed to be small and the vertical motion of the centre of mass may be neglected. Determine the frequency of vibration.

Solution. The restoring torque $M_c = -3S \cos(90 - \varphi) \times r_p = -3Sr_p \sin\varphi$. For a large l and small φ, we may take $\sin\varphi \simeq r_p\theta/l$, and

$\cos \varphi \simeq 1$. Since we assume no vertical motion of the centre of mass, the vertical forces are balanced so that $3S \cos \varphi = Mg$, or $S \simeq Mg/3$.

The torque is now $M_c = -Mg(r_p^2/l)\theta$, and the equation of motion is $I\ddot{\theta} + Mg(r_p^2/l)\theta = 0$, or $\ddot{\theta} + (Mgr_p^2/lI)\theta = 0$. This is SHM with frequency $f = (1/2\pi)\sqrt{(Mgr_p^2/lI)}$ cycles/s.

By measuring the frequency, the moment of inertia I may be determined from this formula. This method is very useful for bodies like ships' propellers where an analytical determination of I is impractical.

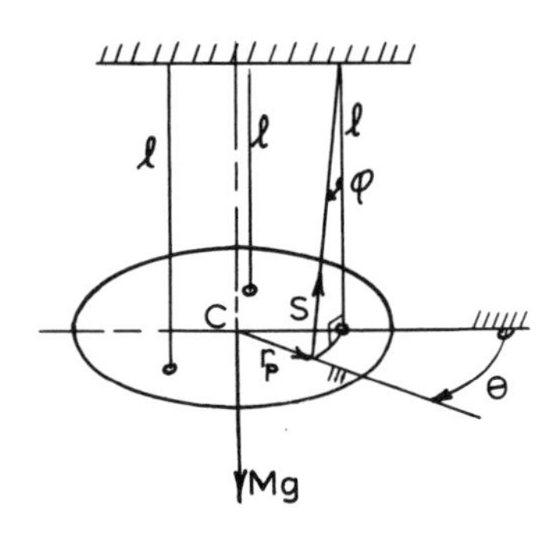

Fig. 7.3

7.3 Alternative form of equation of motion: impulse–momentum equation

The third Euler equation (7.3) of rotational motion $M_0 = I\dot{\omega} = I\,d\omega/dt$ may be integrated directly to give the equation

$$\int_1^2 M_0\,dt = \int_1^2 I\,d\omega = I(\omega_2 - \omega_1) \tag{7.5}$$

assuming that I is constant. The integral $\int_1^2 M_0\,dt$ is called the *angular impulse* of the moment M_0, and $I\omega$ was found before to be the angular momentum about the axis of rotation; the equation then states that the total angular impulse is equal to the total change in angular momentum in the same time, for rotation about a fixed axis. This equation is called *the impulse–momentum equation*. If the moment M_0 *is constant*, the equation takes the form

$$M_0(t_2 - t_1) = I(\omega_2 - \omega_1) \tag{7.6}$$

The moment equation $\mathbf{M} = \dot{\mathbf{H}}$ previously gave the component equation $M_z = \dot{H}_z$ or $M_0 = \dot{H}_z = dH_z/dt = d(I_z\omega_z)/dt = d(I\omega)dt$; this form for the equation of motion may be used when the moment of inertia I about the fixed axis of rotation changes in magnitude due to heating or changes in configuration of the body. If we have *several bodies rotating about the same axis* (Fig. 7.4) where two bodies of inertia I_1 and I_2 are connected by a friction clutch, we have the equation of motion

$$M_0 - M_f = \frac{d(I_1\omega_1)}{dt} \quad \text{and} \quad M_f = \frac{d(I_2\omega_2)}{dt}$$

summing these equations gives

$$M_0 = \frac{d}{dt}(I_1\omega_1 + I_2\omega_2) \tag{7.7}$$

The same type of equation may be established for any number of bodies. It is important to notice that all the bodies must be rotating about the *same axis*. The equation (7.7) is sometimes said to be a statement of *the principle of angular momentum.*

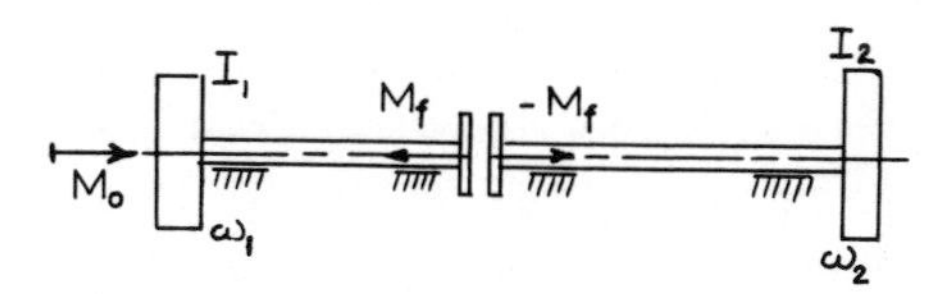

Fig. 7.4

Example 7.6. Figure 7.5 shows a shaft with an inertia $I_1 = 16{\cdot}84\,\text{kg m}^2$; the right-hand end of the shaft is connected through a friction clutch to a shaft which carries an inertia $I_2 = 10{\cdot}5$ kg m^2. The first shaft is originally rotating at 150 rev/min or $\omega_1 = 15{\cdot}70$ rad/s, and the clutch is then engaged to start the other shaft rotating, this being initially at rest; the torque on the driving shaft is removed at the instant the clutch is engaged.

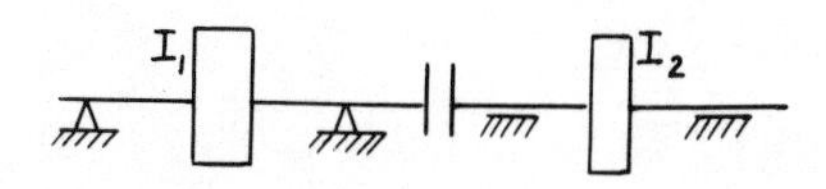

Fig. 7.5

The driving shaft runs in roller bearings with negligible friction, while the driven shaft runs in journal bearings for which the total friction torque may be assumed constant at 13·56 Nm.

The clutch is designed to transmit a maximum torque of 67·8 Nm before it starts to slip.

Determine the angular velocity of the driven shaft just when slipping has ceased, the time of slipping and the number of revolutions of each shaft during slipping.

Solution. The total torque acting on the inertia I_1 during slipping time t is the constant friction torque $M_f = 67{\cdot}8$ Nm; while the total torque on the inertia I_2 is M_f and the resisting friction torque from the journal bearings is 13·56 Nm.

Taking $t_1 = 0$ and $t_2 = t$, and taking the combined speed to be ω_2 when slipping ceases, we find the impulse–momentum equations from (7.6):

$$-67{\cdot}8t = 16{\cdot}84\,(\omega_2 - 15{\cdot}70)$$

$$(67{\cdot}8 - 13{\cdot}56)t = 10{\cdot}5\omega_2$$

The solution is $t = 1{\cdot}71$ s and $\omega_2 = 8{\cdot}84$ rad/s $= 84{\cdot}4$ rev/min. Since M_f is constant, we have $M_f = I\, d\omega/dt$, or $d\omega = (M_f/I)\, dt$, so that $\omega = (M_f/I)t + \omega_1$, showing a linear relationship with time t.

The total rotation is given by

$$\theta = \int_0^t \omega\, dt = \frac{\omega_1 + \omega_2}{2}\, t$$

For I_1 we find $\theta_1 = (15{\cdot}70 + 8{\cdot}84)1{\cdot}71/2 = 20{\cdot}95$ rad $= 3{\cdot}34$ rev. For I_2 the result is $\theta_2 = (8{\cdot}84/2) \times 1{\cdot}71 = 7{\cdot}55$ rad $= 1{\cdot}20$ rev.

If the moment of the external forces with respect to the axis of rotation vanishes, so that $M_0 = 0$, we find from (7.7) that the *total angular momentum* $I_1\omega_1 + I_2\omega_2 + \cdots$ of the system *remains constant*. This is called the *principle of conservation of angular momentum*.

Example 7.7. Determine the angular velocity ω_2 of the system in Example 7.6, if the friction torque from the journal bearings is neglected.

Solution. With no external torque acting, the total angular momentum remains constant. The initial angular momentum is $I_1\omega_1$ and the final value is $(I_1+I_2)\omega_2$; equating these gives $\omega_2=I_1\omega_1/(I_1+I_2)=16{\cdot}84\times 15{\cdot}70/27{\cdot}34=9{\cdot}66$ rad/s. This is somewhat larger than found in Example 7.6, due to the fact that the bearing friction has been neglected in this example.

When the external torque becomes zero, the angular momentum $I\omega$ stays constant; to change the angular velocity ω the moment of inertia I must be changed; this is used by skaters and divers to increase or decrease their angular velocity by changing their moment of inertia.

7.4 Alternative form of equation of motion: work–energy equation

7.4.1 Kinetic energy of a rigid body rotating about a fixed axis

For a rigid body rotating about a fixed axis, each mass particle dm, at a distance r from the axis of rotation, has a velocity $v=r\omega$; the kinetic energy of the particle is then $\frac{1}{2}dm\, v^2=\frac{1}{2}dm\, r^2\omega^2$. Summing up for all particles of the body, we find the total kinetic energy:

$$T = \int \tfrac{1}{2}r^2\omega^2\, dm = \tfrac{1}{2}\omega^2 \int r^2\, dm = \tfrac{1}{2}I\omega^2 \qquad (7.8)$$

this expression holds for any rigid body rotating about a fixed axis: the axis need not be principal and the centre of mass need not be on the axis of rotation.

If the centre of mass is a distance x_C from the axis of rotation, we have $v_C=x_C\omega$. If, for an axis through the centre of mass and *parallel* to the axis of rotation, the moment of inertia is I_C, we have from the parallel-axis theorem that $I=I_C+Mx_C^2$, where M is the total mass of the body. Substituting this in (7.8), we have

$$T = \tfrac{1}{2}I_C\omega^2+\tfrac{1}{2}Mx_C^2\omega^2 = \tfrac{1}{2}I_C\omega^2+\tfrac{1}{2}Mv_C^2 \qquad (7.9)$$

as an alternative form for the kinetic energy.

7.4.2 Work done by a moment or torque

For a couple or moment vector **M**, the two parallel forces forming the couple do no total work during a translation of the couple so that only rotation is important; the work done by a couple during an *infinitesimal* rotation $d\boldsymbol{\theta}$ is defined, in the same way as the work by a force, by the dot product $\mathbf{M}\cdot d\boldsymbol{\theta}$. For rotation from an angle θ_1 to an angle θ_2 the work done is

$$\int_{\theta_1}^{\theta_2} \mathbf{M}\cdot d\boldsymbol{\theta} = \int_1^2 (M_x\, d\theta_x + M_y\, d\theta_y + M_z\, d\theta_z)$$

The *rate of work done* $d(\text{W.D.})/dt$ or the *power* is $\mathbf{M}\cdot d\boldsymbol{\theta}/dt = \mathbf{M}\cdot\boldsymbol{\omega}$, for a couple $\mathbf{M}$.

7.4.3 Work–energy equation

The equation of motion $M_0 = I\, d\omega/dt$ may be integrated as follows:

$$M_0\, d\theta = I\, d\omega\, d\theta/dt = I\, d\omega\, \omega = \tfrac{1}{2}I\, d(\omega^2)$$

so that

$$\text{W.D.} = \int_{\theta_1}^{\theta_2} M_0\, d\theta = \tfrac{1}{2}I\,(\omega_2^2 - \omega_1^2) = T_2 - T_1 \qquad (7.10)$$

where we assume that I is constant.

This equation is called *the work–energy equation*; it states that the work done by the total moment about the axis of rotation is equal to the *total change* in kinetic energy of the body.

If the torque is constant, we find from (7.10) that

$$M_0(\theta_2 - \theta_1) = \tfrac{1}{2}I(\omega_2^2 - \omega_1^2) \qquad (7.11)$$

The equation is particularly useful in problems where time is not involved in the solution.

Example 7.8. In Example 7.6, determine the energy lost in the clutch during slipping, that is the kinetic energy converted to heat in the clutch.

Solution. The total kinetic energy of the system just before the clutch is engaged is $T_1 = \tfrac{1}{2}I_1\omega_1^2 = \tfrac{1}{2}\times 16{\cdot}84 \times 15{\cdot}70^2 = 2075$ Nm.

The total kinetic energy just after slipping ceases is $\tfrac{1}{2}(I_1 + I_2)\omega_2^2 = \tfrac{1}{2}\times 27{\cdot}34 \times 8{\cdot}84^2 = 1067$ Nm. The decrease is 1008 Nm, but part of the loss is due to the friction in the journal bearings, and this part amounts to $13{\cdot}56 \times 7{\cdot}55 = 102$ Nm. The total loss in the clutch is thus 906 Nm.

The result may alternatively be determined as the difference in the work done by the friction torque on the two parts of the system; this is $M_f(\theta_1 - \theta_2) = 67{\cdot}8(20{\cdot}95 - 7{\cdot}55) = 906$ Nm.

Example 7.9. Figure 7.6 shows a pulley of radius r and moment of inertia I. A mass M is connected to the pulley as shown. Neglecting friction, determine the velocity of the mass as a function of the distance moved, if the system is released from rest. Determine also the acceleration of the mass M.

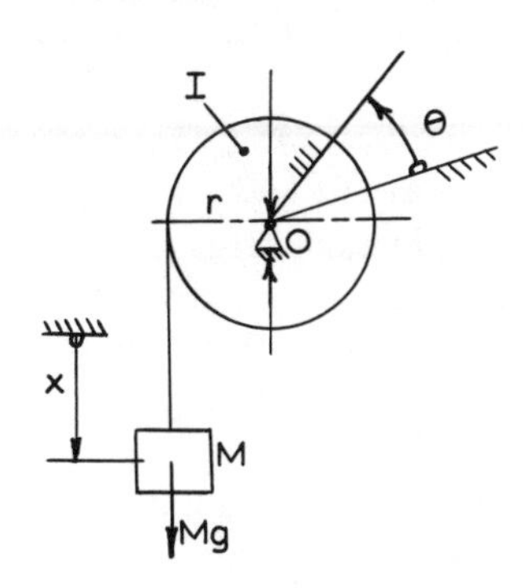

Fig. 7.6

Solution. With the mass at a position x a shown, the total kinetic energy is $T = \frac{1}{2}MV^2 + \frac{1}{2}I\omega^2$; since the kinetic energy at rest is zero, this is also the total change in kinetic energy.

We have now $V^2 = \dot{x}^2$, and $x = r\theta + x_0$, or $\dot{x} = r\dot{\theta} = r\omega$, so that $\omega^2 = \dot{x}^2/r^2$, and $T = \frac{1}{2}(M + I/r^2)\dot{x}^2$.

The only force doing work is the gravity force Mg, the work done is Mgx. The work–energy equation now gives the relation

$$Mgx = \tfrac{1}{2}(M + I/r^2)\dot{x}^2$$

from which

$$\dot{x} = \sqrt{[2Mgx/(M + I/r^2)]}$$

Differentiating the work–energy relationship gives $Mg\dot{x} = [M + (I/r^2)]\dot{x}\ddot{x}$, so that the acceleration is $\ddot{x} = Mg/(M + I/r^2)$.

Example 7.10. Determine the number of revolutions in Example 7.3, by using the work–energy equation.

Solution. The initial angular velocity is $\omega = 20{\cdot}9$ rad/s, so that $T_1 = \frac{1}{2}I\omega^2 = \frac{1}{2} \times 4{\cdot}7 \times 20{\cdot}9^2 = 1026$ Nm. The final kinetic energy $T_2 = 0$.

From the work–energy equation we find $-2 \times \theta = T_2 - T_1 = -1026$, and $\theta = 513$ rad. The number of revolutions are $513/2\pi = 81{\cdot}7$. The difference from the previous result is about 1%, which is due to the use of a slide-rule.

7.5 Principle of conservation of mechanical energy

The principle of conservation of mechanical energy for a particle was stated in Chapter 2, eq. (2.15) as $V + T = \text{constant}$, or the sum of the potential and kinetic energy is constant for a particle moving under the action of conservative forces only.

For a rigid body, the internal forces occur in equal and opposite pairs and produce no work in a summation extended over all particles of the body.

The work–energy eq. (7.10) states that the work done (W.D.) $= T_2 - T_1$; dividing the acting forces into conservative forces doing the work $(\text{W.D.})_c$ and non-conservative forces doing the work $(\text{W.D.})_n$, we have

$$(\text{W.D.}) = (\text{W.D.})_c + (\text{W.D.})_n = T_2 - T_1$$

The potential–energy difference is $V_2 - V_1 = -(\text{W.D.})_c$, so that

$$(\text{W.D.})_n - (V_2 - V_1) = T_2 - T_1 \quad \text{or} \quad (\text{W.D.})_n = (T_2 - T_1) + (V_2 - V_1)$$

Now if all the acting forces are conservative, we have $(\text{W.D.})_n = 0$, or $(T_2 - T_1) + (V_2 - V_1) = 0$, so that $T_2 + V_2 = T_1 + V_1$,

or

$$V + T = \text{constant} \tag{7.12}$$

This is the *principle of conservation of mechanical energy* for a rigid body moving under the action of conservative forces only. The principle is sometimes useful for establishing the equation of motion by differentiation with respect to time, since

$$dV/dt + dT/dt = 0$$

Example 7.11. In Example 7.9, determine the acceleration of the mass M by using the principle of conservation of mechanical energy.

Solution. If we neglect friction on the pulley shaft and air resistance, this is a conservative system.

The reaction from the pulley shaft and the weight force of the pulley do not move and produce no work. The pressure between the string and the pulley and the string tension are internal forces, for which the total work vanishes.

Determining the position of the system by the coordinates x and θ, and taking the datum position at $x=0$, we have $V=-Mgx$. The kinetic energy $T=\frac{1}{2}(M+I/r^2)\dot{x}^2$, as determined in Example 7.9.

Equation (7.12) now states that $\frac{1}{2}(M+I/r^2)\dot{x}^2-Mgx=\text{constant}$, and differentiation gives

$$\left(M+\frac{I}{r^2}\right)\dot{x}\ddot{x}-Mg\dot{x} = 0 \qquad \text{or} \qquad \ddot{x} = \frac{Mg}{(M+I/r^2)}$$

It is always advisable to check a result found in terms of system constants by both a dimensional check and by 'logical extremes'.

A dimensional investigation shows that the right-hand side of the result has the dimension of an acceleration as it should have.

There are four logical extremes: if $M \gg I$, $\ddot{x} \to g$, the pulley has no effect and M is in a free fall. If $M \to 0$, we find $\ddot{x} \to 0$. If $I \gg M$, $\ddot{x} \to 0$, the mass is unable to accelerate the pulley. Finally if $I \to 0$, $\ddot{x} \to g$, the mass M is again in a free fall with acceleration g.

7.6 D'Alembert's principle for rotation. Inertia torque. Dynamic equilibrium

The fundamental moment equation $\mathbf{M}=\dot{\mathbf{H}}$ may be stated in the form $\mathbf{M}+(-\dot{\mathbf{H}})=\mathbf{0}$; this form of the equation was first stated by D'Alembert.

Because of the complexity of the expressions for the components of $\mathbf{H}$, the above equation is convenient in the case of rotation about a fixed axis only when this axis *is a principal axis*. In this case we have the equation $M_0=I\dot{\omega}$, which may be stated in the form

$$M_0+(-I\dot{\omega}) = 0 \tag{7.13}$$

This expresses *D'Alembert's principle* for rotation about a fixed *principal* axis. The term $(-I\dot{\omega})$ has the dimension of a moment or torque and is called *the inertia torque*. If the inertia torque is introduced in a given situation, we say that D'Alembert's principle has been applied to create 'dynamic equilibrium'. The moment equation is then established as in a *statics* case, and may be taken about *any axis* perpendicular to the plane of motion. This may often be an advantage if a point can be found in which unknown forces act, since these forces are eliminated in the moment equation if the point is used as a moment centre.

Example 7.12. Figure 7.7 shows a lift cage of mass M_2 which is raised by a rope passing over a revolving drum, which carries a counterbalance of mass M_1.

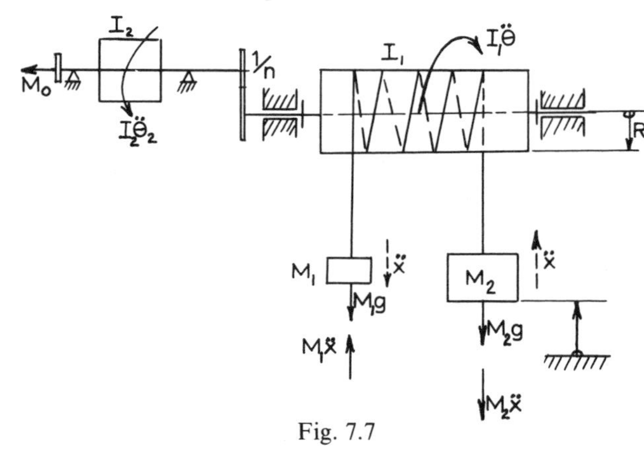

Fig. 7.7

The radius of the drum is R and the moment of inertia about the axis of rotation is I_1; the drum is geared at a ratio $1/n$ to a second rotating inertia I_2 which is driven by an external torque $\mathbf{M}_0$ as shown.

(a) Determine the equation of motion of the total system in terms of the given system constants.

(b) Determine the necessary starting torque if M_2 has to start upwards with an acceleration of 0·610 m/s^2, and $R=0{\cdot}457$ m, $n=5, I_1=75{\cdot}5\,\text{kg}\,\text{m}^2, I_2=8{\cdot}4\,\text{kg}\,\text{m}^2, M_1=3200\,\text{kg}$ and $M_2=6100\,\text{kg}$.

(c) Determine the speed of revolution of the driving torque and the necessary power, if M_2 has to be lifted at a constant speed of 73·2 m/min.

Solution. (a) The system is established in dynamic equilibrium by introducing all the external acting forces and torques: these are the bearing reactions, the gravity forces M_1g and M_2g and the torque M_0. The inertia forces $-M_1\ddot{x}$ and $-M_2\ddot{x}$ and the inertia torques $-I_1\ddot{\theta}$ and $-I_2\ddot{\theta}_2$ are introduced as shown, always opposite to the linear and angular accelerations.

The system may now be treated as a case of statics. Writing a torque balance for the right-hand side of the system, we find the torque M_G on the gear wheel to be

$$M_G = (M_2g+M_2\ddot{x})R+(M_1\ddot{x}-M_1g)R+I_1\ddot{\theta}_1$$

The torque on the left-hand side of the system is now M_G/n, and a torque balance for the left-hand side gives $M_0=I_2\ddot{\theta}_2+M_G/n$. Substituting $\ddot{\theta}_1=\ddot{x}/R$ and $\ddot{\theta}_2=n\ddot{\theta}_1=n\ddot{x}/R$, we find

$$M_0 = \left[\frac{I_1}{Rn}+I_2\frac{n}{R}+(M_1+M_2)\frac{R}{n}\right]\ddot{x}+(M_2-M_1)\frac{R}{n}g$$

or

$$M_0 = \frac{R}{n}\left[C\ddot{x}+(M_2-M_1)g\right]$$

where $C=M_1+M_2+(I_1+n^2I_2)/R^2$.

(b) With the given figures we find $C=10\ 667$ kg, and $M_0=974\ddot{x}+2600$. If $\ddot{x}=0{\cdot}610$ m/s^2, we find $M_0=594+2600=3194$ Nm starting torque.

(c) If $\dot{x}=73{\cdot}2$ m/min constant, $\ddot{x}=0$, so that $M_0=2600$ Nm constant. $\dot{\theta}_2=(n/R)\dot{x}=5/0{\cdot}457\times73{\cdot}2=800$ rad/min. $N=\dot{\theta}_2/2\pi=127{\cdot}2$ rev/min constant. The necessary power is $M_0\dot{\theta}_2=2600\times800/60=34\ 800$ W$=34{\cdot}8$ kW. The power may also be determined as $(M_2-M_1)g\dot{x}=(6100-3200)9{\cdot}81\times73{\cdot}2/60=34\ 800$ W.

7.7 Principle of virtual work in rotation

For a system *in dynamic equilibrium*, we may state that the virtual work vanishes for any virtual displacement of the system. For rotation this may be stated in *D'Alembert's equation*: $(\mathbf{M}-\dot{\mathbf{H}}\cdot\delta\boldsymbol{\theta}=0$.

For the actual solution of problems, this equation is only convenient for *rotation about a principal axis*.

D'Alembert's principle in combination with the principle of virtual work is convenient in many problems; for more complicated problems this method is, however, largely superseded by the method of Lagrange to be discussed in Chapter 9.

Example 7.13. Determine the equation of motion in Example 7.12, by using the principle of virtual work.

Solution. After putting the system in dynamic equilibrium as shown in Fig. 7.7, we give the system a virtual displacement δx. The principal of virtual work now gives the following equation:

$$\delta(\text{W.D.}) = -(M_2 g + M_2 \ddot{x})\,\delta x + (M_1 g - M_1 \ddot{x})\,\delta x - I_1 \ddot{\theta}_1\,\delta\theta_1 - I_2 \ddot{\theta}_2\,\delta\theta_2 + M_0\,\delta\theta_2 = 0$$

Introducing $\delta\theta_1 = \delta x/R$ and $\delta\theta_2 = (n/R)\,\delta x$, we find, by cancelling δx and solving for M_0, the same result for M_0 as previously found in Example 7.12.

The main advantage of the principle of virtual work is that the work of internal forces, fixed reactions and normal forces does not enter the expression for virtual work. If friction forces are present, they must be included with the acting forces.

7.8 Equivalent moment of inertia

Consider the two discs in Fig. 7.8, which are geared together and rotating due to the action of an external torque M_0 as shown. The principle of virtual work gives the equation

$$\delta(\text{W.D.}) = M_0\,\delta\theta_1 - I_1 \ddot{\theta}_1\,\delta\theta_1 - I_2 \ddot{\theta}_2\,\delta\theta_2 = 0$$

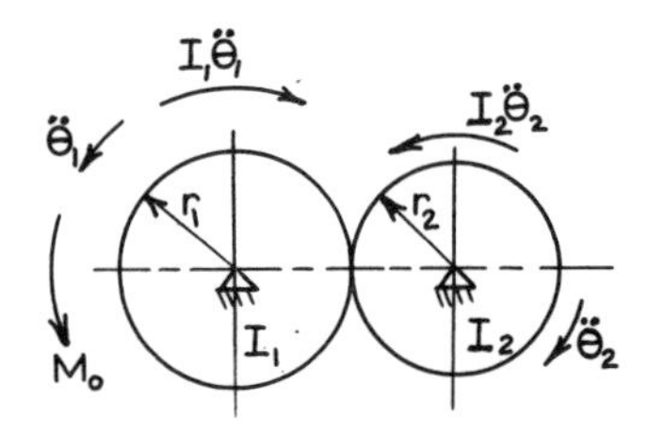

Fig. 7.8

Introducing the equal peripheral speed, we have $r_1\theta_1 = r_2\theta_2$, or

$$\theta_2 = (r_1/r_2)\theta_1 \qquad \delta\theta_2 = (r_1/r_2)\,\delta\theta_1 \qquad \ddot{\theta}_2 = (r_1/r_2)\ddot{\theta}_1$$

Substituting in the virtual work equation leads to

$$M_0 = \left[I_1 + \left(\frac{r_1}{r_2}\right)^2 I_2\right]\ddot{\theta}_1$$

as the equation of motion of the system. The system behaves dynamically as a system with one rotating inertia I_q, so that $M_0 = I_q\ddot{\theta}_1$, and

$$I_q = I_1 + \left(\frac{r_1}{r_2}\right)^2 I_2 = I_1 + \left(\frac{\omega_2}{\omega_1}\right)^2 I_2$$

I_q is called the *equivalent moment of inertia* of the system. The external torque M_0, the angle of rotation θ_1 and the total kinetic energy of the system are unchanged.

The use of an equivalent moment of inertia is very convenient when we are dealing with a series of inertias geared together.

Example 7.14. For the system shown in Fig. 7.9, determine the torque M_0 necessary to give body A an angular acceleration $\dot{\omega}_1$.

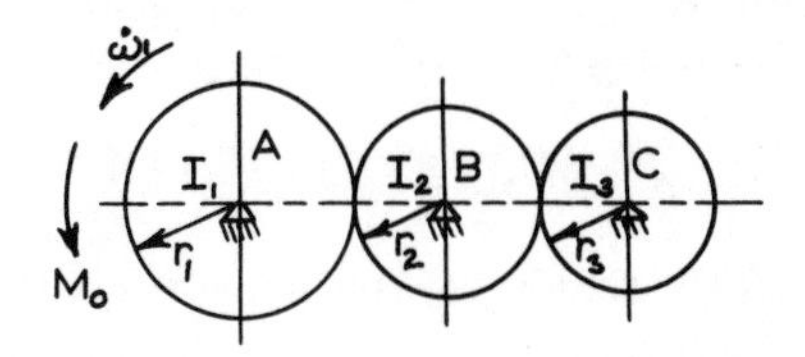

Fig. 7.9

Solution. The equivalent moment of inertia of bodies B and C on the shaft of body B is found to be $I_2 + I_3(r_2/r_3)^2$; combining this with the moment of inertia I_1 on the shaft of body A gives the total equivalent moment of inertia

$$I_q = I_1 + \left[I_2 + I_3\left(\frac{r_2}{r_3}\right)^2\right]\left(\frac{r_1}{r_2}\right)^2 = I_1 + I_2\left(\frac{r_1}{r_2}\right)^2 + I_3\left(\frac{r_1}{r_3}\right)^2$$

and the equation of motion is $M_0 = I_q\dot{\omega}_1$.

7.9 The compound pendulum

A rigid body, free to rotate about a fixed axis due to the action of gravity, is called a compound pendulum, Fig. 7.10.

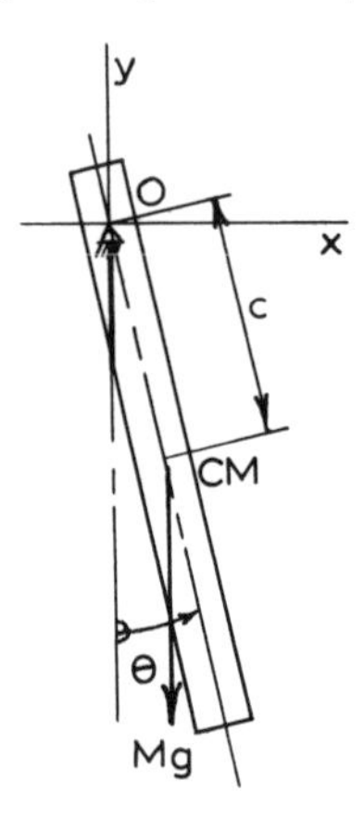

Fig. 7.10

The equation of motion is $M_0 = I_0\dot{\omega}$. The moment about the axis of rotation is the moment of the gravity force Mg, and is in the opposite direction to the angular displacement θ, so that $M_0 = -Mgc \sin\theta$, and the equation of motion is

$$-Mgc \sin\theta = I_0\dot{\omega} = I_0\ddot{\theta}$$

For small values of θ, that is $\theta < 20°$, we may substitute θ for $\sin\theta$ with sufficient accuracy, and the equation takes the form $\ddot{\theta} + (Mgc/I_0)\theta = 0$. Introducing $I_0 = Mr_0^2$, we get $\ddot{\theta} + (gc/r_0^2)\theta = 0$, which indicates a simple harmonic motion with frequency $f = (1/2\pi)\sqrt{(gc/r_0^2)}$. For a simple pendulum of length l, the frequency was found in Example 2.8 to be $f = (1/2\pi)\sqrt{(g/l)}$. Suppose now that we want to determine the length of a simple pendulum so that it has the same frequency as the compound pendulum; equating the frequencies gives $c/r_0^2 = 1/l$, or $l = r_0^2/c$. This length is called the *equivalent length* of the compound pendulum, and determines the distance from the point of suspension to a point P in which the total mass must be assumed concentrated to keep the frequency unchanged.

For a slender uniform bar of length L, suspended at one endpoint, we find that $l = r_0^2/c = (L^2/3)/(L/2) = \frac{2}{3}L$. Introducing $I_0 = I_C + Mc^2$, or $Mr_0^2 = Mr_c^2 + Mc^2$, we have $r_0^2 = r_c^2 + c^2$, so that $l = r_c^2/c + c > c$. The point P is evidently further away from the centre of suspension than is the centre of mass. The point P is called the *centre of percussion*. The theory of the compound pendulum may be used to determine the moment of inertia I_c of a rigid body, as shown in the next example.

Example 7.15. Determine the moment of inertia I_C of the connecting rod shown in Fig. 7.11(a).

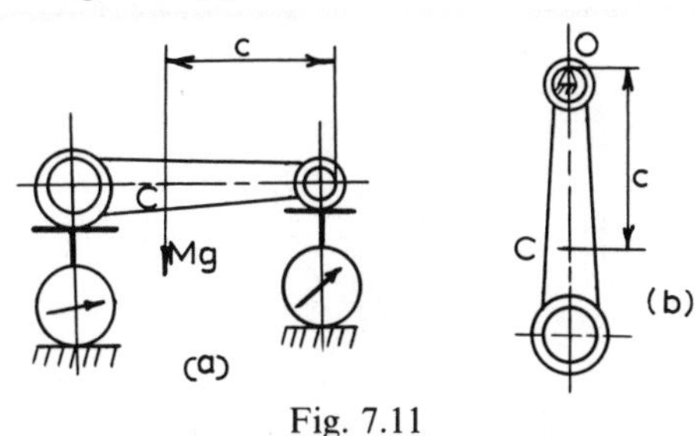

Fig. 7.11

Solution. The rod is arranged, on two scales, with horizontal axis of symmetry, as shown in Fig. 7.11(a). In this way, the total mass M of the rod and the position of the centre of mass C are determined. The distance c as shown is found from statics. Suspending the rod on a knife edge as shown in Fig. 7.11(b), the time for a number of swings of the rod may be measured, and we have $f = (1/2\pi)\sqrt{(cg/r_0^2)}$, which determines r_0^2 and $I_0 = Mr_0^2$. The desired moment of inertia I_C may now be determined from $I_0 = I_C + Mc^2$

7.10 Dynamic equilibrium for the case in which the centre of mass is not on the axis of rotation

Figure 7.12(a) shows a body rotating about a fixed axis under the action of external forces and moments not shown. A coordinate system (x, y, z) is fixed in the body as shown, with the centre of mass C of the body located on the x-axis. D'Alembert's principle and dynamic equilibrium are only convenient if the body is homogeneous and has a *plane of symmetry xy*, in which case $I_{xz} = I_{yz} = 0$, so that the z-axis is principal. We will assume that this is the case here.

The equation of rotational motion is now $M_0 = I_0\dot{\omega}$, where M_0 is the moment about the axis of rotation of *all external forces*.

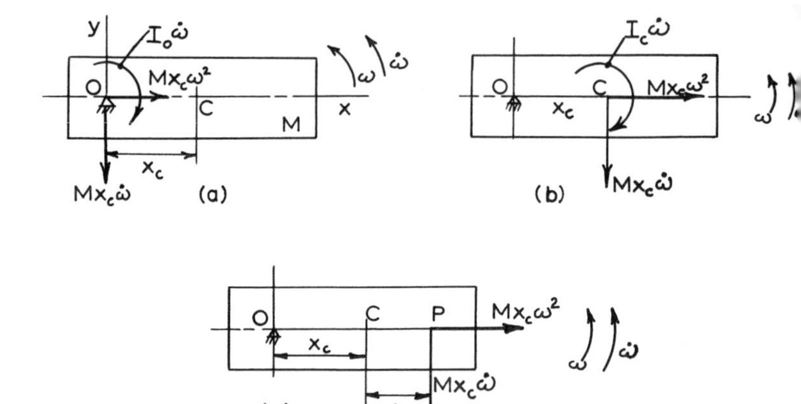

Fig. 7.12

The centre of mass is in a circular motion with radius x_C and acceleration components $x_C\omega^2$ and $x_C\dot{\omega}$. The force equations for the motion of the centre of mass are $F_x = -Mx_C\omega^2$ and $F_y = Mx_C\dot{\omega}$.

Introducing the normal and tangential forces F_N and F_T, we have $F_N = -F_x$ and $F_T = F_y$, so that the equations of motion are $F_N = Mx_C\omega^2$ and $F_T = Mx_C\dot{\omega}$. From D'Alembert's principle, we now have the three equations

$$M_0 + (-I_0\dot{\omega}) = 0 \qquad F_N + (-Mx_C\omega^2) = 0$$

$$F_T + (-Mx_C\dot{\omega}) = 0$$

The system may now be placed in dynamic equilibrium by introducing all acting external forces and moments, together with the *inertia forces* $-Mx_C\omega^2$ and $-Mx_C\dot{\omega}$ and the *inertia torque* $-I_0\dot{\omega}$. The inertia forces and the inertia torque are all opposite to the directions of the accelerations and may be introduced as shown in Fig. 7.12(a). The magnitude of the inertia forces may be calculated by assuming the total mass M of the body concentrated at the centre of mass.

The inertia forces may be moved to the centre of mass, as shown in Fig. 7.12(b). To compensate for the motion of the force $Mx_C\dot{\omega}$ through a distance x_C, the inertia torque about the centre of mass must now be taken as

$$I_0\dot{\omega}-(Mx_C\dot{\omega})x_C = (I_0-Mx_C{}^2)\dot{\omega} = (I_C+Mx_C{}^2-Mx_C{}^2)\dot{\omega} = I_C\dot{\omega}$$

If the inertia forces are moved to a point P a distance a from the centre of mass, as shown in Fig. 7.12(c), the inertia torque must be taken as $I_C\dot{\omega}-(Mx_C\dot{\omega})a=(I_C-Mx_Ca)\dot{\omega}$; equating this to zero, we find $a=I_C/Mx_C=Mr_C^2/Mx_C=r_C^2/x_C$. The total distance of the point P from O is then $l=x_C+r_C^2/x_C$, so that P is the *centre of percussion* discussed in Section 7.9. If the inertia forces are applied at the centre of percussion, no inertia torque is required for dynamic equilibrium.

Example 7.16. Figure 7.13 shows a slender, rigid, uniform homogeneous bar, able to rotate in a vertical plane about a fixed axis through the lower end at point O. The bar is held in a vertical position and released with a slight disturbance so that it rotates as shown. The initial angular velocity may be taken as zero.

Determine the components of the reaction on the bar as functions of the angle of rotation θ.

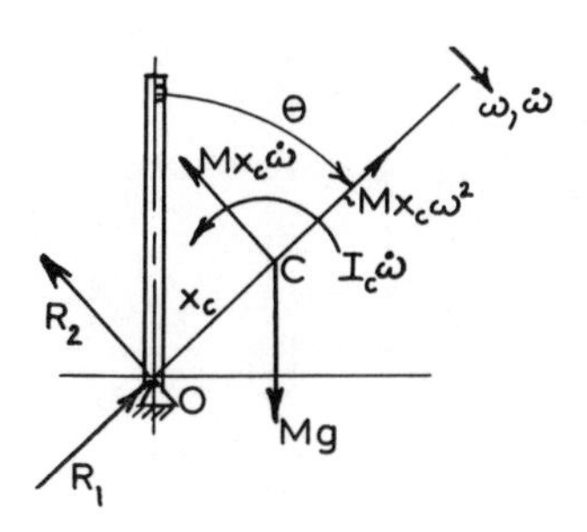

Fig. 7.13

Solution. The components of the reaction at O are introduced as forces R_1 and R_2 along and perpendicular to the bar as shown, the only other external force being the gravity force Mg.

We have dynamic equilibrium after introduction of the inertia forces $Mx_C\dot{\omega}$ and $Mx_C\omega^2$ and the inertia torque $I_C\dot{\omega}$ as shown.

Static equations of equilibrium may now be used on this plane system of forces. Resolving the forces along and perpendicular to the bar gives the equations

$$R_1 - Mg\cos\theta + Mx_C\omega^2 = 0 \qquad \text{(a)}$$

$$R_2 - Mg\sin\theta + Mx_C\dot{\omega} = 0 \qquad \text{(b)}$$

Taking moments about point C gives the result

$$R_2x_C - I_C\dot{\omega} = 0 \qquad \text{(c)}$$

Equation (c) gives $\dot{\omega} = R_2x_C/I_C$, and substituting this in (b) leads to $R_2 = Mg(I_C/I_0)\sin\theta$. The work–energy equation gives W.D.$=$ $Mgx_C(1-\cos\theta) = \frac{1}{2}I_0\omega^2$, or $\omega^2 = (2Mgx_C)/I_0(1-\cos\theta)$, and substituting this in (a) leads to $R_1 = Mg\cos\theta - (2M^2gx_C^2/I_0)(1-\cos\theta)$.

If the bar is of length l, we have $x_C = l/2$, $I_C = \frac{1}{12}Ml^2$ and $I_0 = \frac{1}{3}Ml^2$; substituting this gives

$$R_1 = \tfrac{1}{2}Mg(5\cos\theta - 3) \qquad \text{and} \qquad R_2 = \tfrac{1}{4}Mg\sin\theta$$

Example 7.17. Figure 7.14 shows a compound pendulum at rest. A horizontal force F is suddenly applied in an impact as shown. Determine the distance a so that the horizontal reaction at O vanishes, and show that P must be the centre of percussion. The vertical plane is a plane of symmetry.

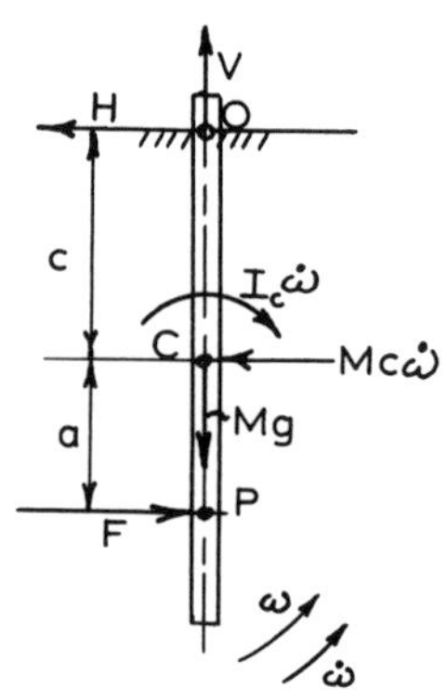

Fig. 7.14

Solution. At the instant considered, the angular velocity $\omega = 0$, while we have a value for the angular acceleration $\dot{\omega}$. The system is in dynamic equilibrium after introduction of the inertia force $Mc\dot{\omega}$ and the inertia torque $I_C\dot{\omega}$ as shown.

Summing the forces vertically and horizontally gives the result $V = Mg$ and

$$F - H - Mc\dot{\omega} = 0 \tag{a}$$

Taking moments about point O gives

$$F(a+c) - Mc^2\dot{\omega} - I_C\dot{\omega} = 0 \tag{b}$$

Equation (b) leads to $\dot{\omega} = F(a+c)/(Mc^2 + I_C)$, and substituting this in (a) leads to

$$H = F\left(\frac{I_C - Mac}{Mc^2 + I_C}\right)$$

For $H=0$, we find $I_C - Mac = 0$, or $a = I_C/Mc = Mr_C^2/Mc = r_C^2/c$. The total distance from point O to the point of impact P is now $l = c + r_C^2/c$, so that P is the centre of percussion.

In some practical cases of impact-testing machines and hammers, the machinery is designed so that the impact force is close to or through the centre of percussion, in order to minimize the effects of the impact forces on the bearings.

7.11 General case of rotation of a rigid body about a fixed axis

Figure 7.15 shows a rigid body rotating about a fixed axis located in two bearings. A rotating coordinate system (x, y, z) has been

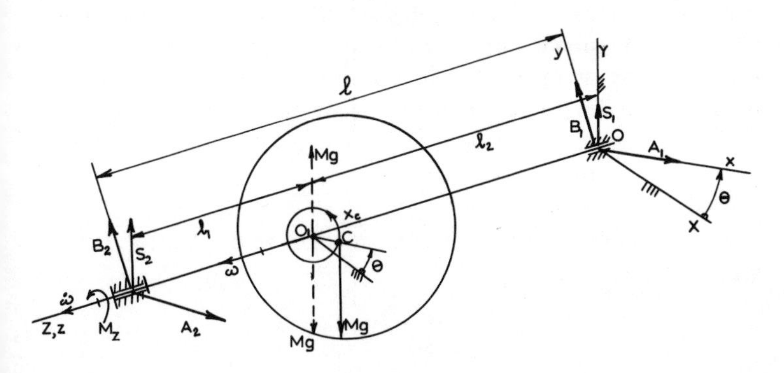

Fig. 7.15

introduced with origin O at the right-hand bearing and the z-axis along the axis of rotation. The centre of mass C is in circular motion with radius x_C; the system (x, y, z) has been located for convenience so that $y_C = 0$. The position of the body is determined by the angle θ as shown.

At the position shown, the gravity force Mg gives a torque about the axis of rotation equal to $Mgx_C \cos\theta$; if the rotor is stationary, this torque must be supplied as an external torque in the opposite direction to keep the rotor in position. Introducing two equal and opposite forces at point O_1, we find from simple statics that the bearing reactions are $S_1 = Mgl_1/l$ and $S_2 = Mgl_2/l$. These static reactions are constant and directed vertically upwards. If the rotor is rotating and we find dynamic reactions at any position, the *total reactions* are determined by adding vectorially the dynamic reactions at this position and the constant vertical static reactions. Keeping this in mind, we may then omit the gravity forces and static reactions in the following discussion of dynamic reactions.

The rotor is located in the longitudinal direction by a thrust bearing at the left-hand bearing as shown.

In any practical system there will generally be longitudinal forces on the rotor, which will give longitudinal bearing reactions, but here we are interested only in dynamic reactions due to the rotation of the rotor. If the rotor has thrust bearings at both bearings, and if it is compressed or extended to fit in the bearings, there will be longitudinal reactions. These can be found only by elasticity and temperature considerations: all we can say from dynamic considerations is that the reactions must be equal and opposite, since the centre of mass has no longitudinal motion. A deformable rotor is in any case outside our discussion, which deals with rigid bodies.

Since the body in Fig. 7.15 has a journal bearing at O, there can be no longitudinal reaction at O, and since the centre of mass has no motion in the z-direction, the longitudinal reaction at the left-hand bearing also vanishes, since there are no external longitudinal forces.

The centre of mass is in circular motion, and the force eqs.(7.1) are

$$F_x = A_1 + A_2 = -Mx_C\omega^2 \quad \text{(a)}$$

$$F_y = B_1 + B_2 = Mx_C\dot{\omega} \quad \text{(b)}$$

The moment eqs. (7.2) are in this case

$$M_x = -B_2 l = I_{yz}\omega^2 - I_{xz}\dot{\omega} \tag{c}$$

$$M_y = A_2 l = -I_{xz}\omega^2 - I_{yz}\dot{\omega} \tag{d}$$

The third Euler equation is $M_z = I\dot{\omega}$; this determines the total external torque about the z-axis (including the gravity torque), to give the angular acceleration $\dot{\omega}$. The discussion of these equations may conveniently be divided into three parts.

Case 1. In this case we will assume that the centre of mass is *not* on the axis of rotation, so that $x_C \neq 0$. The rotor is clearly *statically unbalanced*, since the gravity torque will rotate it until the centre of mass is in the lowest possible position.

If $I_{xz} = I_{yz} = 0$, we find $A_2 = B_2 = 0$; but eqs. (a) and (b) still give values for A_1 and B_1. If $(I_{xz}, I_{yz}) \neq (0, 0)$, we also find values for A_2 and B_2. The system is *dynamically unbalanced.*

Case 2. The centre of mass is *on* the axis of rotation, $x_C = 0$. The system is *statically balanced.* Assuming that $(I_{xz}, I_{yz}) \neq (0, 0)$, eqs. (c) and (d) give values for A_2 and B_2, and (a) and (b) give $A_1 = -A_2$ and $B_1 = -B_2$. The system is *dynamically unbalanced.*

The dynamic reactions may be combined to a couple rotating with the rotor. It is clearly not sufficient to have the centre of mass on the axis of rotation for dynamic balance.

Case 3. $x_C = 0$ and $I_{xz} = I_{yz} = 0$. This means that the axis of rotation is a principal axis at O, and since the centre of mass is on the axis, it is a principal axis for any point on the axis.

Equations (c) and (d) show that $A_2 = B_2 = 0$, and (a) and (b) that $A_1 = B_1 = 0$. In this case there are no dynamic reactions, the system is both *statically and dynamically balanced.*

The bearing reactions consist only of the static reactions due to the weight of the rotor.

7.12 General case of dynamic unbalance of a rotor

Figure 7.16(a) shows a rigid body rotating about a fixed axis. A series of forces, perpendicular to the axis of rotation, are indicated; these forces are due to inhomogeneity of the material, machining tolerances or special parts attached to the rotor. Each mass concentration is in a circular motion with a constant angular velocity. A constant force directed towards the centre of rotation must thus be acting, and the mass elements therefore introduce forces as shown acting *on* the rotor.

We may reduce the force field to a resultant force **R** at an arbitrary point E along the axis of rotation, by moving all forces to this point and adding them vectorially; in doing this we introduce for each force a moment acting in a plane which passes through the centre line AB and through the force. The moment vectors may be combined to a resultant moment vector at E; this vector is again perpendicular to AB. A couple may be substituted for the moment vector, and one of the couple vectors combined with **R** to give a force **P** at point E, as shown in Fig. 7.16(b), the other couple vector being **Q**. The force system is thus reduced to two forces **P** and **Q** perpendicular to AB.

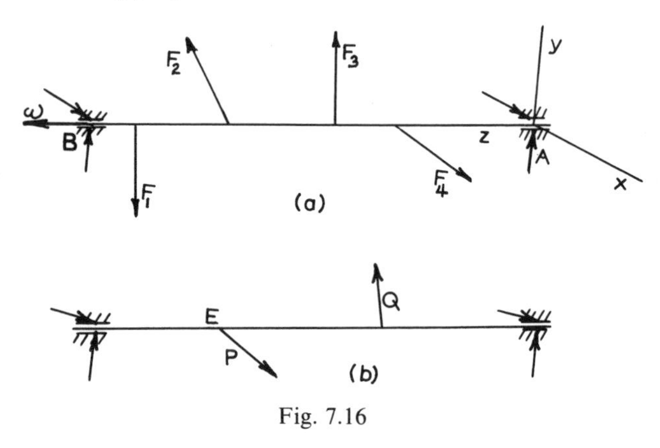

Fig. 7.16

Dynamic balance of the rotor may now be obtained by placing two *correction weights* in two arbitrary planes, the correction planes, perpendicular to the axis of rotation, in such a way that the centre of mass of the system falls on the axis of rotation, and the weights set up *equal and opposite* dynamic reactions to those already present.

The position and magnitude of the unbalance of a rotor is generally determined in a balancing machine.

Example 7.18. Figure 7.17 shows a rigid rotor for which the dynamic unbalance has been found to be equivalent to two masses M_1 and M_2 with location as shown.

The system is to be balanced by placing two correction weights of mass M_3 and M_4 in the two correction planes A and B as shown at radial distances r_3 and r_4.

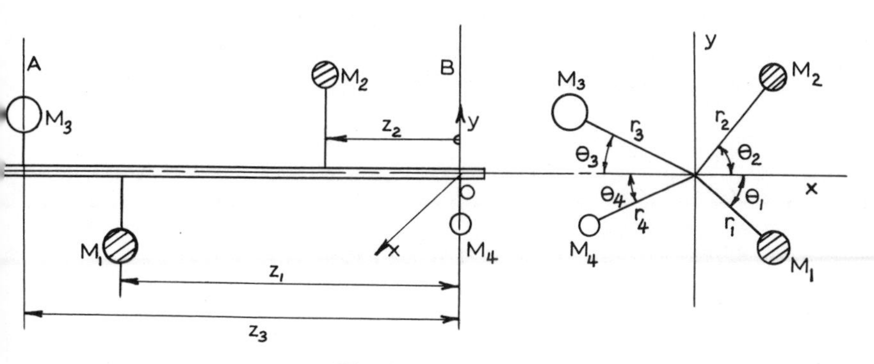

Fig. 7.17

Solution. Placing a coordinate system in the rotor with the z-axis as the axis of rotation, and origin at O, the following system constants are available: M_1, M_2, r_1, r_2, θ_1, θ_2, z_1, z_2. The correction weights are to be placed at known radial distances r_3 and r_4, and at longitudinal positions given by z_3 and, since O is in the correction plane, $z_4=0$. The problem now consists in finding the four constants M_3, M_4, θ_3 and θ_4.

We will assume that the correction weights are so small that contributions to I_{xz} and I_{yz} may be calculated as for concentrated masses.

For dynamical balance of the rotor, the axis of rotation must be a principal axis through the centre of mass, which gives the conditions:

$$I_{xz} = 0 \qquad \text{or} \qquad \sum_{1}^{4} M_n x_n z_n = 0 \tag{a}$$

which gives the equation

$$M_1(r_1 \cos\theta_1)z_1 + M_2(r_2 \cos\theta_2)z_2 - M_3(r_3 \cos\theta_3)z_3 = 0 \tag{a$'$}$$

$$I_{yz} = 0 \qquad \text{or} \qquad \sum_{1}^{4} M_n y_n z_n = 0 \tag{b}$$

which gives the result

$$-M_1(r_1 \sin\theta_1)z_1 + M_2(r_2 \sin\theta_2)z_2 + M_3(r_3 \sin\theta_3)z_3 \doteq 0 \tag{b$'$}$$

these two equations determine M_3 and θ_3.

The condition that the centre of mass must be on the z-axis means that $x_C=0$ and $y_C=0$, or $x_C M=\int x\,dm=0$ and $y_C M=\int y\,dm=0$.

Since the centre of mass of the system without $M_1,\ldots, M_4$ is already on the z-axis, we need only state that the centre of mass of $M_1,\ldots, M_4$ must be on the z-axis; this gives the equations

$$\sum_1^4 M_n x_n = 0 \tag{c}$$

or

$$M_1 r_1 \cos\theta_1 + M_2 r_2 \cos\theta_2 - M_3 r_3 \cos\theta_3 - M_4 r_4 \cos\theta_4 = 0 \tag{c$'$}$$

and

$$\sum_1^4 M_n y_n = 0 \tag{d}$$

or

$$-M_1 r_1 \sin\theta_1 + M_2 r_2 \sin\theta_2 + M_3 r_3 \sin\theta_3 - M_4 r_4 \sin\theta_4 = 0 \tag{d$'$}$$

These two equations determine M_4 and θ_4.

The theory of balancing works well for short rigid rotors. For long flexible rotors, it cannot be applied, since such rotors can generally only be balanced at one particular speed and temperature condition. Balancing is done in this case by 'trial and error' methods under field conditions.

7.13 Euler's equations for principal axis at the centre of mass

So far we have developed Euler's equations (7.2) for a set of axes (x, y, z) fixed in the body, with the origin at a fixed point about which the body was rotating.

For cases of rotation as shown in Fig. 5.6(e), it is convenient to develop Euler's equations for a set of axes (x, y, z) fixed in the body with origin at the centre of mass. The equations only take a simple form if these axes are *principal axes* at the centre of mass, so that $I_{xy}=I_{xz}=I_{yz}=0$, and we shall limit the development to this case.

The moment of momentum components from (5.5) are $H_x=I_x\omega_x$, $H_y=I_y\omega_y$ and $H_z=I_z\omega_z$, where I_x, I_y and I_z are the *principal moments* of inertia at the centre of mass.

With respect to the centre of mass, we have $\mathbf{M} = \dot{\mathbf{H}} = \dot{\mathbf{H}}_r + \boldsymbol{\omega} \times \mathbf{H}$. Taking $\mathbf{H} = H_x\mathbf{i} + H_y\mathbf{j} + H_z\mathbf{k}$ and $\boldsymbol{\omega} = \omega_x\mathbf{i} + \omega_y\mathbf{j} + \omega_z\mathbf{k}$, we find

$$\dot{\mathbf{H}}_r = \dot{H}_x\mathbf{i} + \dot{H}_y\mathbf{j} + \dot{H}_z\mathbf{k} = I_x\dot{\omega}_x\mathbf{i} + I_y\dot{\omega}_y\mathbf{j} + I_z\dot{\omega}_z\mathbf{k}$$

For the cross product $\boldsymbol{\omega} \times \mathbf{H}$ we find

$$\begin{aligned} \boldsymbol{\omega} \times \mathbf{H} &= (\omega_y H_z - \omega_z H_y)\mathbf{i} - (\omega_x H_z - \omega_z H_x)\mathbf{j} + (\omega_x H_y - \omega_y H_x)\mathbf{k} \\ &= (I_z - I_y)\omega_y\omega_z\mathbf{i} - (I_z - I_x)\omega_x\omega_z\mathbf{j} + (I_y - I_x)\omega_x\omega_y\mathbf{k} \end{aligned}$$

Combining the **i**-, **j**- and **k**-terms for **M**, we find the moment equations:

$$\left.\begin{aligned} M_x &= I_x\dot{\omega}_x - (I_y - I_z)\omega_y\omega_z \\ M_y &= I_y\dot{\omega}_y - (I_z - I_x)\omega_z\omega_x \\ M_z &= I_z\dot{\omega}_z - (I_x - I_y)\omega_x\omega_y \end{aligned}\right\} \qquad (7.15)$$

these equations are Euler's equations for a *central principal axis*. The development of these equations may also be done by taking a *fixed point, about which the body is rotating*, as a moment centre for the equation $\mathbf{M} = \dot{\mathbf{H}}$, and taking a coordinate system with (x, y, z) fixed in the body and along *principal axes* at the fixed point.

Equations (7.15) are therefore valid for all the cases 5.6(a)–(e), as long as the axes (x, y, z) are *principal axes fixed* in *the body* with *origin at the centre of mass* or at a *fixed point* about which the body is rotating (this includes *any* point on a *fixed axis of rotation*).

Equations (7.15) are easy to remember, since the first term in each is the usual simple equation from elementary physics, relating a moment or torque about an axis to the product of the moment of inertia and the angular acceleration; all signs put down are negative, and the subscripts are a cyclic rotation of xyz, that is xyz for the first equation, yzx for the second and zxy for the third.

If the body is rotating in such a way that $\boldsymbol{\omega}$ is a given function of time that can be differentiated, the necessary moments to produce the motion may be found by direct substitution in (7.15); if, however, the moments are given functions of time, and $\boldsymbol{\omega}$ is wanted, the *general solution* of (7.15) is not known. Numerical solutions may, of course, always be obtained by computer, but the solution cannot be stated in a general formula (except for the cases where no forces are acting or only gravity is acting or for one other special case).

Since the equations are valid at all times, we may use them at any specific moment in time to find *instantaneous* values of $\boldsymbol{\omega}$ or $\mathbf{M}$ for any particular position of the body.

The moments in eqs. (7.15) are moments, at any instant, about the x-, y- or z-axis of all acting forces and torques on the body. In calculating moments of a force about a line or axis, it is usually most convenient to use components of the force, which may be components in the rotating system (x, y, z) or in an inertial system. If the force components are found in the system (x, y, z) they may be found, at the same instant, by projection from the x, y, z axes onto the X, Y, Z axes of an inertial system to give the components in the fixed inertial system.

Example 7.19. Figure 7.18 shows a right circular disc attached obliquely to a horizontal shaft in bearings A and B. The shaft is rotated by a torque $\mathbf{T}$ as shown and at the instant shown the angular velocity and acceleration is $\boldsymbol{\omega}$ and $\dot{\boldsymbol{\omega}}$. The shaft is secured in the longitudinal direction by a thrust bearing at A, while bearing B is open. The mass of the disc is M and its radius r. Neglecting the mass of the shaft and friction, determine the total bearing reactions and the torque $\mathbf{T}$ in terms of $\boldsymbol{\omega}$, $\dot{\boldsymbol{\omega}}$ and the system constants.

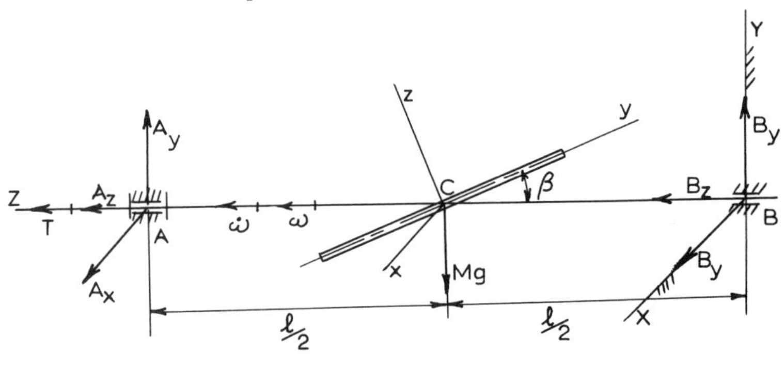

Fig. 7.18

Solution. Introducing components of the reactions as shown in the inertial system (X, Y, Z) we find from Newton's second law that the forces are balanced in any direction, since the centre of mass C has no acceleration:

$$A_z + B_z = 0 \tag{a}$$

Since $B_z = 0$, we find $A_z = 0$,

$$A_y + B_y = Mg \tag{b}$$

$$A_x + B_x = 0 \tag{c}$$

Remembering that forces parallel to or intersecting a line have no moment about that line, we find the following equations, by taking moments of all forces and torques acting about the set of *principal axis* (x, y, z) *fixed in the body at* C:

$$M_x = -A_y l/2 + B_y l/2 \tag{d}$$

$$M_y = A_x(l/2)\sin\beta - B_x(l/2)\sin\beta - T\cos\beta \tag{e}$$

$$M_z = A_x(l/2)\cos\beta - B_x(l/2)\cos\beta + T\sin\beta \tag{f}$$

To use Euler's eqs.(7.15), we establish the following information:

$$\omega_x = 0 \qquad \omega_y = -\omega\cos\beta \qquad \omega_z = \omega\sin\beta$$

$$\dot{\omega}_x = 0 \qquad \dot{\omega}_y = -\dot{\omega}\cos\beta \qquad \dot{\omega}_z = \dot{\omega}\sin\beta$$

$$I_x = I_y = \tfrac{1}{4}Mr^2 \qquad I_z = \tfrac{1}{2}Mr^2$$

Euler's equations now give the equations

$$M_x = -\tfrac{1}{4}Mr^2\omega^2\sin\beta\cos\beta \tag{g}$$

$$M_y = -\tfrac{1}{4}Mr^2\dot{\omega}\cos\beta \tag{h}$$

$$M_z = \tfrac{1}{2}Mr^2\dot{\omega}\sin\beta \tag{i}$$

We can now eliminate M$_x$, M_y, and M_z between eqs.(d)–(i), so that we now have 5 equations (including (b) and (c)) with 5 unknowns A_x, B_x, A$_y$, B_y and T:

$$A_y + B_y = Mg \tag{b}$$

$$A_x + B_x = 0 \tag{c}$$

$$A_y - B_y = \frac{Mr^2\omega^2}{2l}\sin\beta\cos\beta \tag{j}$$

$$A_x\sin\beta - B_x\sin\beta - \frac{2T}{l}\cos\beta = -\frac{Mr^2\dot{\omega}}{2l}\cos\beta \tag{k}$$

$$A_x\cos\beta - B_x\cos\beta + \frac{2T}{l}\sin\beta = \frac{Mr^2\dot{\omega}}{l}\sin\beta \tag{l}$$

Equations (b) and (j) may be solved directly to give

$$A_y = \frac{Mg}{2} + \frac{Mr^2\omega^2}{8l} \sin 2\beta$$

$$B_y = \frac{Mg}{2} - \frac{Mr^2\omega^2}{8l} \sin 2\beta$$

The first part in each expression is due to the static gravity force.

Substituting $B_x = -A_x$ from (c) into (k) and (l) and solving these gives

$$A_x = -B_x = \frac{Mr^2\dot{\omega}}{8l} \sin 2\beta$$

$$T = \frac{Mr^2\dot{\omega}}{4}(1+\sin^2 \beta)$$

A_x and B_x vanish if ω is constant, otherwise they form a couple rotating with the disc. All dynamic reactions disappear if $\sin 2\beta = 0$, that is if $\beta = 0°$ or $\beta = 90°$; in both cases the rotation is about a principal axis. The dynamic reactions are maximum for $\sin 2\beta = 1$ or $\beta = 45°$.

Since this example is rotation of a rigid body about a *fixed axis*, the torque T may also be determined from the elementary formula $T = I_z\dot{\omega}$. From Example 6.3, $I_z = \frac{1}{4}Mr^2\ (1+\sin^2\ \beta)$, so that this gives the same expression for T as found above.

It must be kept in mind that Euler's equations give the forces as they act *on* the body in question; if the forces *on* the bearings are wanted, these are determined by reversing the forces from Euler's equations.

Example. 7.20. Figure 7.19 shows a slender, rigid, uniform bar DB of length $l = 3$ m and mass $M = 20$ kg. The bar is rotating at a constant 100 rev/min about the axis DA. The top end B of the bar is connected to the axis AD by a string AB.

Determine the tension in the string AB (a) by using Euler's equations, (b) by introducing dynamic equilibrium. Assume that the connection at D can only transmit forces.

Solution. (a) A rotating Euler system may be introduced with origin at the centre of mass C, or with origin at any point on the axis of rotation DA. We shall use a system with origin at D, and with axes fixed in the body along principal axes at D as shown in the figure.

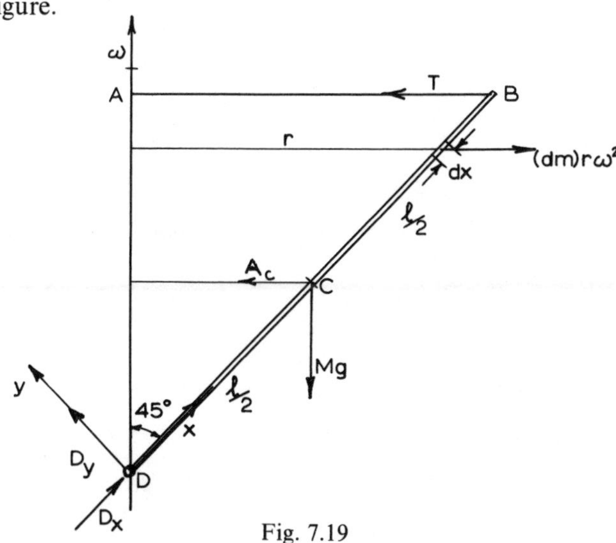

Fig. 7.19

The angular velocity is $\omega = 2\pi \cdot 100/60 = 10{\cdot}48$ rad/s.

The centre of mass C is in a circular motion with acceleration $\mathbf{A}_C$ as shown. The magnitude of $\mathbf{A}_C$ is $A_C = r\omega^2$, and $r = (l/2)\sin 45 = 1{\cdot}06$ m, so that $A_c = 116{\cdot}2$ m/s^2.

The force equations for the centre of mass now give the equations

$$D_x - 196\cos 45 - T\cos 45 = -20\times 116{\cdot}2\cos 45 \qquad \text{(a)}$$

$$D_y - 196\cos 45 + T\cos 45 = 20\times 116{\cdot}2\cos 45 \qquad \text{(b)}$$

$$D_z = 0 \qquad \text{(c)}$$

Taking moments of the acting forces give the equations

$$M_x = 0 \qquad M_y = 0 \qquad M_z = 3T\cos 45 - 196\cdot(1{\cdot}5)\sin 45$$

For Euler's eqs. (7.15), we have $\dot{\omega}_x = \dot{\omega}_y = \dot{\omega}_z = 0$, and $\omega_x = \omega\cos 45 = 7{\cdot}40$ rad/s, $\omega_y = \omega\cos 45 = 7{\cdot}40$ rad/s, $\omega_z = 0$. Substituting this in Euler's equations gives the equations

$$M_x = 0 \qquad \text{(d)}$$

$$M_y = 0 \qquad \text{(e)}$$

$$M_z = (I_y - I_x)\omega_x\omega_y \qquad \text{(f)}$$

Equations (d) and (e) are in agreement with the result of the moments of the acting forces. We have now

$$I_x \simeq 0 \qquad I_y = \tfrac{1}{3}Ml^2 = \tfrac{1}{3} \times 20 \times 3^2 = 60 \text{ kg m}^2$$

so that

$$M_z = 60 \times 7{\cdot}40 \times 7{\cdot}40 = 3280 \text{ Nm} \tag{g}$$

Introducing the moments of the acting forces we have

$$3T \cos 45 - 196 \times 1{\cdot}5 \sin 45 = 3280 \tag{h}$$

or $T = 1648$ N. D_x and D_y may now be determined from eqs. (a) and (b).

(b) Since $\dot{\omega} = 0$, each element dx of the bar may be put in dynamic equilibrium by introducing the inertia force $dm\ r\omega^2$, as shown in Fig. 7.19.

Taking moments about the z-axis, we find

$$Tl \sin 45 - Mg\ l/2 \sin 45 - \textstyle\int_0^l r\omega^2 x \sin 45\ dm = 0$$

introducing $r = x \sin 45$ and $dm = (M/l)\ dx$, we get

$$Tl - Mg\frac{l}{2} - \frac{M}{l}\ \omega^2 \sin 45 \int_0^l x^2\ dx = 0$$

and

$$T = \frac{Mg}{2} + \frac{Ml}{3}\ \omega^2 \sin 45 = 98 + 1550 = 1648 \text{ N}$$

Of this tension, 98 N is due to gravity and 1550 N due to rotation.

Problems

7.1 Derive the equation of motion for the system in Fig. 7.20. Find the frequency for small vibrations.

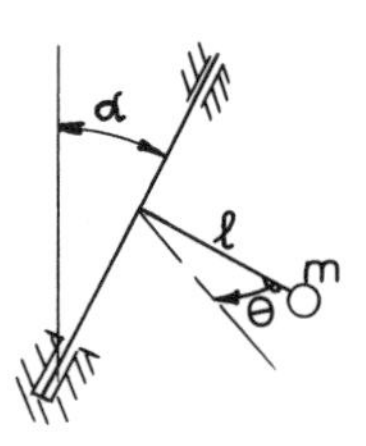

Fig. 7.20

7.2 A rotor with moment of inertia I_1 (Fig. 7.21) is driven at a constant angular velocity ω_1. It is brought into contact with a second rotor I_2 which is initially at rest. The contact pressure is a constant normal force P between the rotors and the coefficient of friction between the rotors is μ. Find the time of slipping.

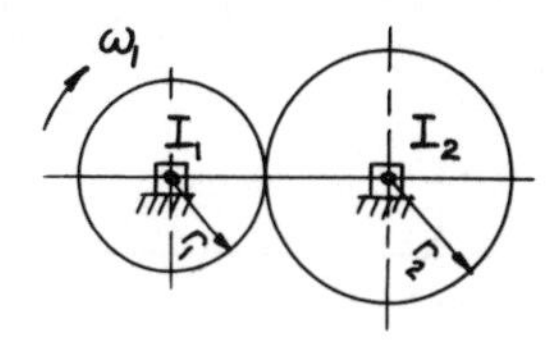

Fig. 7.21

If the outside torque is removed from I_1 at the moment of contact, what is the new time of slipping and the final angular velocity of I_1? Neglect bearing friction.

7.3 The system in Fig. 7.22 consists of a body of mass $M = 12{\cdot}7$ kg which is connected to a rotating drum as shown. The drum has a moment of inertia about its axis of rotation of $16{\cdot}84$ kg m^2.

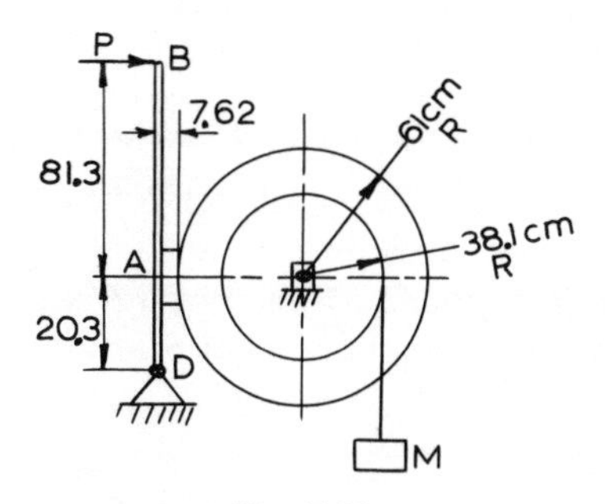

Fig. 7.22

The drum may be slowed down by applying a force P to the brake handle DB, which connects the brake shoe A to the drum. The coefficient of friction between the brake shoe and the drum is 0·40. Friction on the drum shaft may be neglected.

(a) Determine the necessary constant horizontal force P on the brake handle to stop the mass M in 2 s, if the mass is initially moving downwards with a velocity of 4·88 m/s.

(b) Determine the horizontal and vertical reactions on the brake lever at point D during this motion.

7.4 Figure 7.23 shows a small rigid rotor in two bearings A and B. The rotor has been examined in a balancing machine, and the unbalance has been found to be equivalent to two masses M_1 and M_2. The mass $M_1 = 113$ g is at a radial distance $r_1 = 12{\cdot}7$ cm from the rotor centre line, and is located as shown, in the plane of the figure. $M_2 = 28{\cdot}4$ g at $r_2 = 15{\cdot}2$ cm, and M_2 is behind the plane of the paper in Fig. 7.23.

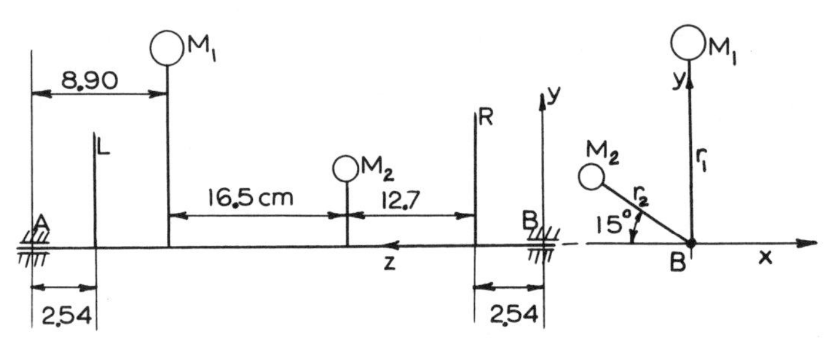

Fig. 7.23

The rotor is to be dynamically balanced by placing two balance masses M_L and M_R in the balance planes L and R shown in Fig. 7.23; the radial distance is to be $r_L = r_R = 8$ cm. Determine the mass of M_L and M_R and the angular position of each.

7.5 A slender rigid uniform bar of length l and mass M is rigidly attached at its midpoint to a horizontal shaft in bearings A and B as shown in Fig. 7.24. Shaft AB may be assumed massless and is secured in the longitudinal direction at bearing A only.

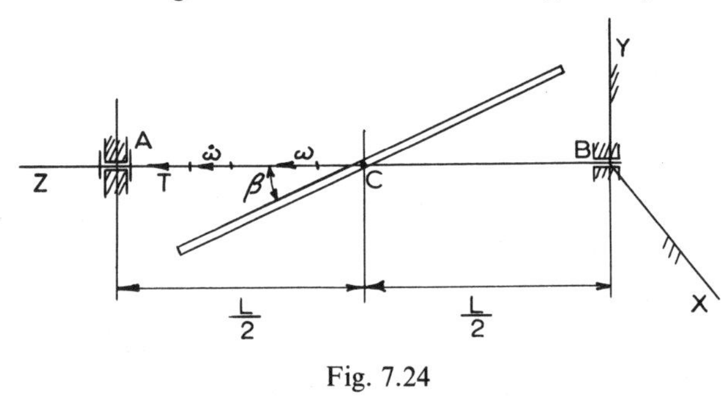

Fig. 7.24

A torque $\mathbf{T}$ is applied to shaft AB as shown, and at the instant considered the angular velocity of the shaft is $\boldsymbol{\omega}$.

(a) Determine the components of the reactions on the shaft at A and B in the inertial system shown, expressed as functions of the given system constants and ω and $\dot{\omega}$.

(b) Find an expression for the torque $\mathbf{T}$ as a function of the system constants and $\dot{\omega}$.

7.6 In Fig. 7.25 a thin rectangular plate weighing 111 N rotates about axis AB. The plate is restrained in the Z-direction at bearing A only. If a torque of 40·7 Nm in the direction of rotation is applied to the shaft on which the plate rotates, what is the angular acceleration at the instant of application of the torque? Find the reactions *on the bearings* at this instant. The plate is in the XZ-plane when the torque is applied and has an angular velocity of 20 rad/s.

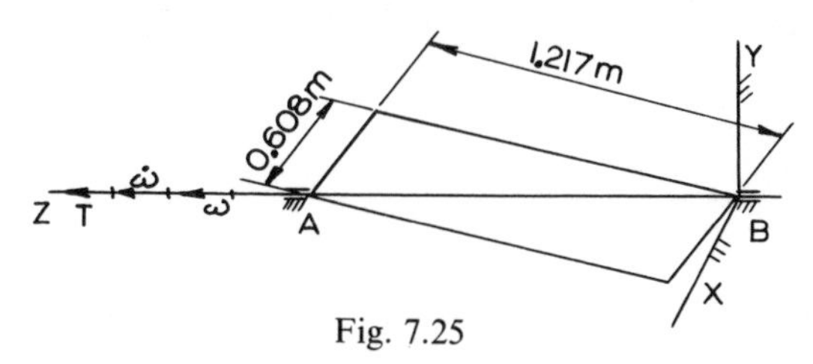

Fig. 7.25

7.7 Figure 7.26 shows a horizontal shaft in two bearings A and B. The bearing at A is a thrust bearing securing the system in the direction AB. Two square plates are welded onto the shaft as shown. The total mass of the plates is M, while the mass of the shaft may be neglected.

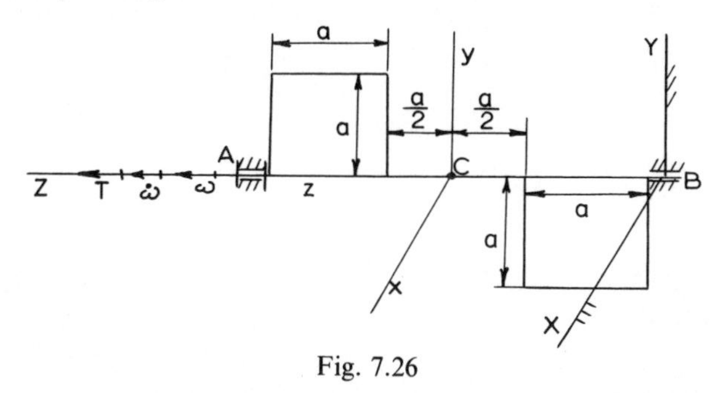

Fig. 7.26

An external torque $\mathbf{T}$ rotates the system, and at the instant shown the plates are in a vertical plane and rotating with angular velocity ω as shown. Wind resistance and friction may be neglected.

(a) Write Newton's second law for the centre of mass C.

(b) Determine the product moments I_{xz} and I_{yz} and the moment of inertia I_z of the system with respect to the Euler system (x, y, z) shown at the centre of mass.

(c) Determine the angular acceleration $\dot{\omega}$ and the components of the bearing reactions in terms of the given system constants T, ω, a and M.

CHAPTER EIGHT

Plane Motion of a Rigid Body

Plane motion of a rigid body is defined as motion in which a certain plane of the body always remains in a fixed plane. Any point of the body then moves in a plane parallel to the fixed plane. Some examples of plane motion were given in Chapter 3. The case of rotation about a fixed axis discussed in Chapter 7 is clearly a special case of plane motion.

Plane motion is perhaps the most important type of motion encountered in engineering Dynamics.

8.1 Equations of plane motion

Figure 8.1 shows a body in plane motion in an inertial system (X, Y, Z). Because of the importance of the centre of mass C of the body, the plane of motion, the XY-plane, has been taken through the centre of mass C.

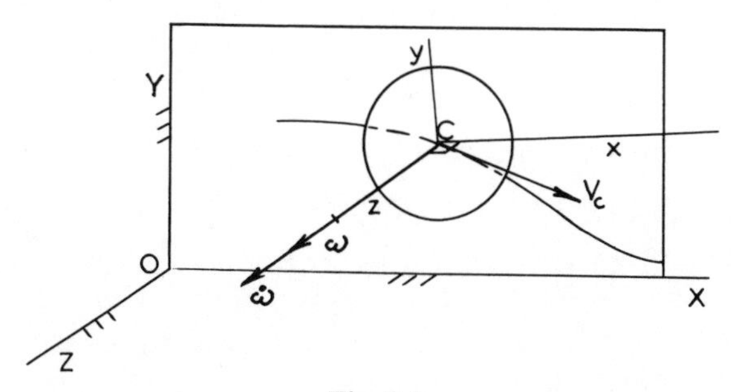

Fig. 8.1

A body in plane motion has three degrees of freedom: the position of the body may be given, for instance, by the coordinates (X_C, Y_C) of the centre of mass and an angle θ between the X-axis and a line fixed in the body in the XY-plane.

The centre of mass C moves along a curve as shown in the XY-plane, so it is a point in plane curvilinear motion. The angular velocity vector $\boldsymbol{\omega}$ and the angular acceleration vector $\dot{\boldsymbol{\omega}}$ of the body have been located for convenience at the centre of mass, and are both always perpendicular to the plane of motion.

The scalar magnitudes of the acceleration components of the centre of mass in the inertial system are $\ddot{X}_C$ and $\ddot{Y}_C$, while $\ddot{Z}_C = 0$. If M is the total mass of the body, Newton's second law gives the force equations of motion

$$F_X = M\ddot{X}_C \qquad F_Y = M\ddot{Y}_C \qquad F_Z = 0 \tag{8.1}$$

The first two equations of (8.1) describes the motion of the centre of mass in the XY-plane.

Fixing a coordinate system (x, y, z) in the body as shown with origin at the centre of mass and the z-axis perpendicular to the plane of motion, we have

$$\omega_x = \omega_y = 0 \qquad \text{and} \qquad \boldsymbol{\omega} = \omega_z \mathbf{k}$$

In the same way

$$\dot{\omega}_x = \dot{\omega}_y = 0 \qquad \text{and} \qquad \dot{\boldsymbol{\omega}} = \dot{\omega}_z \mathbf{k}$$

Equations (5.5) for the moment of momentum vector components are in this case $H_x = -I_{xz}\omega_z$, $H_y = -I_{yz}\omega_z$ and $H_z = I_z\omega_z$. Using the equation $\mathbf{M} = \dot{\mathbf{H}}$, which is valid for the moving centre of mass, we have $\mathbf{M} = \dot{\mathbf{H}}_r + \boldsymbol{\omega} \times \mathbf{H}$ from eq. (4.1).

Since $(\mathbf{i}, \mathbf{j}, \mathbf{k})$ are in a fixed direction in the system (x, y, z) we find

$$\dot{\mathbf{H}}_r = \dot{H}_x\mathbf{i} + \dot{H}_y\mathbf{j} + \dot{H}_z\mathbf{k}$$

The vector cross product is

$$\begin{aligned}\boldsymbol{\omega} \times \mathbf{H} &= \omega_z\mathbf{k} \times (H_x\mathbf{i} + H_y\mathbf{j} + H_z\mathbf{k}) \\ &= \omega_z H_x \mathbf{k} \times \mathbf{i} + \omega_z H_y \mathbf{k} \times \mathbf{j} + \omega_z H_z \mathbf{k} \times \mathbf{k} = \omega_z H_x \mathbf{j} - \omega_z H_y \mathbf{i}\end{aligned}$$

We have now

$$\mathbf{M} = (\dot{H}_x - \omega_z H_y)\mathbf{i} + (\dot{H}_y + \omega_z H_x)\mathbf{j} + \dot{H}_z\mathbf{k}$$

introducing

$$\dot{H}_x = -I_{xz}\dot{\omega}_z \qquad \dot{H}_y = -I_{yz}\dot{\omega}_z \qquad \dot{H}_z = I_z\dot{\omega}_z$$

leads to the scalar moment equations

$$\begin{aligned} M_x &= -I_{xz}\dot{\omega}_z + I_{yz}\omega_z^2 \\ M_y &= -I_{yz}\dot{\omega}_z - I_{xz}\omega_z^2 \\ M_z &= I_z\dot{\omega}_z \end{aligned} \tag{8.2}$$

The development of these equations and the result is identical to the eqs. (7.2) for rotation about a fixed axis. The difference lies in the rotating coordinate system applied: for plane motion we take as origin the moving centre of mass, whereas for rotation about a fixed axis the origin is on the axis of rotation.

Equations (8.2) are Euler's equations for plane motion. It may be pointed out here that the Euler equations (7.15) for principal axes at the centre of mass are valid for all types of motion and therefore may also be applied in plane motion; in general it will be found, however, that eqs. (8.2) are better suited to the treatment of plane motion than (7.15), since (8.2) takes advantage of the fact that the axis of rotation is in a fixed direction in plane motion.

If the z-axis *is a principal axis*, we have $I_{xz} = I_{yz} = 0$; and eqs. (8.2) take the simpler form

$$M_x = 0 \qquad M_y = 0 \qquad M_z = I_z\dot{\omega}_z \tag{8.3}$$

The most important case of this is *a homogeneous body with the XY-plane as a plane of symmetry*.

The third Euler equation in (8.2) describes the rotation of the body, while the first two Euler equations in (8.2) and the force eqs. (8.1) determine the forces necessary to keep a cross section of the body in the plane of motion.

The fact that the equation of rotational motion may be taken exactly as if the body were rotating about a fixed axis through the centre of mass is sometimes called the *principle of independence of translation and rotation in plane motion*. This principle holds only if the axis of rotation is taken through the centre of mass.

8.2 Translation of a rigid body

Translation of a rigid body is defined as a motion in which all lines in the body remain parallel to their original directions. This means that the body has no rotation, so the angular velocity and acceleration are always zero.

Translation is the simplest form of rigid-body motion. Although it need not be plane motion, we shall consider only translation in a plane. The equations of motion in translation are the force equations (8.1) and the moment equations (8.2); since we have $\boldsymbol{\omega}=\dot{\boldsymbol{\omega}}=\mathbf{0}$, the moment equations take the simple form

$$M_x = 0 \qquad M_y = 0 \qquad M_z = 0 \tag{8.4}$$

An example will illustrate the general procedure for the solution of problems of this nature.

Example 8.1. Figure 8.2 shows a sliding door of mass M_1 which is carried on two small rollers at A and B. The rollers run on a horizontal track as shown. The door is opened by releasing a mass M_2 which is connected to the door by a string.

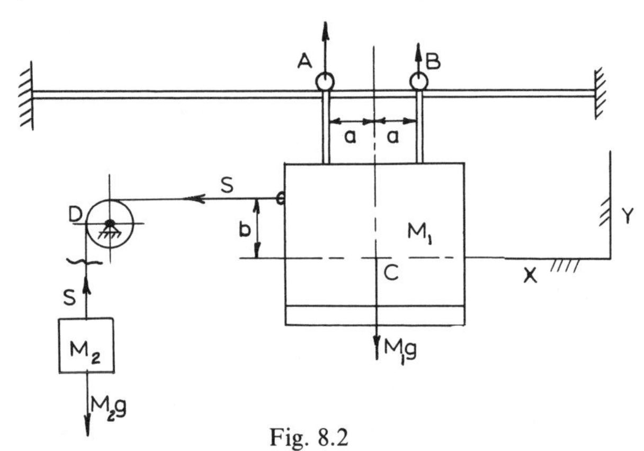

Fig. 8.2

Determine the acceleration of the door, the reactions at A and B and the string tension. The inertia of the rollers and the pulley at D and all friction may be neglected.

Solution. All the acting forces on the door and on the falling mass M_2 have been shown on the figure.

Applying the force equations (8.1) to the door gives the equations

$$F_X = S = M_1\ddot{X} \tag{a}$$

$$F_Y = A+B-M_1g = M_1\ddot{Y} = 0 \tag{b}$$

The third Euler equation from (8.4) is

$$M_z = Aa - Ba - Sb = 0 \tag{c}$$

The force equation for the falling weight is

$$M_2 g - S = M_2 \ddot{X} \tag{d}$$

Substituting $S = M_1 \ddot{X}$ from (a) into (d) gives the result

$$\ddot{X} = \frac{M_2 g}{M_1 + M_2}$$

which result is correct dimensionally, and $\ddot{X} < g$ as it should be, since this is also the acceleration of the falling mass. The result from (a) is now

$$S = \frac{M_1 M_2}{M_1 + M_2} g$$

Again this result is correct dimensionally, and $S < M_2 g$. Equations (b) and (c) give the reactions

$$A = \frac{M_1 ag + Sb}{2a} \quad \text{and} \quad B = \frac{M_1 ag - Sb}{2a}$$

8.3 General plane motion

When the body in plane motion is translating and rotating at the same time, the equations of motion are (8.1) and (8.2). If the z-axis is principal the moment equations are (8.3) instead of (8.2). As an example on general plane motion consider the following.

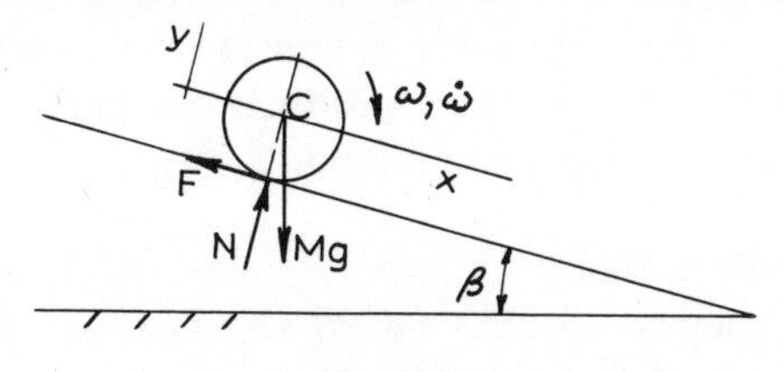

Fig. 8.3

Example 8.2. Figure 8.3 shows a homogeneous body of revolution, which is rolling down an inclined plane. The mass of the body is M, the radius of gyration of the body with respect to its geometric axis is r_C, and the rolling radius is R. Determine the acceleration of the centre of mass of the body, the angular acceleration of the body and the friction force.

Solution. To describe the motion of the body, we introduce a fixed coordinate system (x, y) in the most suitable position, with the x-axis through the centre of mass and parallel to the inclined plane; the motion of the centre C is then rectilinear motion along the x-axis.

The forces acting are the gravity force Mg and the reaction from the plane, and this reaction is taken for convenience in two components, a normal force N and a friction force F as shown. As long as there is no relative motion between the body and the plane, the friction force is of unknown magnitude and $F<\mu N$, where μ is the coefficient of friction between the body and the plane. If the body is sliding on the plane, the friction force is at a maximum $F = \mu N$.

The force equations (8.1) are

$$F_y = N - Mg \cos \beta = M\ddot{y} = 0 \qquad \text{therefore} \qquad N = Mg \cos \beta \tag{a}$$

$$F_x = Mg \sin \beta - F = M\ddot{x} \tag{b}$$

Since the geometric axis is a principal axis the third Euler equation from (8.3) is

$$|M_z| = FR = I_z\dot{\omega} \qquad \text{where} \qquad I_z = Mr_C^2 \tag{c}$$

We have two equations with three unknown quantities $\ddot{x}$, F and $\dot{\omega}$. To get a fourth equation we must consider the constraints on the system, so we divide the investigation in two parts, depending on whether the body is in pure rolling or sliding.

A. Rolling without sliding

In this case the relationship $x = R\theta + x_0$ is valid, where x is the distance moved of the centre C, and θ is the angle of rotation; x_0 is a constant which is the value of x when $\theta = 0$; differentiating this equation twice gives an equation

$$\ddot{x} = R\ddot{\theta} = R\dot{\omega} \tag{d}$$

The equations (b), (c) and (d) may now be solved with the result that

$$\ddot{x} = \frac{g \sin \beta}{1+(r_C/R)^2} \qquad \dot{\omega} = \frac{\ddot{x}}{R} \qquad F = \frac{Mg \sin \beta}{1+(R/r_C)^2}$$

The acceleration and the friction force are constant. The acceleration is reduced by a factor $\sin \beta/[1+(r_C/R)^2]$; this fact was used by Galileo (about 1500) in studying the motion of bodies rolling down a plane; by using a body as indicated in Fig. 8.4, the ratio $(r_C/R)^2$ may be made larger than 10, with a resultant lowering of the acceleration of the centre of the body. Table 8.1 gives the acceleration for various types of bodies rolling down a plane without slipping.

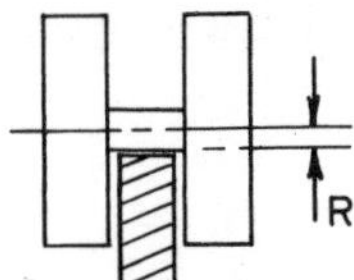

Fig. 8.4

TABLE 8.1

Type of body	r_C^2	$\dfrac{\ddot{x}}{g \sin \beta}$
Solid sphere	$\frac{2}{5}R^2$	0·714
Solid cylinder	$\frac{1}{2}R^2$	0·667
Spherical shell	$\frac{2}{3}R^2$	0·600
Cylindrical shell or hoop	R^2	0·500

The bodies have been arranged in the order in which they would reach the end of an inclined plane, independent of size or material, assuming that they roll without sliding, are homogeneous and that wind resistance may be neglected. It may be seen that all solid spheres would get ahead of all cylinders.

B. Rolling with sliding

In this case the equations of motion (a)–(c) are still valid, but (d) no longer holds. The friction force is now a maximum, and a new equation (d′) is then:

$$F = \mu N = \mu Mg \cos \beta \qquad \text{(d}'\text{)}$$

The solution in this case is

$$\ddot{x} = g(\sin\beta - \mu\cos\beta) \qquad \dot{\omega} = \frac{Rg\mu}{r_C^2}\cos\beta \qquad F = \mu Mg\cos\beta$$

In any given problem it is evidently necessary to investigate whether there is pure rolling or sliding. Pure rolling requires

$$F = \frac{Mg\sin\beta}{1+(R/r_C)^2} < F_{\max} = \mu Mg\cos\beta \qquad \text{or}$$

$$\tan\beta < \mu[1+(R/r_C)^2]$$

otherwise we have rolling with sliding. For instance for a solid cylinder $r_C^2 = R^2/2$, and we have pure rolling if $\tan\beta < \mu(1+2) = 3\mu$; if $\mu = 1/3$, then the cylinder is in pure rolling if $\tan\beta < 1$, or $\beta < 45°$. The cylinder will roll and slide for this value of μ if $\beta > 45°$.

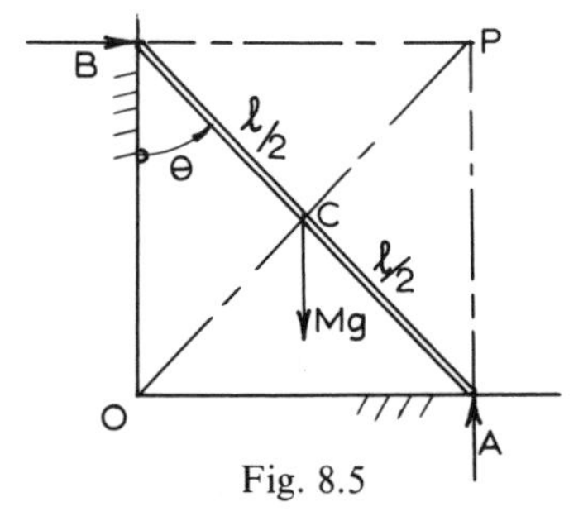

Fig. 8.5

Example 8.3. Figure 8.5 shows a uniform slender rigid bar of length l and mass M. The bar moves in the vertical plane with the ends of the bar sliding on frictionless walls as shown. Given that the bar starts from rest in the vertical position, determine the reactions A and B as functions of the angle θ. Determine also the angle at which the bar leaves the vertical wall.

Solution. The distance OC is equal to $l/2$ for any position of the bar in contact with both walls. The centre of mass C of the bar then moves on a circle with centre O and radius $l/2$. The acceleration of C is thus $\frac{1}{2}l\dot{\theta}^2$ towards O, and $\frac{1}{2}l\ddot{\theta}$ perpendicular to OC. The equations of motion of C are thus in the normal and tangential direction respectively:

$$Mg\cos\theta - B\sin\theta - A\cos\theta = \tfrac{1}{2}Ml\dot{\theta}^2 \qquad \text{(a)}$$

$$Mg\sin\theta + B\cos\theta - A\sin\theta = \tfrac{1}{2}Ml\ddot{\theta} \qquad \text{(b)}$$

The point P shown on Fig. 8.5 is the instantaneous centre of rotation, this point is also accelerating towards C and may therefore be used as a moment centre. Hence

$$\tfrac{1}{2}Mgl \sin\theta = I_P\ddot{\theta} \qquad I_P = \tfrac{1}{3}Ml^2$$

so that

$$\ddot{\theta} = (3g/2l)\sin\theta$$

We may write

$$d\theta\,\frac{d\dot{\theta}}{dt} = \dot{\theta}\,d\dot{\theta} = \tfrac{1}{2}d\dot{\theta}^2 = \frac{3g}{2l}\sin\theta\,d\theta$$

so that

$$d(\dot{\theta}^2) = \frac{3g}{l}\sin\theta\,d\theta \qquad \dot{\theta}^2 = -\frac{3g}{l}\cos\theta + C$$

When $\theta=0$, $\dot{\theta}=0$; therefore $C=3g/l$ and $\dot{\theta}^2=(3g/l)\,(1-\cos\theta)$. Substituting the expressions for $\ddot{\theta}$ and $\dot{\theta}^2$ in eqs. (a) and (b) gives two equations in A and B. The solution is

$$A = \frac{Mg}{4}(3\cos\theta-1)^2 \qquad \text{and} \qquad B = 3\frac{Mg}{4}(3\cos\theta-2)\sin\theta$$

This result is valid only as long as the bar is in contact with both walls. The bar loses contact with the vertical wall when $B=0$; therefore $3\cos\theta-2=0$, or $\cos\theta=\tfrac{2}{3}$, $\theta=48{\cdot}19°$. The above expressions for A and B are then valid for $0\leqslant\theta\leqslant 48{\cdot}19°$.

8.4 Alternative form of equations of motion: impulse–momentum equations

The first two of the force equations (8.1) and the third Euler equation (8.2) may be integrated directly to give an alternative form of the equations of motion.

Taking the coordinates of the centre of mass as (x, y), introducing $\omega_z=\omega$ and assuming that the mass M and moment of inertia I_z are constants, we find

$$\left.\begin{aligned} \int_1^2 F_x\,dt &= \int_1^2 M\,d\dot{x} = M(\dot{x}_2-\dot{x}_1) \\ \int_1^2 F_y\,dt &= \int_1^2 M\,d\dot{y} = M(\dot{y}_2-\dot{y}_1) \\ \int_1^2 M_z\,dt &= \int_1^2 I_z\,d\omega = I_z(\omega_2-\omega_1) \end{aligned}\right\} \qquad (8.5)$$

These equations are the impulse–momentum equations in plane motion. The two linear impulse–momentum equations state that the impulse of a force in a certain time and direction is equal to the total *change* in linear momentum of the body in the same time and direction.

The third equation states that the total angular impulse about the z-axis through the centre of mass is equal to the total *change* in angular momentum about that axis in the same time.

Equations (8.5) are simply a different form of the original equations of motion, and therefore contain the same information as these. The new equations are, however, convenient in certain problems.

If the force components and the moment M_z are constants during a time t, eqs. (8.5) take the simpler form:

$$\left.\begin{aligned} F_x t &= M(\dot{x}_2 - \dot{x}_1) \\ F_y t &= M(\dot{y}_2 - \dot{y}_1) \\ M_z t &= I_z(\omega_2 - \omega_1) \end{aligned}\right\} \qquad (8.6)$$

If there is no impulse in the x-direction in a certain time interval, we have $\int_1^2 Fx\, dt = 0$, and $M\dot{x}_2 = M\dot{x}_1$, or the linear momentum in that direction is *conversed* in that time interval. In the same manner, if the angular impulse in a certain time is $\int_1^2 M_z\, dt = 0$, $I_z\omega =$ constant, so that the angular momentum is conserved during that time; this is called *the principle of conservation of angular momentum.*

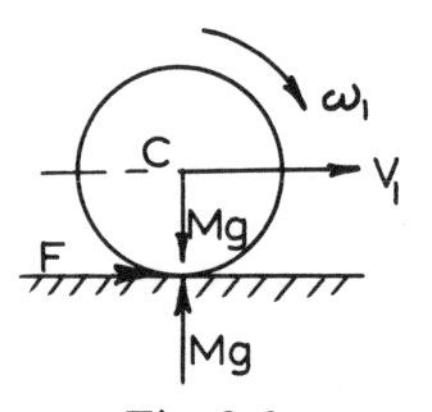

Fig. 8.6

Example 8.4. Figure 8.6 shows a solid, homogeneous right circular cylinder of mass M and radius r. The cylinder is initially rotating about its axis with an angular velocity ω_1 directed as shown, while the linear velocity of the centre C is $V_1 = 0$. The cylinder is suddenly placed on a horizontal plane with coefficient of friction μ between the cylinder and the plane. Determine the angular velocity of the cylinder when pure rolling starts, and the time of slipping and the distance moved by the centre C in this time.

Solution. As long as slipping occurs, the friction force is $F=\mu Mg$. This force accelerates the centre C and slows down the rotation; pure rolling occurs when the angular velocity is ω_2 and the linear velocity is $V_2=r\omega_2$. If the slipping time is t, the impulse–momentum equations (8.6) give the following result:

$$Ft = M(V_2-V_1) \qquad \text{(a)}$$

$$-(Fr)t = I(\omega_2-\omega_1) = \tfrac{1}{2}Mr^2(\omega_2-\omega_1) \qquad \text{(b)}$$

Substituting $F=\mu Mg$, $V_1=0$ and $V_2=r\omega_2$ leads to

$$\mu Mgt = Mr\omega_2 \qquad \text{(a}'\text{)}$$

$$\mu Mgrt = \tfrac{1}{2}Mr^2(\omega_1-\omega_2) \qquad \text{(b}'\text{)}$$

from which $r\omega_2=\frac{1}{2}r(\omega_1-\omega_2)$, or $\omega_2=\frac{1}{3}\omega_1$. The linear velocity when pure rolling starts is $V_2=r\omega_2=\frac{1}{3}r\omega_1$. The time of slipping from (a′) is $t=r\omega_2/\mu g=r\omega_1/3\mu g$. Since the friction force is constant, the distance moved by the centre is

$$x = \tfrac{1}{2}(V_1+V_2)t = \tfrac{1}{2}V_2t = \tfrac{1}{6}r\omega_1 t = r^2\omega_1^2/18\mu g$$

8.5 Kinetic energy of a rigid body in plane motion

Figure 8.7 shows a rigid body in plane motion. The total kinetic energy is $T=\frac{1}{2}\int v^2\,dm$. Using the centre of mass C as a pole, we have the velocity V of any particular mass element expressed by $\mathbf{V}=\mathbf{V}_C+\boldsymbol{\omega}\times\mathbf{r}$.

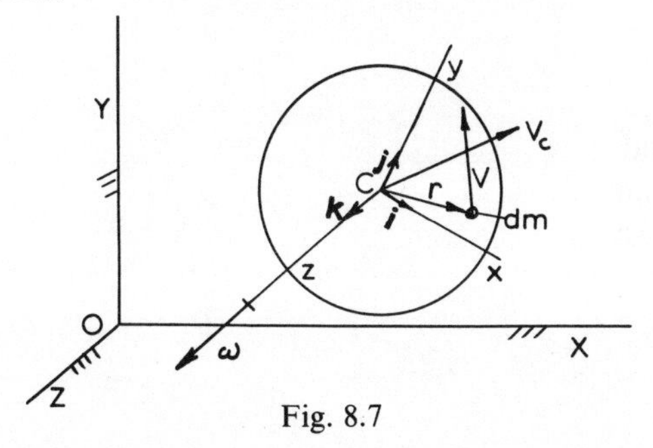

Fig. 8.7

All points of the body have velocities parallel to the XY-plane, and we may write

$$\mathbf{V}_C = V_{Cx}\mathbf{i}+V_{Cy}\mathbf{j} \qquad \boldsymbol{\omega}=\omega\mathbf{k} \qquad \mathbf{r}=x\mathbf{i}+y\mathbf{j}+z\mathbf{k}$$

We have now

$$\boldsymbol{\omega}\times\mathbf{r} = \omega\mathbf{k}\times(x\mathbf{i}+y\mathbf{j}+z\mathbf{k})$$
$$= x\omega\mathbf{k}\times\mathbf{i}+y\omega\mathbf{k}\times\mathbf{j}+z\omega\mathbf{k}\times\mathbf{k} = x\omega\mathbf{j}-y\omega\mathbf{i}$$

so that

$$\mathbf{V} = (V_{Cx}-y\omega)\mathbf{i}+(V_{Cy}+x\omega)\mathbf{j}$$

and

$$V^2 = (V_{Cx}-y\omega)^2+(V_{Cy}+x\omega)^2$$
$$= V_C^2+(x^2+y^2)\omega^2+2(xV_{Cy}-yV_{Cx})\omega$$

The total kinetic energy is thus

$$T = \tfrac{1}{2}\int V_C^2\,dm+\tfrac{1}{2}\int(x^2+y^2)\omega^2\,dm+\int(xV_{Cy}-yV_{Cx})\omega\,dm$$
$$= \tfrac{1}{2}V_C^2\int dm+\tfrac{1}{2}\omega^2\int(x^2+y^2)\,dm+V_{Cy}\omega\int x\,dm-V_{Cx}\omega\int y\,dm$$

Since $\int x\,dm=\int y\,dm=0$ according to (2.7), we have finally

$$T = \tfrac{1}{2}MV_C^2+\tfrac{1}{2}I_z\omega^2 \tag{8.7}$$

The first term in (8.7) represents the kinetic energy of translation of the total mass with the instantaneous velocity V_C of the centre of mass, and the second term represents the kinetic energy of rotation about the centre of mass. To use (8.7) we must use the centre of mass of the body, otherwise the expression takes a more complicated form.

The formula (8.7) corresponds to the formula (7.9) for rotation about a fixed axis.

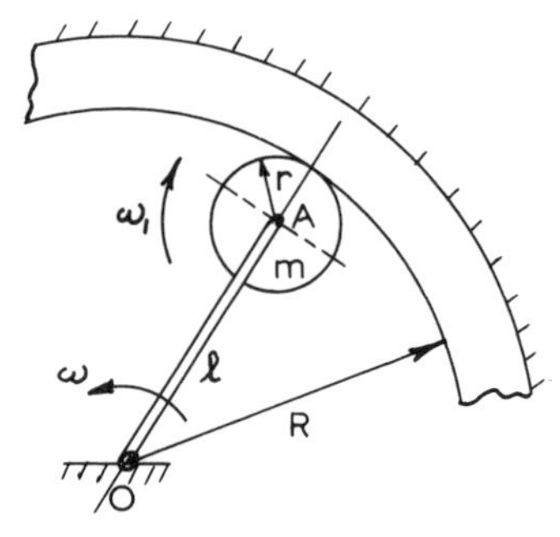

Fig. 8.8

Example 8.5. Figure 8.8 shows a circular disc of radius r and mass m which rolls without slipping on a cylindrical surface of radius R. The disc is pin-connected to a rigid, slender uniform bar OA of length l and mass M. The instantaneous angular velocity of OA is ω. Determine the total kinetic energy of the system in terms of the system constants l, ω, M and m. Calculate the numerical value of the kinetic energy if $l=0{\cdot}40$ m, $M=m=6{\cdot}5$ kg and $\omega=10$ rad/s.

Solution. The velocity of the centre A of the disc is $V_A=l\omega=r\omega_1$, so $\omega_1=(l/r)\omega$. The kinetic energy of the bar is $T_B=\frac{1}{2}I_O\omega^2=\frac{1}{2}(\frac{1}{3}Ml^2)\omega^2=\frac{1}{6}Ml^2\omega^2$. For the disc we get

$$T_D = \tfrac{1}{2}mV_A^2+\tfrac{1}{2}I_A\omega_1^2 = \tfrac{1}{2}ml^2\omega^2+\tfrac{1}{2}(\tfrac{1}{2}mr^2)\frac{l^2}{r^2}\omega^2 = \tfrac{3}{4}ml^2\omega^2$$

so that

$$T = \tfrac{1}{6}Ml^2\omega^2+\tfrac{3}{4}ml^2\omega^2 = \frac{l^2\omega^2}{12}(2M+9m)$$

The numerical value is

$$T = \frac{0{\cdot}40^2\times 10^2}{12}\times 11\times 6{\cdot}5 = 95{\cdot}2J$$

Example 8.6. Figure 8.9 shows a slender rigid bar of length l_1 and mass M_1. The bar is rotating in a horizontal plane about a fixed vertical axis through O. A second bar of length l_2 and mass M_2 is hinged to the first bar at B and is performing vibrations in the horizontal plane relative to the first bar. Using the coordinates φ and θ as shown, determine the total kinetic energy of the system in terms of the system constants.

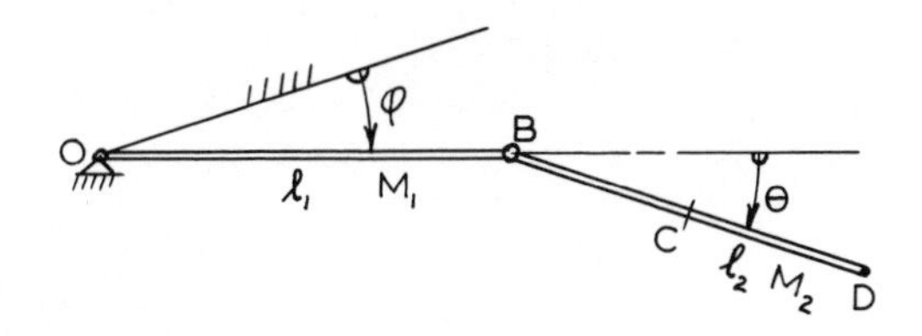

Fig. 8.9

Solution. For the bar OB, the angular velocity is $\omega_1=\dot{\varphi}$ and the moment of inertia about the axis of rotation is $I_O=\frac{1}{3}M_1l_1^2$, so that the kinetic energy is $T_1=\frac{1}{2}I_O\omega_1^2=\frac{1}{6}M_1l_1^2\dot{\varphi}^2$.

The angular velocity in formula (8.7) is the absolute angular velocity; for the bar OD we have $\omega_2 = \dot{\varphi} + \dot{\theta}$, and $I_C = \frac{1}{12} M_2 l_2^2$, where C is the centre of mass of OD.

The velocity of C is

$$\mathbf{V}_C = \mathbf{V}_B + \mathbf{V}_{CB} = l_1 \dot{\varphi} \downarrow + \frac{l_2}{2} (\dot{\varphi} + \dot{\theta}) \text{ at } 270 - \theta°$$

therefore

$$V_C^2 = l_1^2 \dot{\varphi}^2 + \frac{l_2^2}{4} (\dot{\varphi} + \dot{\theta})^2 + l_1 l_2 \dot{\varphi} (\dot{\varphi} + \dot{\theta}) \cos\theta$$

The kinetic energy of BD is thus $T_2 = \frac{1}{2} M_2 V_C^2 + \frac{1}{2} I_C \omega_2^2$, and the total kinetic energy of the system is $T = T_1 + T_2$.

8.6 Alternative form of equations of motion: work–energy equation for a rigid body in plane motion

It is shown in statics that any force system may be reduced to a resultant force and a couple acting about any particular point. The forces and couples acting on a rigid body may, therefore, always be reduced to a *resultant force* $\mathbf{F}$ and a *resultant moment* $\mathbf{M}_C$ acting at the centre of mass C of the body.

If the instantaneous velocity of C is $\mathbf{V}_C$ and the instantaneous angular velocity is $\boldsymbol{\omega}$, the displacement of the body in a time dt is $\mathbf{V}_C \cdot dt$ and the rotation is $\boldsymbol{\omega}\, dt$. The total work done on the body in the time dt is thus $\mathbf{F} \cdot \mathbf{V}_C\, dt + \mathbf{M}_C \cdot \boldsymbol{\omega}\, dt$. Hence the total work done in moving from position 1 to position 2 is

$$\text{W.D.} = \int_1^2 \mathbf{F} \cdot \mathbf{V}_C\, dt + \int_1^2 \mathbf{M}_C \cdot \boldsymbol{\omega}\, dt$$

Introducing Newton's second law $\mathbf{F} = M d\mathbf{V}_C/dt$, we find that the work done by the resultant force is

$$\int_1^2 \mathbf{F} \cdot \mathbf{V}_C\, dt = \int_1^2 M \frac{d\mathbf{V}_C}{dt} \cdot \mathbf{V}_C\, dt = M \int_1^2 \mathbf{V}_C \cdot d\mathbf{V}_C$$

$$= \tfrac{1}{2} M \int_1^2 d(\mathbf{V}_V \cdot \mathbf{V}_C)$$

$$= \tfrac{1}{2} M [V_C^2]_1^2 = \tfrac{1}{2} M (V_{C2}^2 - V_{C1}^2)$$

For the couple or moment $\mathbf{M}_C$, we have $\mathbf{M}_C = M_{Cx}\mathbf{i} + M_{Cy}\mathbf{j} + M_{Cz}\mathbf{k}$. Since $\boldsymbol{\omega} = \omega\mathbf{k}$, we find $\mathbf{M}_C \cdot \boldsymbol{\omega} = M_{Cz}\omega$; introducing $M_{Cz} = I_z \dot{\omega}$ from (8.2) we have

$$\int_1^2 \mathbf{M}_C \cdot \boldsymbol{\omega}\, dt = \int_1^2 I_z \dot{\omega} \omega\, dt = I_z \int_1^2 \omega\, d\omega = \tfrac{1}{2} I_z (\omega_2^2 - \omega_1^2)$$

We have then the total work done

$$\text{W.D.} = \tfrac{1}{2}M(V_{C2}^2 - V_{C1}^2) + \tfrac{1}{2}I_z(\omega_2^2 - \omega_1^2)$$

$$= \tfrac{1}{2}(MV_{C2}^2 + I_z\omega_2^2) - \tfrac{1}{2}(MV_{C1}^2 + I_z\omega_1^2) = T_2 - T_1 \quad (8.8)$$

This is the work–energy equation for a rigid body in plane motion. The equation states that the *total work* done on the body in the motion from a position 1 to a position 2 is equal to the *total change* in the kinetic energy of the body between the two positions.

Example 8.7. Consider Example 8.2. Determine the velocity $\dot{x}$ of the centre of mass as a function of the distance moved x, if the body starts from rest. Determine also the acceleration $\ddot{x}$. Use the work–energy equation.

Solution. Taking the position $x=0$ as the first position, we have $T_1=0$, since the body is at rest. For any position x we have

$$T_2 = \tfrac{1}{2}M\dot{x}^2 + \tfrac{1}{2}I\omega^2 = \frac{M}{2}(\dot{x}^2 + r_C^2\omega^2)$$

since

$$x = R\theta + x_0 \qquad \dot{x} = R\dot{\theta} = R\omega$$

so that

$$T_2 = \frac{M}{2}\left[\dot{x}^2 + \left(\frac{r_C}{R}\right)^2 \dot{x}^2\right] = \frac{M}{2}\dot{x}^2\left[1 + \left(\frac{r_C}{R}\right)^2\right] = T_2 - T_1$$

The normal force **N** does no work, and the friction force **F** is always directed through the instantaneous centre and does no work. The only work done is by the gravity force component along the inclined plane, so that W.D. $= Mg(\sin\beta)x$.

The work–energy equation now gives the result

$$\text{W.D.} = Mg(\sin\beta)x = T_2 - T_1 = \frac{M}{2}\dot{x}^2\left[1 + \left(\frac{r_C}{R}\right)^2\right]$$

therefore

$$\dot{x} = \left(\frac{2g\sin\beta}{1 + (r_C/R)^2}\, x\right)^{1/2}$$

Differentiating the equation gives

$$g(\sin\beta)\dot{x} = \dot{x}\ddot{x}\left[1+\left(\frac{r_C}{R}\right)^2\right] \qquad \text{or} \qquad \ddot{x} = \frac{g\sin\beta}{1+(r_C/R)^2}$$

as found in Example 8.2.

Example 8.8. Consider Example 8.3. By using the work–energy equation, determine the angular velocity and acceleration of the bar as a function of the angle θ if the bar starts from rest.

Solution. Since the bar starts from rest, $T_1 = 0$. The motion of the bar may be considered pure rotation about the instantaneous centre P (Fig. 8.5):

$$I_P = \tfrac{1}{3}Ml^2 \qquad \text{and} \qquad T_2 = \tfrac{1}{2}I_P\omega^2 = \tfrac{1}{6}Ml^2\dot{\theta}^2 = T_2 - T_1$$

Only the gravity force does work, so that

$$\text{W.D.} = \tfrac{1}{2}Mgl(1-\cos\theta) \qquad \text{and} \qquad \tfrac{1}{6}Ml^2\dot{\theta}^2 = \tfrac{1}{2}Mgl(1-\cos\theta)$$

This gives the result $\omega^2 = \dot{\theta}^2 = (3g/l)\,(1-\cos\theta)$, and by differentiation $2\dot{\theta}\ddot{\theta} = (3g/l)\,(\sin\theta)\dot{\theta}$, or $\ddot{\theta} = \dot{\omega} = (3g/2l)\sin\theta$. Both results are in agreement with the results found in Example 8.3.

8.7 Principle of conservation of mechanical energy in plane motion

The principle of conservation of mechanical energy for a particle was stated in eq. (2.15) as V + T = constant, that is the sum of the potential and kinetic energy is constant for a particle moving under the action of conservative forces only.

For a rigid body, the principle was stated in (7.12) with the form

$$V + T = \text{constant} \tag{8.9}$$

This is the principle of conservation of mechanical energy for a rigid body under the action of conservative external forces only. This principle is not as powerful as the work–energy equation, since that equation also includes friction forces and all non-conservative forces.

The principle (8.9) is sometimes useful for establishing the equation of motion by differentiation, since $dV/dt + dT/dt = 0$.

Example 8.9. Figure 8.10 shows a *rocking pendulum*, which consists of a body with mass M and moment of inertia with respect to an axis perpendicular to the plane of motion through the centre of mass $I_C = Mr_C^2$. The body rolls without slipping on a horizontal plane. Determine the equation of motion of the body by using the principle of conservation of mechanical energy. Determine also the frequency of small-displacement rocking.

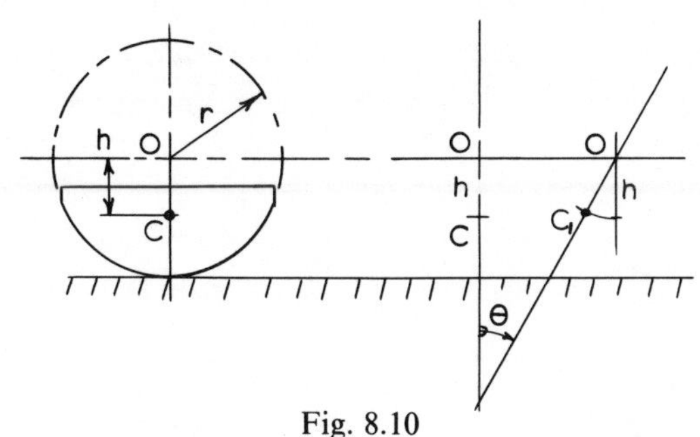

Fig. 8.10

Solution. This is a conservative system, since only the gravity force does any work during motion of the body. For a rotation θ of the body, the centre of mass increases its height above the plane by a distance $h(1-\cos\theta)$, so that

$$V = Mgh(1-\cos\theta)$$

The velocities involved are given by

$$\mathbf{V}_C = \mathbf{V}_O + \mathbf{V}_{CO} \qquad \mathbf{V}_O = r\dot{\theta} \rightarrow \qquad \mathbf{V}_{CO} = h\dot{\theta} \text{ at } 180-\theta^\circ$$

from which $V_C^2 = (r^2+h^2-2rh\cos\theta)\dot{\theta}^2$. The kinetic energy is

$$T = \tfrac{1}{2}MV_C^2 + \tfrac{1}{2}I_C\omega^2 = \tfrac{1}{2}M(r^2+h^2-2rh\cos\theta + r_C^2)\dot{\theta}^2$$

We have now

$$Mgh(1-\cos\theta) + \tfrac{1}{2}M(r^2+h^2-2rh\cos\theta+r_C^2)\dot{\theta}^2 = \text{constant}$$

Differentiating and cancelling M and $\dot{\theta}$ gives *the equation of motion*

$$(r^2+h^2-2rh\cos\theta+r_C^2)\ddot{\theta} + (rh\sin\theta)\dot{\theta}^2 + gh\sin\theta = 0$$

For small displacements we may take $\sin\theta \sim \theta$, $\cos\theta \sim 1$ and also $\dot{\theta}$ small, so that the term $(rh\sin\theta)\dot{\theta}^2 \sim rh\theta\dot{\theta}^2$ may be neglected. The equation thus takes the form

$$\ddot{\theta} + \left(\frac{gh}{(r-h)^2 + r_C^2}\right)\theta = 0$$

This represents simple harmonic motion with

$$f = \frac{1}{2\pi}\sqrt{\frac{gh}{(r-h)^2 + r_C^2}}$$

Example 8.10. Figure 8.11 shows a *rolling pendulum*. This consists of a cylinder of radius r and mass M rolling without slipping on a cylindrical surface.

Determine the equation of motion of the cylinder and the frequency for small displacements φ from the vertical line OA.

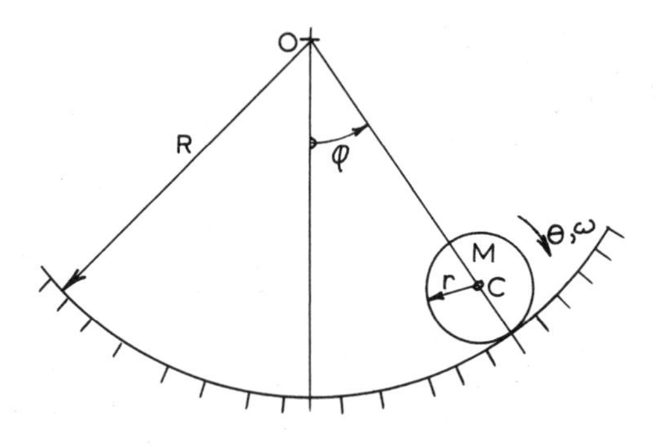

Fig. 8.11

Solution. Only gravity does work on the motion. The potential energy is $V = Mg(R-r)(1-\cos\varphi)$.

If the angle of rotation of the cylinder is θ, the velocity of the centre C is $V_C = (R-r)\dot{\varphi} = r\dot{\theta}$, so that $\dot{\theta} = \omega = [(R-r)/r]\dot{\varphi}$. With $I_C = Mr^2/2$, the kinetic energy is

$$T = \tfrac{1}{2}M(R-r)^2\dot{\varphi}^2 + \tfrac{1}{2}\left(\frac{Mr^2}{2}\right)\frac{(R-r)^2}{r^2}\dot{\varphi}^2 = \tfrac{3}{4}M(R-r)^2\dot{\varphi}^2$$

We have then $Mg(R-r)(1-\cos\varphi) + \tfrac{3}{4}M(R-r)^2\dot{\varphi}^2 = \text{constant}$; differentiating and cancelling the factors M, $(R-r)$ and $\dot{\varphi}$ gives the equation of motion

$$\ddot{\varphi} + \frac{2g}{3(R-r)}\sin\varphi = 0$$

For small values of φ the equation is

$$\ddot{\varphi} + \frac{2g}{3(R-r)}\varphi = 0$$

This represents simple harmonic motion with frequency

$$f = \frac{1}{2\pi}\sqrt{\frac{2g}{3(R-r)}}$$

8.8 D'Alembert's principle. Dynamic equilibrium in plane motion

The equations of motion for a rigid body in plane motion, with the z-axis as a principal axis, were eqs. (8.1) and (8.3), that is $F_X = M\ddot{X}$, $F_Y = M\ddot{Y}$ and $M_z = I_z\dot{\omega}$, where X and Y are the co-ordinates of the centre of mass. We may now apply D'Alembert's principle to these equations, that is we write the equations in terms of the inertia forces, inertia torques and applied external forces and moments as follows:

$$\left.\begin{aligned} F_X + (-M\ddot{X}) &= 0 \\ F_Y + (-M\ddot{Y}) &= 0 \\ M_z + (-I_z\dot{\omega}) &= 0 \end{aligned}\right\} \qquad (8.10)$$

This is D'Alembert's form of the equations of motion. Because of the complexity of the first two moment equations in eqs. (8.2) when the z-axis is not principal, we shall not apply D'Alembert's principle in such cases, but use the principle only in the form (8.10), in which the z-axis is a principal axis at the centre of mass. The forces and moments acting on the body, together with the inertia forces and torques, may now be taken as a system of plane forces in equilibrium, so that the body is in dynamic equilibrium and equations of statics may be used for the equilibrium conditions. The most important of these is the condition that moments may be taken about any axis perpendicular to the plane of motion, or more briefly about any point in the plane, while in Euler's equations the moment centres were limited to the centre of mass or a fixed point about which the body was rotating.

Example 8.11. Determine all the forces acting between the bars of the mechanism in Fig. 3.20, and the external torque at O at the instant considered in Example 3.3. The bars are all uniform, slender, rigid bars with masses $M_1 = 0{\cdot}18$ kg, $M_2 = 0{\cdot}62$ kg and $M_3 = 0{\cdot}39$ kg for bar OA, AB and CB respectively. The mechanism moves in a horizontal plane and friction may be neglected.

Solution. Since the mechanism is horizontal, the constant vertical forces at the hinges due to gravity may be found from statics and added vectorially to the dynamic forces.

A general kinematic analysis of this mechanism was carried out in Examples 3.3 and 3.4. The result was

$$\omega_1 = 12 \qquad \omega_2 = 3{\cdot}48 \circlearrowleft \qquad \omega_3 = 7{\cdot}10 \circlearrowleft \text{ rad/s}$$

$$\dot{\omega}_1 = 0 \qquad \dot{\omega}_2 = 4 \circlearrowleft \qquad \dot{\omega}_3 = 32{\cdot}1 \text{ rad/s}^2 \circlearrowright$$

$$A_A = 21{\cdot}6 \searrow \text{m/s}^2$$

Since the bars are uniform, the centre of mass of each bar is at the middle of the bar, and the analysis may easily be extended to determine the acceleration of each centre of mass, with the result

$$A_{C_1} = 10{\cdot}8 \searrow \qquad A_{C_2} = 20{\cdot}3 \searrow + 0{\cdot}62 \nearrow \qquad A_{C_3} = 5{\cdot}24 \searrow + 8{\cdot}21 \swarrow \text{ m/s}^2$$

The moments of inertia for the bars are

$$I_{C_2} = 13{\cdot}8 \times 10^{-3} \text{ kg m}^2 \qquad \text{and} \qquad I_{C_3} = 3{\cdot}45 \times 10^{-3} \text{ kg m}^2$$

The inertia forces and torques may now be calculated for each bar:

Bar OA: inertia force $= 1{\cdot}94 \nwarrow$ N, no inertia torque
Bar AB: inertia force $= 12{\cdot}6 \nwarrow + 0{\cdot}384 \swarrow$ N
inertia torque $= 55{\cdot}1 \times 10^{-3} \circlearrowright$ Nm
Bar CB: inertia force $= 2{\cdot}04 \nwarrow + 3{\cdot}20 \nearrow$ N
inertia torque $= 111 \times 10^{-3} \circlearrowleft$ Nm

All these forces and torques are in the opposite direction to the accelerations.

The bar AB may now be arranged in dynamic equilibrium as shown in Fig. 8.12(a).

Summing forces along and perpendicular to the bar and taking moments about point B gives the equations

$$F_1 - F_3 - 0{\cdot}384 = 0 \qquad \text{(a)}$$

$$F_2 + F_4 - 12{\cdot}6 = 0 \qquad \text{(b)}$$

$$-F_2 \times 0{\cdot}517 + 12{\cdot}6 \times 0{\cdot}259 + 55{\cdot}1/10^3 = 0 \qquad \text{(c)}$$

Equation (c) gives $F_2 = 6{\cdot}41$ N, and (b) gives $F_4 = 6{\cdot}19$ N. Forces F_1 and F_3 cannot be determined from this bar alone.

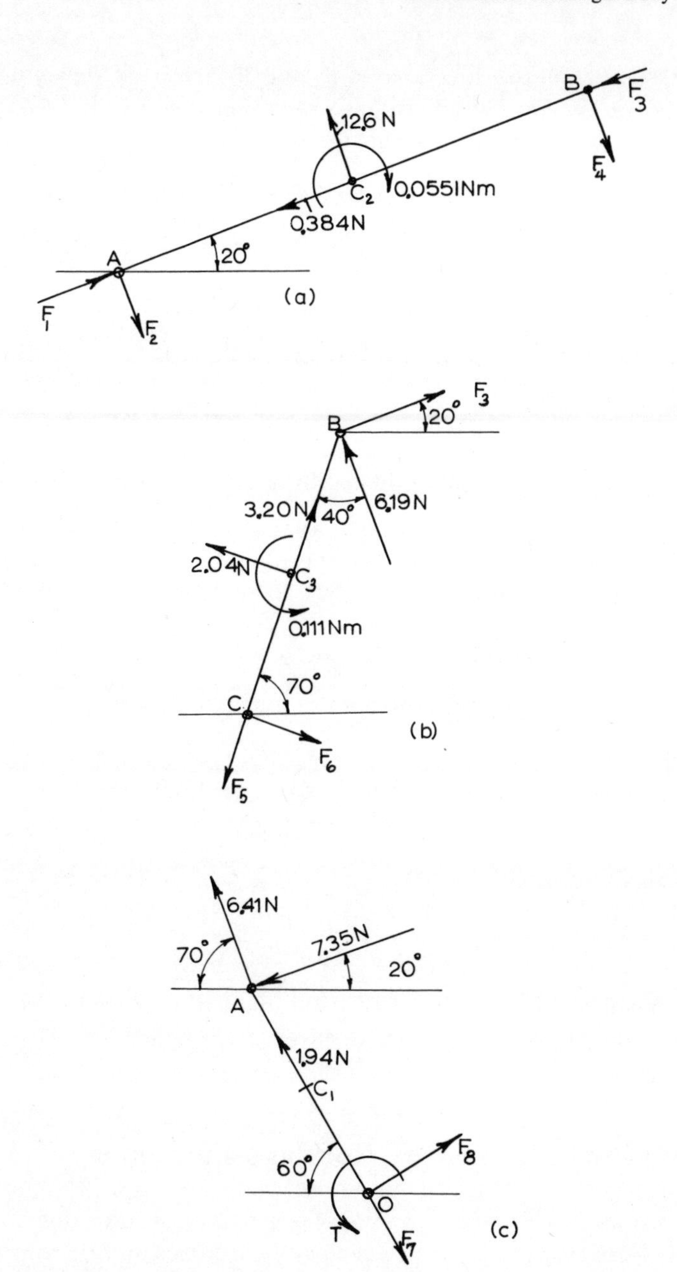
B
F_3
12.6 N
F_4
C_2
0.0551Nm
0.384N
A
20°
(a)
F_1
F_2
F_3
20°
B
3.20N
40°
6.19N
2.04N
C_3
0.111Nm
70°
C
(b)
F_6
F_5
6.41N
70°
7.35N
20°
A
1.94N
C_1
F_8
60°
O
T
F_7
(c)

Fig. 8.12

By reversing the directions of F_3 and F_4, the bar CB may now be put in dynamic equilibrium as shown in Fig. 8.12(b). Taking moments about point C, we get the equation

$$0{\cdot}111+2{\cdot}04\times 0{\cdot}163+6{\cdot}19\,(\sin 40)\times 0{\cdot}326-F_3\,(\sin 50)\times 0{\cdot}326 = 0 \tag{d}$$

from which $F_3=6{\cdot}97$ N.

Taking moments about point B gives the result

$$-F_6\times 0{\cdot}326-0{\cdot}111+2{\cdot}04\times 0{\cdot}163 = 0 \tag{e}$$

whence $F_6=0{\cdot}678$ N.

Summing forces along the bar, we obtain

$$-F_5+3{\cdot}20+6{\cdot}19\cos 40+F_3\cos 50 = 0 \tag{f}$$

so $F_5=12{\cdot}4$ N. Returning to eq. (a) and substituting $F_3=6{\cdot}97$ N gives $F_1=7{\cdot}35$ N. The bar OA may now be drawn in dynamic equilibrium as shown in Fig. 8.12(c), by reversing the direction of the forces F_1 and F_2.

Summing forces along the bar gives

$$F_7-1{\cdot}94-6{\cdot}41\cos 10-7{\cdot}35\cos 80 = 0 \tag{g}$$

whence $F_7=9{\cdot}54$ N.

Summing forces perpendicular to the bar, there results

$$F_8+6{\cdot}41\sin 10-7{\cdot}35\sin 80 = 0 \tag{h}$$

giving $F_8=6{\cdot}12$ N.

Finally, by taking moments about point A, we obtain

$$F_8\times 0{\cdot}15+T = 0 \tag{i}$$

from which $T=-0{\cdot}917$ N m. The instantaneous external torque T is in a direction opposite to that shown in the figure.

8.9 Principle of virtual work in plane motion

Virtual displacements $\delta\mathbf{r}$ of a particle were defined in Section 2.14. Any force system acting on a rigid body in statics may be reduced to a single resultant force $\mathbf{F}$ and a couple vector $\mathbf{M}$ at any chosen point. If the body is given a virtual displacement $\delta\mathbf{r}$ and a virtual rotation $\delta\boldsymbol{\theta}$, the work done by all acting forces is the virtual work $\delta(\text{W.D.}) = \mathbf{F}\cdot\delta\mathbf{r} + \mathbf{M}\cdot\delta\boldsymbol{\theta}$; if the force system is in equilibrium, it will also be in equilibrium during this virtual displacement according to the definition of virtual displacements, and we have consequently $\delta(\text{W.D.}) = 0$ for any virtual displacement. This is the *principle of virtual work* in statics.

In dynamics the acting forces are generally not in equilibrium, but if D'Alembert's principle is applied we have dynamic equilibrium and the principle of virtual work may then be applied as in a static case.

If the force system is reduced to a resultant force **F** at the centre of mass C and a resultant couple $\mathbf{M}_C$ about C, the principle of virtual work in dynamics for a rigid body may be expressed as follows:

$$(\mathbf{F} - M\ddot{\mathbf{r}}_C)\cdot\delta\mathbf{r} = 0 \qquad (\mathbf{M}_C - \dot{\mathbf{H}}_C)\cdot\delta\boldsymbol{\theta} = 0 \qquad (8.11)$$

These equations are called D'Alembert's equations and they are valid for general three-dimensional motion of a rigid body; they are also of fundamental importance in further developments in dynamics.

For the actual practical solutions of problems in plane motion, the second of eqs. (8.11) is convenient only for a body *rotating about a principal axis*.

Although D'Alembert's principle combined with the principle of virtual work is a powerful method in many problems, it has generally been superseded by the more useful Lagrange method to be discussed in Chapter 9. The principle retains its importance, however, since it is the foundation for Lagrange's equations.

Example 8.12. Figure 8.13 shows a two-step pulley of total mass M, and radius of gyration with respect to a central axis through C equal to r_C. The pulley carries a mass M_2 as shown. A second hub and vertical string on the far side makes the vertical plane a plane of symmetry. The mass M_2 is accelerating upwards.

Determine the acceleration of the mass M_2 by applying the principle of virtual work.

Solution. The system is shown in dynamic equilibrium with all inertia forces and external forces and torques.

Giving the centre C a virtual displacement δx_1 as shown, we have the virtual work

$$\delta(\text{W.D.}) = (M_1 g - M_1 \ddot{x}_1)\,\delta x_1 - I_C \ddot{\theta}\,\delta\theta - (M_2 g + M_2 \ddot{x}_2)\,\delta x_2 = 0$$

Substituting

$$\delta\theta = \frac{\delta x_1}{r_1} \qquad \delta x_2 = r_2\,\delta\theta - r_1\,\delta\theta = (r_2 - r_1)\frac{\delta x_1}{r_1}$$

and cancelling δx_1, we obtain

$$(M_1 g - M_1 \ddot{x}_1) - M_1 r_C^2 \frac{\ddot{\theta}}{r_1} - (M_2 g + M_2 \ddot{x}_2)\frac{r_2 - r_1}{r_1} = 0$$

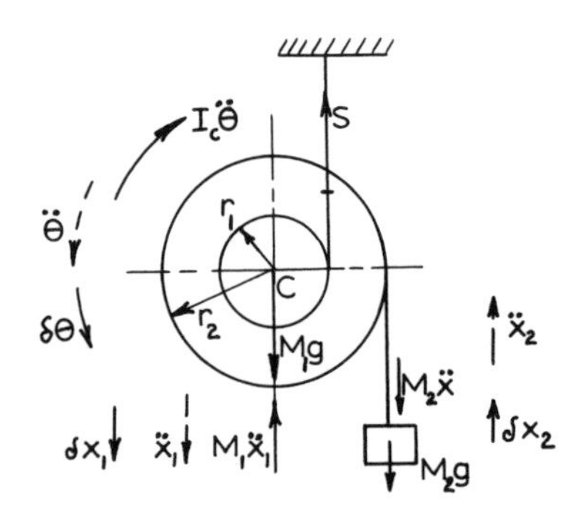

Fig. 8.13

Substituting

$$\ddot{x}_1 = \frac{r_1}{r_2 - r_1}\ddot{x}_2 \qquad \ddot{\theta} = \frac{\ddot{x}_1}{r_1} = \frac{\ddot{x}_2^2}{r_2 - r_1}$$

and solving for $\ddot{x}_2$ gives the result

$$\ddot{x}_2 = \frac{[(M_1 + M_2)r_1 - M_2 r_2](r_2 - r_1)g}{M_1(r_1^2 + r_C^2) + M_2(r_2 - r_1)^2}$$

The great advantage of the principle of virtual work is, as in statics, that the work of internal forces, fixed reactions and normal forces does not enter the expression for virtual work. If friction forces are present, they must be included with the other acting forces.

8.10 Impact in plane motion

For impact between rigid bodies in plane motion, we apply the same rules as for impact between particles (Section 2.12). These rules are

1. The impact forces are so great that other forces like gravity, friction, etc. may be neglected during impact.
2. The time of impact is so short that no appreciable motion can take place. The configuration just after impact may therefore be assumed to be the same as just before impact.

Since the impact forces and the impact time are generally unknown, equations of motion during impact cannot be established, but an *approximate* solution may be obtained by applying the above rules, together with the concept of *moment of momentum.*

The transfer formula (5.2) for the moment of momentum $\mathbf{H}_P$ of a rigid body in a fixed or moving point P is

$$\mathbf{H}_P = \mathbf{r}_{PC} \times M\mathbf{V}_C + \mathbf{H}_C$$

For a rigid body in plane motion (Fig. 8.14), we find, with respect to the fixed point O, that

$$\mathbf{H}_O = \mathbf{r}_C \times M\mathbf{V}_C + \mathbf{H}_C$$

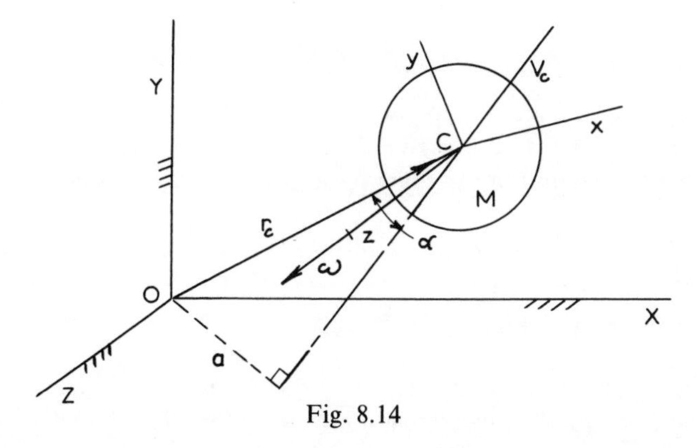

Fig. 8.14

We shall now restrict the discussion to bodies for which *the z-axis* through the centre of mass *is a principal axis*, which includes all homogeneous bodies with the plane of motion as a plane of symmetry. We have in these cases $I_{xz}=I_{yz}=0$, and since $\omega_x=\omega_y=0$, we find $H_x=H_y=0$ and $H_z=I_z\omega_z=I\omega$, so that $\mathbf{H}_C=I\omega\mathbf{k}$. The vectors $\mathbf{r}_C$ and $\mathbf{V}_C$ are in the xy-plane, so that $\mathbf{r}_C\times M\mathbf{V}_C$ is directed along the z-axis; the magnitude of this vector is $|\mathbf{r}_C|M|\mathbf{V}_C|\sin\alpha= aMV_C$. This is the scalar moment of MV_C about the z-axis. We find in these cases that vector notation is unncecessary, and the moment of momentum equation may be stated in scalar form:

$$H_O = (MV_C)a+I\omega \tag{8.12}$$

It is sufficient to take either the clockwise or anti-clockwise direction of rotation as positive, and use this for both terms in eq. (8.12).

For a fixed point O, we have the equation $\mathbf{M}_O=\dot{\mathbf{H}}_O$. If $\mathbf{M}_O=\mathbf{0}$ during a certain time, we have $\dot{\mathbf{H}}_O=\mathbf{0}$, or $\mathbf{H}_O$ is a vector in a fixed direction and with constant magnitude during the same time. In impact problems we consider only the impact force acting during impact; by taking moments about the point of impact O, we have then $\mathbf{M}_O=\mathbf{0}$, so that $\mathbf{H}_O$ is constant during impact, which means that H_O is the same just before and after impact.

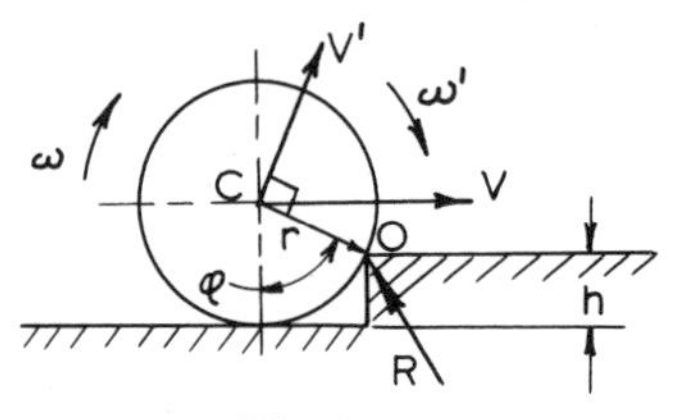

Fig. 8.15

Example 8.13. Figure 8.15 shows a right circular cylinder rolling without slipping on a horizontal plane. The velocity of the centre of mass C is V, and the angular velocity is ω. The cylinder is shown at the moment of impact with a step of height h. Assuming that there is no slip or rebound, determine the velocity of C just after impact. Investigate the conditions under which the cylinder will continue along the higher plane.

Solution. The moment of inertia of the cylinder is $I=Mr^2/2$, and we have $V=r\omega$. Taking moments about point O, and considering only the impact force R, we have $\mathbf{M}_O=\mathbf{0}=\dot{\mathbf{H}}_O$, so that H_O is constant during impact. Just before impact we find

$$H_O = (MV)r\cos\varphi+I\omega = MVr(\cos\varphi+\tfrac{1}{2})$$

Just after impact, the point O becomes the new instantaneous centre, so that V' is directed as shown. We find then $H'_O = MV'r + I\omega'$, with $V' = r\omega'$. This becomes $H'_O = \frac{3}{2}MV'r$, and so taking $H'_O = H_O$, we find $V' = \frac{1}{3}V(1+2\cos\varphi)$, which determines the motion just after impact. If $\varphi = 0$, there is no step and we get $V' = V$; if $\varphi = 90°$, the result is $V' = V/3$.

The work–energy equation may be used to determine whether the cylinder will continue along the upper plane. The kinetic energy after impact is

$$T_1 = \tfrac{1}{2}MV'^2 + \tfrac{1}{2}I\omega'^2$$

while the work done in moving up on the step is W.D. $= -Mgh$, so that

$$T_2 - T_1 = -Mgh \qquad \text{or} \qquad T_2 = T_1 - Mgh$$

If this expression is positive, the cylinder will roll along the upper plane; the velocity in this motion may be determined from $T_2 = \frac{1}{2}MV_2^2 + \frac{1}{2}I\omega_2^2$.

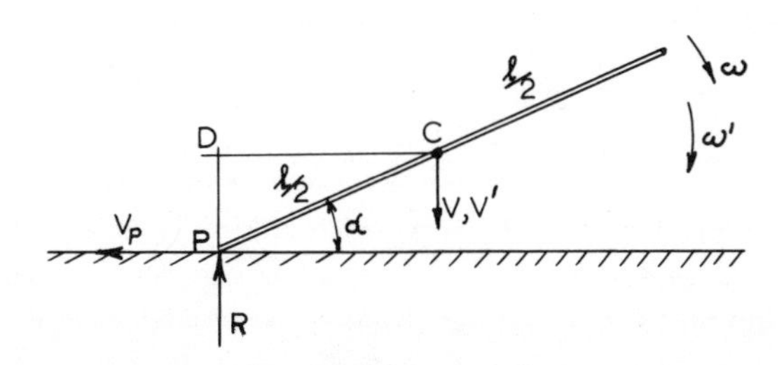

Fig. 8.16

Example 8.14. A slender rigid uniform bar of length l and mass M is falling without rotation in a vertical plane, as shown in Fig. 8.16. The lower end of the bar comes into contact with a smooth horizontal plane, and the velocity of the centre of mass C at this instant is V. Determine the motion of the bar after impact. Assume that there is no rebound at the impact.

Solution. Before, during, and after the impact, all forces acting are vertical; there is therefore no horizontal acceleration of the centre of mass C, and the horizontal component of the velocity of C, which is zero before impact, must remain zero; this means that

C moves down in a vertical line, so that the velocity V' after impact is vertical.

During the impact $\mathbf{M}_P = \mathbf{0} = \dot{\mathbf{H}}_P$ or $H_P = \text{constant}$
Before impact $H_P = \frac{1}{2}MVl \cos\alpha + I\omega = \frac{1}{2}MVl \cos\alpha$
After impact $H'_P = \frac{1}{2}MV'l \cos\alpha + I\omega' = \frac{1}{2}MV'l \cos\alpha + \frac{1}{12}Ml^2\omega'$
Equating the two expressions leads to the result

$$\omega' = (6/l)(V - V')\cos\alpha$$

Since this expression contains two unknown quantities ω' and V', the conservation of angular momentum is not sufficient to solve the problem. A second equation may be established kinematically by using the instantaneous centre D after impact; this gives the relationship $V' = (\frac{1}{2}l \cos\alpha)\omega'$, and the solution is

$$V' = \frac{3V\cos^2\alpha}{1 + 3\cos^2\alpha} \qquad \omega' = \frac{6V\cos\alpha}{l(1 + 3\cos^2\alpha)}$$

If $\alpha = 90°$, the bar is vertical, and $V' = \omega' = 0$. The solution does not apply for $\alpha = 0$, since then the bar would be in impact along its length; for $\alpha \to 0$ the result is $V' \to \frac{3}{4}V$, $\omega' \to \frac{3}{2}V/l$.

8.11 Simple gyroscopic motion

We shall consider only the most elementary type of gyroscopic motion, and consequently we *define a gyroscope* as a homogeneous body of revolution, spinning about its geometrical axis. This is then rotation about a central principal axis. In a gyroscope this axis of spin is itself rotating or changing direction in space. We shall here consider only the cases where this change of direction of the spin axis takes place in a plane, so that the spin axis rotates about an axis perpendicular to the spin axis. Such a situation is shown in Fig. 8.17(a), which shows a homogeneous right circular cylinder in two bearings A and B fixed on a horizontal platform. The bearing A is a thrust bearing which holds the cylinder in the direction AB.

In the position shown, the cylinder is rotating about its axis with angular velocity $\boldsymbol{\omega}_S$ and angular acceleration $\dot{\boldsymbol{\omega}}_S$ relative to the platform. The platform is rotating about the fixed vertical Z-axis with angular velocity $\boldsymbol{\omega}_P$ and angular acceleration $\dot{\boldsymbol{\omega}}_P$ as shown.

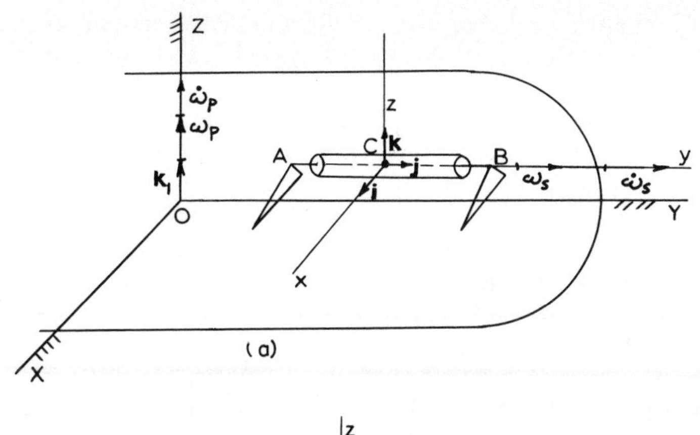

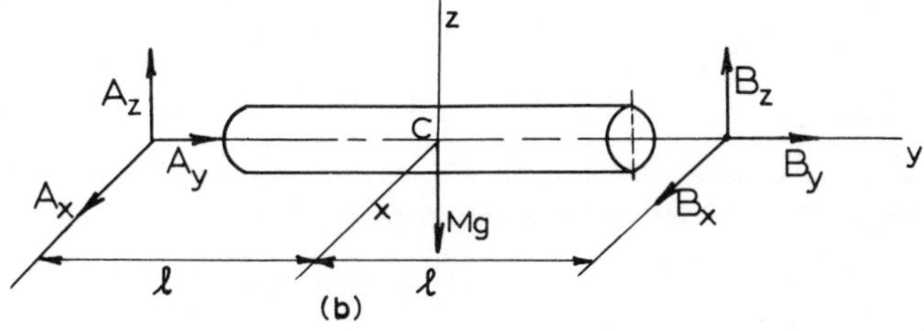

Fig. 8.17

A rotating Euler system (x, y, z) is shown fixed in the cylinder with origin at the centre of mass C and axes along principal axes of the cylinder. Euler's equations in this system are eqs. (7.15):

$$M_x = I_x\dot{\omega}_x-(I_y-I_z)\omega_y\omega_z$$
$$M_y = I_y\dot{\omega}_y-(I_z-I_x)\omega_z\omega_x$$
$$M_z = I_z\dot{\omega}_z-(I_x-I_y)\omega_x\omega_y$$

The total angular velocity of the cylinder, at the instant shown, is

$$\boldsymbol{\omega} = \boldsymbol{\omega}_S+\boldsymbol{\omega}_P = \omega_S\mathbf{j}+\omega_P\mathbf{k}$$

so that $\omega_x=0$, $\omega_y=\omega_S$ and $\omega_z=\omega_P$. The expression for $\boldsymbol{\omega}$ is, however, *only valid at this instant*: a moment later the x-axis will be pointing downward and the angular velocity $\boldsymbol{\omega}_P$ will then have a component along the x-axis; ω_x is therefore only zero at the instant shown, and we would expect a value for $\dot{\omega}_x$ at this instant. Since the expression for $\boldsymbol{\omega}$ is valid only instantaneously, it cannot be differentiated. To find an expression for $\boldsymbol{\omega}$ that is valid in general, we introduce the unit vector $\mathbf{k}_1$ on the fixed Z-axis; we have then $\boldsymbol{\omega}=\omega_S\mathbf{j}+\omega_P\mathbf{k}_1$.

To determine $\dot{\boldsymbol{\omega}}$, we apply the general vector formula (4.1) $\dot{\mathbf{r}}=\dot{\mathbf{r}}_r+\boldsymbol{\omega}\times\mathbf{r}$ to the vector $\boldsymbol{\omega}$, giving $\dot{\boldsymbol{\omega}}=\dot{\boldsymbol{\omega}}_r+\boldsymbol{\omega}\times\boldsymbol{\omega}=\dot{\boldsymbol{\omega}}_r$, so that *the angular acceleration is the same in the inertial system and the rotating system.* By differentiating a vector function for $\boldsymbol{\omega}$, we find $\dot{\boldsymbol{\omega}}_r$, and taking components of this vector along the x-, y- and z-axes gives us the values of $\dot{\omega}_x$, $\dot{\omega}_y$ and $\dot{\omega}_z$ for Euler's equations.

By differentiation of the general expression above for $\boldsymbol{\omega}$, we find

$$\dot{\boldsymbol{\omega}}=\dot{\boldsymbol{\omega}}_r = \dot{\omega}_S\mathbf{j}+\omega_S\dot{\mathbf{j}}+\dot{\omega}_P\mathbf{k}_1+\omega_P\dot{\mathbf{k}}_1$$

The vector $\mathbf{j}$ is a constant-length vector fixed in a rotating rigid body, so that

$$\dot{\mathbf{j}} = \boldsymbol{\omega}\times\mathbf{j} = (\omega_S\mathbf{j}+\omega_P\mathbf{k}_1)\times\mathbf{j} = \omega_P\mathbf{k}_1\times\mathbf{j}$$

Since $\mathbf{k}_1$ is of constant length and in a fixed direction, we have $\dot{\mathbf{k}}_1=\mathbf{0}$, and the general vector expression for $\dot{\boldsymbol{\omega}}$ is

$$\dot{\boldsymbol{\omega}} = \dot{\omega}_S\mathbf{j}+\omega_P\omega_S\mathbf{k}_1\times\mathbf{j}+\dot{\omega}_P\mathbf{k}_1$$

At the instant considered we have $\mathbf{k}_1=\mathbf{k}$, so that

$$\begin{aligned}\dot{\boldsymbol{\omega}} &= \dot{\omega}_S\mathbf{j}+\omega_P\omega_S\mathbf{k}\times\mathbf{j}+\dot{\omega}_P\mathbf{k}\\ &= -\omega_P\omega_S\mathbf{i}+\dot{\omega}_S\mathbf{j}+\dot{\omega}_P\mathbf{k} = \dot{\omega}_x\mathbf{i}+\dot{\omega}_y\mathbf{j}+\dot{\omega}_z\mathbf{k}\end{aligned}$$

The components of the angular acceleration are then $\dot{\omega}_x=-\omega_P\omega_S$, $\dot{\omega}_y=\dot{\omega}_S$ and $\dot{\omega}_z=\dot{\omega}_P$. Substituting the expressions for the components of $\boldsymbol{\omega}$ and $\dot{\boldsymbol{\omega}}$ in Euler's equations, gives the result for M_x:

$$M_x = -I_x\omega_P\omega_S-(I_y-I_z)\omega_S\omega_P$$

however $I_x=I_z$ for the cylinder and for any homogeneous body of revolution with the y-axis as geometrical axis; the Euler equations now take the following form, with the notations $I_y=I_s$ and $I_z=I_p$:

$$M_x = -I_S\omega_P\omega_S \qquad M_y = I_s\dot{\omega}_S \qquad M_z = I_P\dot{\omega}_P$$

The moment of the forces acting (Fig. 8.17(b)) gives $M_y=0$, since all forces intersect or are parallel to the y-axis. The torque $M_y=I_s\dot{\omega}_s$ must then be supplied from outside by a driving motor (not shown), to give the angular acceleration $\dot{\omega}_S$, otherwise ω_S must be constant, assuming no friction.

The moment about the z-axis is $M_z = A_x l - B_x l = I_P \dot{\omega}_P$. This torque is supplied by the *horizontal bearing reactions*.

The moment of the acting forces about the x-axis is $M_x = B_z l - A_z l = -I_S \omega_P \omega_S$. This torque is in the negative x-direction as a vector, it acts in the *vertical plane* and is supplied by the *vertical bearing reactions*. The torque M_x in the *vertical* plane necessary to rotate the spin axis (the y-axis) in the *horizontal* plane is called a *gyroscopic torque*. In vector notation the magnitude and direction of the gyroscopic torque may be given in the formula

$$\mathbf{T} = I_S \boldsymbol{\omega}_P \times \boldsymbol{\omega}_S \tag{8.13}$$

The formula (8.13) is particularly easy to remember, since it relates a torque to a moment of inertia multiplied by the product of two angular velocities which has the dimension of angular acceleration. The angular velocities ω_P and ω_S are also arranged in alphabetic order.

It must be kept in mind that (8.13) *gives the torque as it acts on the spinning body*; if the *bearing reactions* are required, the direction of $\mathbf{T}$ must be reversed. The formula (8.13) holds for ω_P and ω_S constant or variable, and no assumption needs to be made concerning the relative magnitude of ω_P and ω_S.

The rotation of the spin axis about the z-axis is called *precessional motion* of the spin axis in the horizontal plane; ω_S is the angular spin velocity, ω_P the angular precessional velocity and $\omega_P \omega_S$ is the gyroscopic acceleration.

The formula (8.13) is really the first Euler equation in vectorial form; to determine all the acting forces it is generally necessary to apply also the other two Euler equations, and the three Newton equations for the motion of the centre of mass.

Example 8.15. Derive the formula (8.13) for the gyroscopic torque in simple gyroscopic motion, by the vector formula $\mathbf{M}_C = \dot{\mathbf{H}}_C$ directly, without using Euler's equations.

Solution. Figure 8.18 shows the general situation, a coordinate system (x, y, z) moving with the body, with origin at C and with a z-axis that is always vertical and a y-axis always along the axis of spin, has been introduced. With respect to the moving centre of mass C, we have $\mathbf{M}_C = \dot{\mathbf{H}}_C$, where the differentiation is absolute. Now the absolute value of $\mathbf{H}_C$ is equal to its relative value, and since the x-, y- and z-axes are principal axes, $\mathbf{H}_C = I_S\boldsymbol{\omega}_S + I_P\boldsymbol{\omega}_P$ at all times.

Now

$$d\mathbf{H}_C/dt = I_S\dot{\boldsymbol{\omega}}_S + I_P\dot{\boldsymbol{\omega}}_P = \mathbf{M}_C$$

The change in $\boldsymbol{\omega}_S$ is a change in magnitude, and a change due to the rotation $\boldsymbol{\omega}_P$; from eq. (4.1) we find that

$$\dot{\boldsymbol{\omega}}_S = \dot{\boldsymbol{\omega}}_{S_r} + \boldsymbol{\omega}_P \times \boldsymbol{\omega}_S = \dot{\omega}_S\mathbf{j} + \omega_P\mathbf{k} \times \omega_S\mathbf{j} = \dot{\omega}_S\mathbf{j} - \omega_P\omega_S\mathbf{i}$$

$$\dot{\boldsymbol{\omega}}_P = \frac{d}{dt}(\omega_P\mathbf{k}) = \dot{\omega}_P\mathbf{k} + \omega_P\dot{\mathbf{k}} = \dot{\omega}_P\mathbf{k}$$

since $\mathbf{k}$ is always vertical. The result is thus

$$\mathbf{M}_C = -I_S\omega_P\omega_S\mathbf{i} + I_S\dot{\omega}_S\mathbf{j} + I_P\dot{\omega}_P\mathbf{k}$$

so that M_x in vector form is $\mathbf{T} = I_S\boldsymbol{\omega}_P \times \dot{\boldsymbol{\omega}}_S$ as in (8.13).

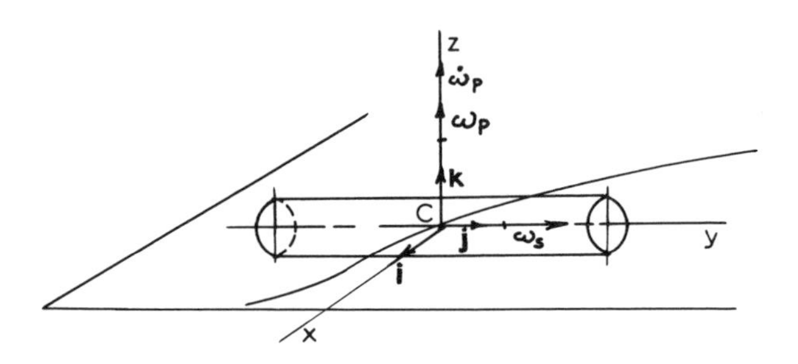

Fig. 8.18

The situation is particularly simple if $\boldsymbol{\omega}_P$ and $\boldsymbol{\omega}_S$ are of constant magnitude: then $\boldsymbol{\omega}_S$ is a constant-length vector fixed in a rigid body rotating at $\boldsymbol{\omega}_P$, so that $\dot{\boldsymbol{\omega}}_S = \boldsymbol{\omega}_P \times \boldsymbol{\omega}_S = -\omega_P\omega_S\mathbf{i}$, and $\dot{\boldsymbol{\omega}}_P = \mathbf{0}$. We have then $\mathbf{M}_C = \dot{\mathbf{H}}_C = I_S\dot{\boldsymbol{\omega}}_S = -I_S\omega_P\omega_S\mathbf{i} = \mathbf{T}$.

If $\boldsymbol{\omega}_S$ and $\boldsymbol{\omega}_P$ are of constant magnitude and C *is fixed*, the constant-length vector $\boldsymbol{\omega}_S$ rotates in the plane about C with angular velocity ω_P, we find in this case

$$\frac{d\boldsymbol{\omega}_S}{dt} = -\omega_S\frac{d\theta}{dt}\mathbf{i} = -\omega_S\omega_P\mathbf{i} \qquad \text{and} \qquad \dot{\boldsymbol{\omega}}_P = \mathbf{0}$$

$$\mathbf{H}_C = I_S\boldsymbol{\omega}_S + I_P\boldsymbol{\omega}_P \qquad \text{and} \qquad \mathbf{M}_C = \dot{\mathbf{H}}_C = I_S\dot{\boldsymbol{\omega}}_S + I_P\dot{\boldsymbol{\omega}}_P = I_S\dot{\boldsymbol{\omega}}_S$$

$$= -I_S\omega_S\omega_P\mathbf{i}$$

Example 8.16. An aircraft (Fig. 8.19(a)) is flying at a constant speed of 707 km/h in a horizontal circle with radius 3050 m. The propeller and rotating parts of the engine have a moment of inertia of 26·2 kg m^2, and are rotating at a constant 3200 rev/min in a clockwise direction viewed from the rear of the plane. Determine the effect of the gyroscopic couple on the engine bearings.

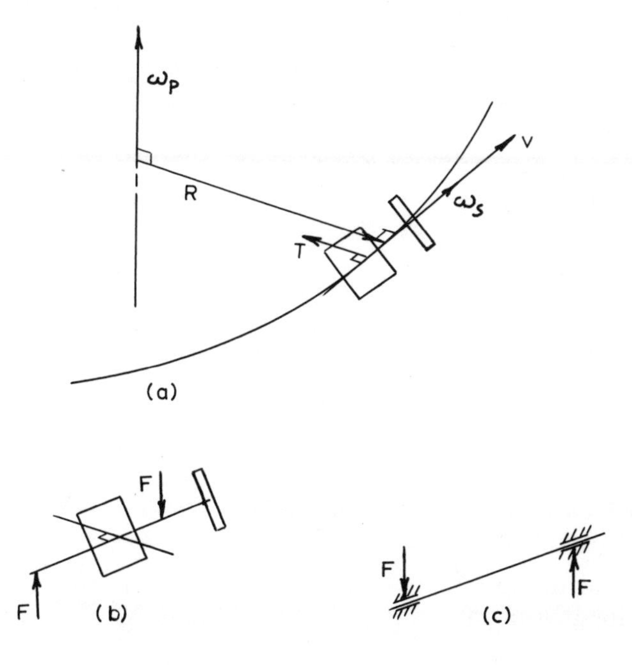

Fig. 8.19

Solution. The gyroscopic torque is $\mathbf{T} = I_S \boldsymbol{\omega}_P \times \boldsymbol{\omega}_S$. We have $I_S = 26{\cdot}2\ \text{kg m}^2$, $\omega_S = (\pi/30) \times 3200 = 335$ rad/s, $\omega_P = V/R = (707 \times 1000)/(3600 \times 3050) = 0{\cdot}0644$ rad/s.

T acts on the spinning body as shown in Fig. 8.19(a). The magnitude of **T** is

$$T = I_S \omega_P \omega_S = 26{\cdot}2 \times 0{\cdot}0644 \times 335 = 566 \text{ N m}$$

The action of **T** on the spinning body is shown in Fig. 8.19(b). The bearing forces **F** adjust themselves to produce the torque **T**. Figure 8.19(c) shows the forces *on* the bearings. It is evident that the nose of the aircraft tends to move upwards.

8.12 The spinning top

It is well known that the spinning top shown in Fig. 8.20 is able to perform a rotational motion about the vertical line through the point of support O. Assuming that the top is a homogeneous body of revolution, we introduce a coordinate system (x, y, z) with origin at O and fixed in the body with axes along the principal axes as shown.

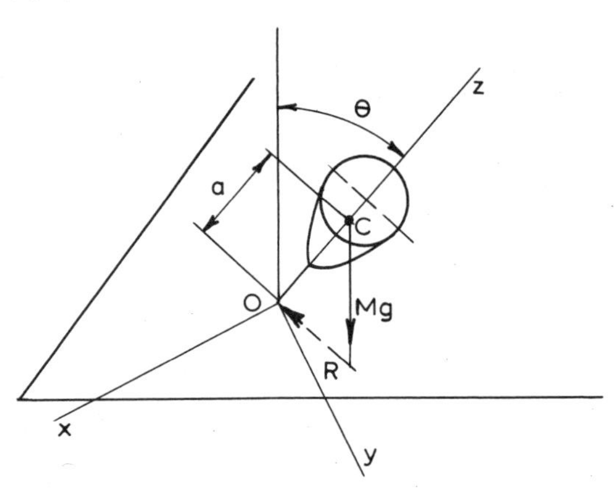

Fig. 8.20

Equations (5.5) for the components of the moment of momentum vector at O are then

$$H_x = I_x\omega_x \qquad H_y = I_y\omega_y \qquad H_z = I_z\omega_z$$

To obtain a *simplified* solution, we now assume that ω_z is large compared to ω_x and ω_y, and that I_z is not small compared to I_x and I_y; thus $H_z = I_z\omega_z$ becomes large compared to H_x and H_y. We will ignore the latter components and take $\mathbf{H}_O = \mathbf{H}_z = I_z\boldsymbol{\omega}_z$. Introducing the notation $I_z = I_S$ and $\omega_z = \omega_S$, we have $\mathbf{H}_O = I_S\boldsymbol{\omega}_S$ along the z-axis. We will also assume that $\boldsymbol{\omega}_S$ is of constant magnitude, and that friction and wind resistance may be ignored.

We have now the equation $\mathbf{M}_O = \dot{\mathbf{H}}_O$, with respect to the fixed point O. Referring to Fig. 8.21, we have $\mathbf{M}_O = \mathbf{r}_C \times \mathbf{M}g$; the magnitude of $\mathbf{M}_O$ is $|\mathbf{M}_O| = |\mathbf{r}_C||\mathbf{M}g| \sin\theta = aMg \sin\theta$, and the direction of $\mathbf{M}_O$ is perpendicular to the vertical plane OAB, so that $\mathbf{M}_O$ is horizontal as shown.

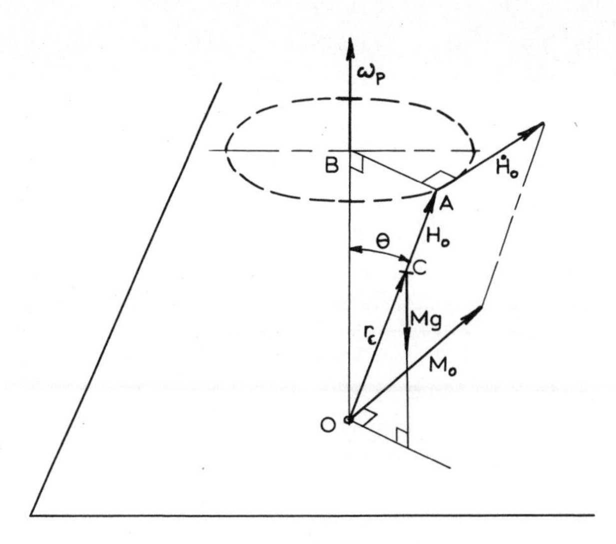

Fig. 8.21

Since $\mathbf{M}_O = \dot{\mathbf{H}}_O = I_S\dot{\boldsymbol{\omega}}_S$, the vector $\dot{\mathbf{H}}_O$ may be introduced as shown, tangential to the horizontal circle with centre B and radius BA. The magnitude of $\dot{\mathbf{H}}_O$ is $aMg \sin \theta$.

The constant-length vector $\mathbf{H}_O$ is a position vector for the point A, so its derivative $\dot{\mathbf{H}}_O$ is the velocity $\mathbf{V}$ of point A. The endpoint of the $\mathbf{H}_O$ vector thus describes a horizontal circle with centre B and radius $BA = I_S\omega_S \sin \theta$, which means that the spin axis describes a circular cone of angle 2θ; the angular velocity of $\mathbf{H}_O$ about the vertical axis is $\boldsymbol{\omega}_P$, where $|\mathbf{V}| = |\dot{\mathbf{H}}_O| = AB\omega_P$, so that

$$\omega_P = \frac{|\dot{\mathbf{H}}_O|}{AB} = \frac{aMg \sin \theta}{I_S\omega_S \sin \theta} = \frac{aMg}{I_S\omega_S}$$

This is constant and independent of θ for $\theta \neq 0$. This is the angular velocity of *precession* of the cone.

For $a > 0$ as shown in the figure, with O below C, ω_P is in the same direction as ω_S. For $a < 0$, which means that O is above C, the precession is in the opposite direction of the spin. If $a = 0$, O coincides with C, and the centre of mass is a fixed point, in which case $\omega_P = 0$, so that the gyroscope is stationary in a fixed direction θ in space. This fact is of great importance in directional gyroscopic indicators, in the spinning effects of bullets, etc.

If $\theta=90°$, the spin axis is horizontal and $\sin\theta=1$. The torque is then of magnitude $Mga=M_O=T$, and $\omega_P=I/I_S\omega_S$, or $T=I_S\omega_P\omega_S$ as found before for this case.

Problems

8.1 For the system shown in Fig. 8.22, determine the acceleration of the falling mass M_1 and the string tension. The cylinder rolls without slip and the fixed pulley is frictionless.

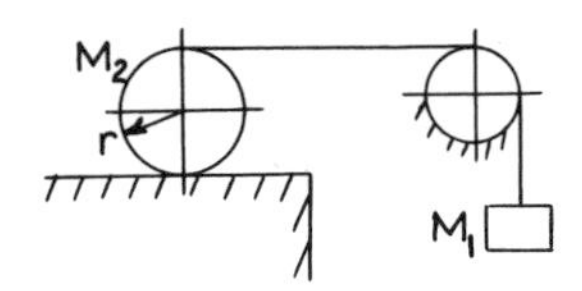

Fig. 8.22

8.2 Figure 8.23 shows a homogeneous spool consisting of two cylindrical discs of radius R rigidly connected by a cylindrical section of radius r. The spool rolls without slipping on a horizontal plane due to the pull P in a string which is wrapped around the spool.

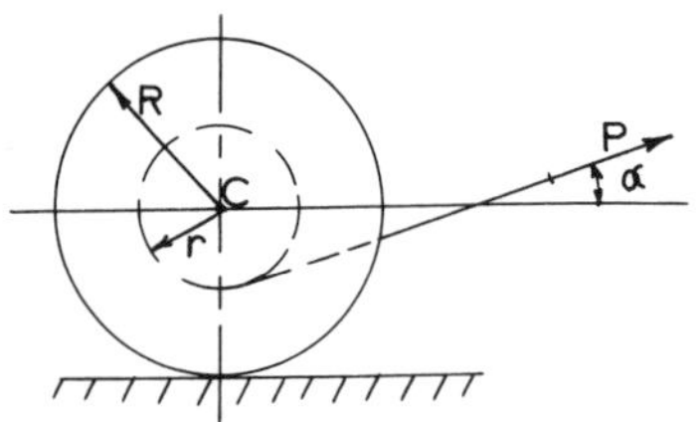

Fig. 8.23

The force P acts in the vertical plane of symmetry of the spool and is inclined at an angle α to the horizontal direction.

The total mass of the spool is M and the moment of inertia with respect to the geometrical axis is $I=Mr_C^2$, where r_C is the radius of gyration.

(a) Determine the acceleration of the centre of mass of the spool in terms of the given system constants and the force P.

(b) Determine the limits of the angle α for the spool to roll to the right or to the left, if $r=R/2$.

(c) Determine the magnitude of the acceleration of the centre of mass of the spool, if $P=44{\cdot}5$ N; $M=9{\cdot}1$ kg; $R=15{\cdot}2$ cm; $r=R/2$; $r_C=R/3$ and $\alpha=30°$.

(d) Determine the minimum value of the coefficient of friction between the spool and the plane, for the conditions given under (c).

8.3 A thin steel hoop of mass M and radius r starts from rest at A (Fig. 8.24) and rolls down along a circular cylindrical surface of radius a. Determine the angle θ_0 defining the position B where the hoop starts to slip, if the coefficient of friction is $\mu=\frac{1}{3}$.

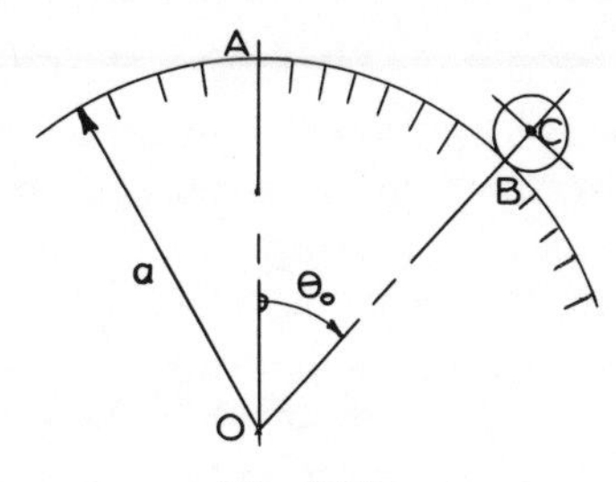

Fig. 8.24

8.4 Find the period of oscillation of a homogeneous right semicircular cylinder of radius 30·5 cm for small amplitudes of rolling without slip on a horizontal plane. Repeat for a homogeneous hemisphere of radius 30·5 cm.

8.5 The thin cylindrical shell in Fig. 8.25 approaches the inclined plane with velocity V, rolls up this plane and eventually rolls back to the left. Find the velocity of the shell to the left. Assume no slip or rebound.

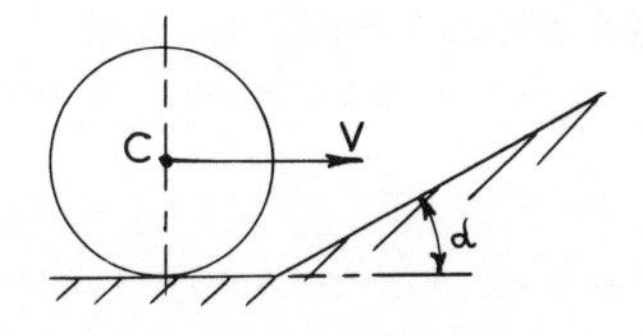

Fig. 8.25

8.6 An arrow of length 76·2 cm travelling with a velocity $V=$ 15·25 m/s strikes a smooth hard wall (Fig. 8.26). Assuming that the end A slides downward without friction or rebound, find the angular velocity of the arrow just after impact if $\alpha = 30°$. Determine also the velocity of the centre of mass just after impact.

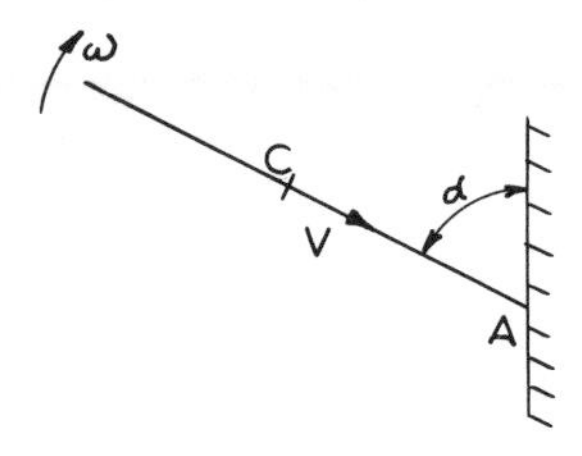

Fig. 8.26

8.7 The armature of the motor of an electric car weighs 2670 N and rotates in a direction opposite to the rotation of the car wheels. The distance between its bearings is 0·610 m and its radius of gyration is 15·24 cm. The motor makes 4 revolutions to 1 revolution of the car wheels which have diameters of 0·84 m. If the car is moving forward around a curve of 30·5 m radius with a velocity of 6·10 m/s, find the total pressures on the bearings of the armature if the centre of the curve is to the right.

The centre-line of the armature is parallel to the wheel centre-line.

8.8 The low-pressure turbine rotor of a cargo steamer weighs 84·8 kN and has a radius of gyration of 0·381 m. The speed of rotation is 3300 rev/min in a clockwise sense when viewed from the stern of the ship. If the ship is pitching with an amplitude of 6° above and below the horizontal with a period of 10 s in simple harmonic motion, calculate the maximum torque which is transmitted to the turbine bed plates. State the direction of the torque in relation to the pitching motion of the ship.

The centre-line of the turbine is on the longitudinal centre-line of the ship.

8.9 An aircraft landing gear assembly is shown in Fig. 8.27. After taking off the landing gear is retracted into the wing by rotation about the axis AA. The wheel continues to spin as it is being retracted.

Calculate the magnitude of the gyroscopic couple due to the spinning wheel if the weight of the wheel is 667 N, the radius of gyration about the axis CC is 0·549 m, the take-off speed is 290 km/h and the maximum speed of retraction ω_P is 3 rad/s.

Indicate the direction of the gyroscopic couple and its action on the strut AC.

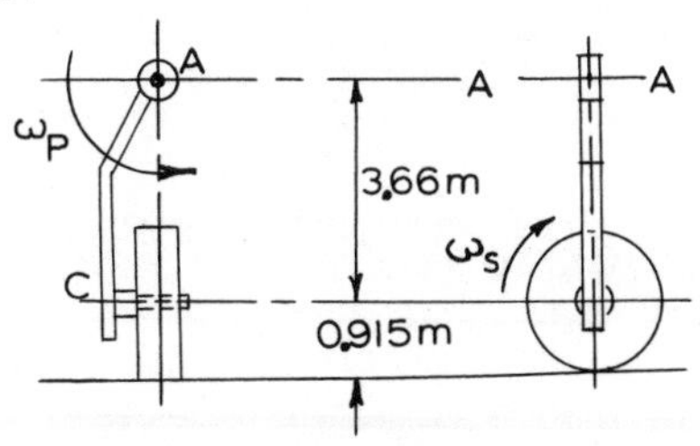

Fig. 8.27

8.10 A right circular disc of mass $M = 7$ kg and radius $r = 8$ cm, is attached to a shaft AB as shown in Fig. 8.28. The disc rotates at an angular velocity $\omega_2 = 100$ rad/s constant relative to the bearings. The bearings are mounted on a horizontal platform which rotates at a constant angular velocity $\omega_1 = 20$ rad/s about the axis OZ. The mass of the shaft AB may be neglected, and it may be assumed that bearing A alone retains the system in the Y-direction. For the instant shown in Fig. 8.28, compute the components of the reactions on the shaft at A and B.

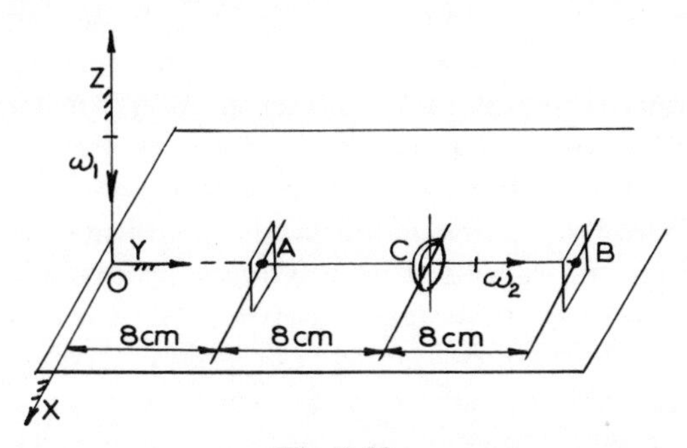

Fig. 8.28

CHAPTER NINE

Lagrange's Equations

9.1 Degrees of freedom. Equations of constraints

The number of degrees of freedom n of a body was defined as the number of independent coordinates necessary to specify the position of the body at any time in an inertial reference frame or absolute coordinate system.

For a particle, the degrees of freedom and equations of constraints were discussed in Section 2.15. The maximum value of n was $n=3$.

For a rigid body, $n=0$ if the body is fixed. For *rectilinear translation* or *rotation about a fixed axis* $n=1$; for a combination of these $n=2$, unless there is a fixed relationship between the rotational and longitudinal motion, as in a screw motion, when $n=1$. For a body in *plane curvilinear translation*, with one point following a curve $y=f(x)$, $n=1$. If the body is free to translate in a plane, $n=2$. For *curvilinear translation* with one point following a space curve, $n=1$. For general curvilinear translation in space $n=3$.

A rigid body in *plane motion* has $n=3$. For *rotation about a fixed point*, $n=3$; in this case the coordinates of two points other than the fixed point may be given, which determines a triangle fixed in the body and gives 6 coordinates. However, the length of each side of the triangle is constant, giving three equations of constraint of the form

$$(x_2-x_1)^2+(y_2-y_1)^2+(z_2-z_1)^2 = a^2$$

Only three of the six coordinates may be chosen arbitrarily and $n=3$.

A rigid body *free in space* has $n=6$. The coordinates may be taken as the three coordinates of a point in the body, together with three angles of rotation of the body about three axes through the point. Any *elastic body* has an infinite number of degrees of freedom.

Practically all dynamics systems in engineering are systems with constraints. Machines may usually be considered as systems of connected rigid bodies with various constraints like gears, bearings, foundations, cylinders, etc. These constraints limit the number of degrees of freedom of the total system.

Equations of constraint sometimes involve the time t and derivatives of time (moving constraints); in most cases these problems are much more complicated than problems where time is not involved explicitly. Problems with moving constraints are outside the scope of this book.

9.2 Generalized coordinates

In most problems the position of a body may be given in rectangular coordinates. In general these are not independent but connected by equations of constraints, and to avoid the complications of these equations, it is a great advantage if some *independent quantities* can be found to describe the system. These quantities are called *generalized coordinates*, and there are just enough of them to specify the position of the system. The number of generalized coordinates is therefore equal to the number of *degrees of freedom* of the system.

In most cases several different sets of generalized coordinates may be found; often the usual rectangular coordinates and angles of rotation may be used. Other quantities than lengths or angles may also be used: for instance the pressure in a cylinder may be taken as generalized coordinate for the piston, if it uniquely determines the position of the piston.

Because different sets of generalized coordinates may be found, it is customary to use the notation $(q_1, q_2, \ldots, q_n)$ for the generalized coordinates of a system with n degrees of freedom.

The usual rectangular coordinates (x, y, z) may be expressed in terms of the generalized coordinates by certain functions:

$$\left.\begin{aligned} x_i &= G_i(q_1, \ldots, q_n) \\ y_i &= H_i(q_1, \ldots, q_n) \\ z_i &= K_i(q_1, \ldots, q_n) \end{aligned}\right\} \tag{9.1}$$

As an example, consider the system shown in Fig. 9.1. This is a double pendulum swinging in the vertical plane. The system has two degrees of freedom. The position of the two bobs A and B may be given by rectangular coordinates x_1, y_1, x_2 and y_2; these are not independent, but connected by the equations of constraint $x_1^2+y_1^2=l_1^2$ and $(x_2-x_1)^2+(y_2-y_i)^2=l_2^2$. Only two of the coordinates may be used as independent generalized coordinates. A simpler set of generalized coordinates are the angles θ and φ as shown, which are *absolute* coordinates. We may also use the angles θ and φ_1 measured as shown on the figure, and in this case φ_1 is a *relative* coordinate. A mixture of absolute and relative coordinates is often most useful for use as generalized coordinates.

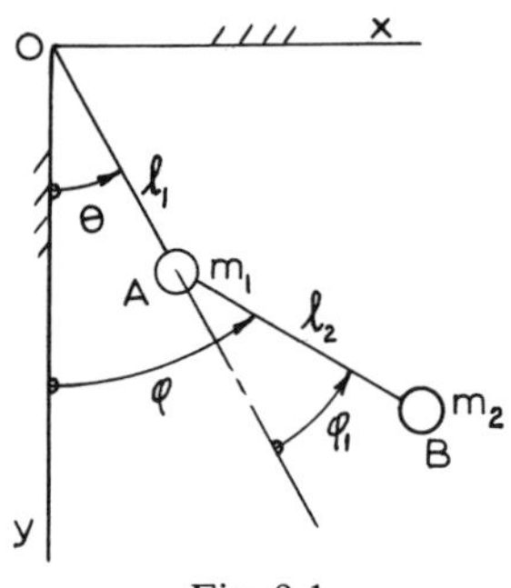

Fig. 9.1

Using θ and φ as coordinates, we find for the functions (9.1) in this case:

$$x_1 = G_1(\theta,\varphi) = l_1 \sin\varphi \qquad x_2 = G_2(\theta, \varphi) = l_1 \sin\theta + l_2 \sin\varphi$$

$$y_1 = H_1(\theta, \varphi) = l_1 \cos\theta \qquad y_2 = H_2(\theta, \varphi) = l_1 \cos\theta + l_2 \cos\varphi$$

9.3 Generalized forces

9.3.1 Definition of generalized forces

Consider a general case of a system with n degrees of freedom. The position of the system may be given by a set of generalized coordinates $(q_1, \ldots, q_n)$, and these must all be independent, which means that we may give a small increment δq_i to the coordinate q_i, without changing any of the other coordinates.

To this change δq_i corresponds a certain displacement of the system, and the forces acting will move through certain distances, dependent on δq_i, and perform an amount of work δ(W.D.).

We now express the work done as $\delta(\text{W.D.}) = \delta q_i[Q_i]$, where Q_i is an expression involving the forces acting. The expression Q_i is called the *generalized force corresponding to the generalized coordinate* q_i. The generalized force is thus that quantity by which we must multiply δq_i, to obtain the work done by all the forces acting on a displacement δq_i of the system, with all the other coordinates kept constant.

As an example, consider the double pendulum in Fig. 9.1. Using the angles θ and φ as generalized coordinates, we find the generalized force Q_φ corresponding to the coordinate φ by keeping θ constant and give a small increment $\delta\varphi$ to φ as shown in Fig. 9.2.

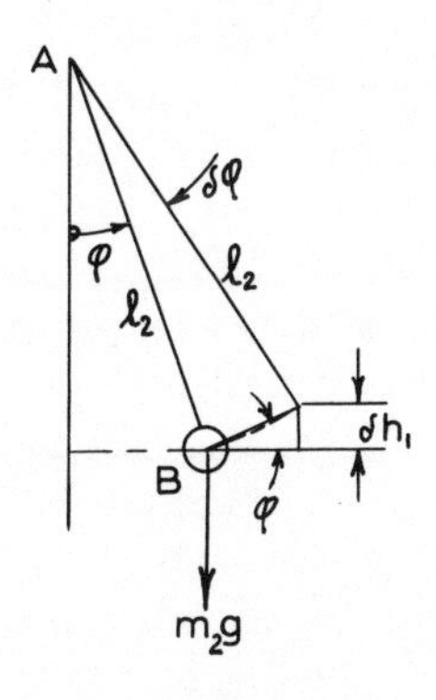

Fig. 9.2

The point B moves up a distance $\delta h_1 = l_2\ \delta\varphi \sin\varphi$, and the total work done is $\delta(\text{W.D.}) = -m_2 g l_2\ \delta\varphi \sin\varphi = \delta\varphi[Q_\varphi]$. We thus find $Q_\varphi = -m_2 g l_2 \sin\varphi$; this has the dimension of a moment (N m), which agrees with the fact that φ is an angle, and $Q_\varphi\ \delta\varphi$ must have the dimension of work.

To find the generalized force Q_θ corresponding to θ, we keep φ constant and give a small increment $\delta\theta$ to θ. The point A moves up a distance $\delta h_2 = l_1\ \delta\theta \sin\theta$, and since φ is unchanged, the line

AB moves parallel to itself, so that point B also moves a distance δh_2. The total work done is thus

$$\delta(\text{W.D.}) = -(m_1 g + m_2 g)\delta h_2 = -(m_1 + m_2) g l_1 \, \delta\theta \sin\theta$$
$$= \delta\theta \, [Q_\theta]$$

so that $Q_\theta = -(m_1 + m_2) g l_1 \sin\theta$, which again has the dimensions of a moment.

In general only the external forces are involved in the expressions for generalized forces. Of these external forces, fixed reactions and normal forces do no work and are not involved; internal forces occur in equal and opposite pairs and again do no work in a summation of work done. There are also cases where no external forces do any work on δq_i, and $Q_i = 0$.

9.3.2 Generalized forces and potential energy

The generalized forces may be found in a different, and usually simpler, way for forces with a *potential*. The two most common conservative forces are gravity and elastic forces.

If the position of a *conservative system* is determined by the generalized coordinates $(q_1, \ldots, q_n)$, the potential energy V is a function of the coordinates $V = f(q_1, \ldots, q_n)$. If the coordinate q_i is given an increment δq_i, while all the other coordinates are kept constant, the incremental change in V is $\delta V = (\partial V/\partial q_i)\, \delta q_i$. By the definition of potential energy, this change is also the negative of the work done by all the potential forces, so that $\delta V = (\partial V/\partial q_i)\delta q_i = -\delta(\text{W.D.})$. The generalized force Q_i corresponding to q_i is defined by $Q_i \delta q_i = \delta(\text{W.D.})$, thus $Q_i \delta q_i = -\delta V = -(\partial V/\partial q_i)\, \delta q_i$, or

$$Q_i = -\partial V/\partial q_i \tag{9.2}$$

For conservative systems, the generalized forces may thus be found by partial differentiation of the potential energy-function expressed in generalized coordinates. We shall find that this way of determining the generalized forces generally results in less labour than the definition of generalized forces in Section 9.31.

As an example, consider the generalized forces for the system in Fig. 9.1. The potential-energy function is

$$V = m_1 g l_1 (1 - \cos\theta) + m_2 g [l_1(1 - \cos\theta) + l_2(1 - \cos\varphi)]$$

where the datum position with $V=0$ has been taken with both pendulums in the vertical line. Application of (9.2) directly gives the result

$$Q_\varphi = -\frac{\partial V}{\partial \varphi} = -m_2 g l_2 \sin \varphi$$

and

$$Q_\theta = -\frac{\partial V}{\partial \theta} = -(m_1 + m_2) g l_1 \sin \theta$$

as found before.

9.3.3 Determination of generalized forces from force functions

A third method exists for the determination of generalized forces. Consider a plane system with the functions (9.1): $x_i = G_i(q_1, \ldots, q_n)$, and $y_i = H_i(q_1, \ldots, q_n)$. We find the differential

$$dx_i = \frac{\partial x_i}{\partial q_1} dq_1 + \cdots + \frac{\partial x_i}{\partial q_n} dq_n$$

and the virtual displacement

$$\delta x_i = \frac{\partial x_i}{\partial q_1} \delta q_1 + \cdots + \frac{\partial x_i}{\partial q_n} \delta q_n$$

with a similar expression for δy_i.

Since the generalized coordinates are independent, we may give an increment δq_i to q_i, while keeping the other coordinates constant; with $\delta q_1 = \delta q_2 = \cdots = 0$, we have then

$$\delta x_i = \frac{\partial x_i}{\partial q_i} \delta q_i \qquad \text{and} \qquad \delta y_i = \frac{\partial y_i}{\partial q_i} \delta q_i$$

The total work done on this increment is, by summation over all particles,

$$\delta(\text{W.D.}) = \sum (F_{ix} \delta x_i + F_{iy} \delta y_i) = \sum \left(F_{ix} \frac{\partial x_i}{\partial q_i} + F_{iy} \frac{\partial y_i}{\partial q_i} \right) \delta q_i$$

The total work done is also by definition equal to $(Q_i)\delta q_i$; equating the two expressions gives the result:

$$Q_i = \sum \left(F_{ix} \frac{\partial x_i}{\partial q_i} + F_{iy} \frac{\partial y_i}{\partial q_i} \right) \tag{9.3}$$

This result may be extended directly to three-dimensional motion.

If the functions $F_{ix}(x, y)$ and $F_{iy}(x, y)$, together with the functions $x_i = G_i(q_1, \ldots, q_n)$ and $y_1 = H_i(q_1, \ldots q_n)$ are known for a system, the generalized forces may be determined from (9.3); it is only necessary to express F_{ix} and F_{iy} as functions of the generalized coordinates.

As an example, consider the determination of Q_θ and Q_φ for the system in Fig. 9.1. We have previously found the functions $x_1 = l_1 \sin\theta$, $x_2 = l_1 \sin\theta + l_2 \sin\varphi$; $y_1 = l_1 \cos\theta$ and $y_2 = l_1 \cos\theta + l_2 \cos\varphi$ for this system. The forces acting are the gravity forces $m_1 g$ and $m_2 g$, for which we have $F_{1x} = 0$, $F_{1y} = m_1 g$, $F_{2x} = 0$ and $F_{2y} = m_2 g$. We find from this that

$$\frac{\partial y_1}{\partial \varphi} = 0 \qquad \frac{\partial y_2}{\partial \varphi} = -l_2 \sin\varphi \qquad \frac{\partial y_1}{\partial \theta} = -l_1 \sin\theta$$

$$\frac{\partial y_2}{\partial \theta} = -l_1 \sin\theta$$

Equation (9.3) now gives the results

$$Q_\varphi = F_{1x}\frac{\partial x_1}{\partial \varphi} + F_{2x}\frac{\partial x_2}{\partial \varphi} + F_{1y}\frac{\partial y_1}{\partial \varphi} + F_{2y}\frac{\partial y_2}{\partial \varphi} = -m_2 g l_2 \sin\varphi$$

$$Q_\theta = -(m_1 + m_2) g l_1 \sin\theta$$

as found before. This method is not generally as convenient as the two previous methods for the determination of Q_i.

9.3.4 Generalized forces for viscous friction forces

Consider the system shown in Fig. 9.3. The system consists of a mass m suspended on a spring of constant K. A dashpot with damping coefficient c is attached as shown, and the mass is vibrated by an external exciting force $F = F_0 \sin\omega t$.

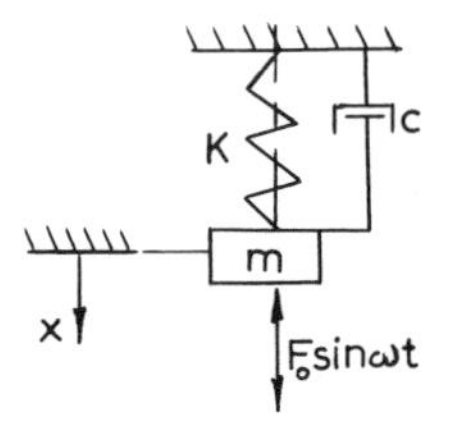

Fig. 9.3

The generalized coordinate may be taken as the distance x measured from the static equilibrium position. The viscous damping force is of magnitude $c\dot{x}$ and is in the opposite direction to the velocity $\dot{x}$.

Giving the system an incremental displacement δx, we may consider all forces constant through this small displacement, and the total work done is then

$$\begin{aligned}\delta(\text{W.D.}) &= mg\,\delta x + F\,\delta x - (mg+kx)\,\delta x - c\dot{x}\,\delta x \\ &= [F_0 \sin \omega t - kx - c\dot{x}]\,\delta x = [Q_x]\,\delta x\end{aligned}$$

The generalized force is thus $Q_x = F_0 \sin \omega t - kx - c\dot{x}$. In this example the viscous friction force $-c\dot{x}$ presented no difficulties, but in other cases the situation may be more complicated, and it is an advantage to determine an expression for the generalized forces corresponding to the viscous friction forces.

By analogy with the kinetic energy $\frac{1}{2}m\dot{x}^2$ for the system in Fig. 9.3, we define a *dissipation function* $D = \frac{1}{2}c\dot{x}^2$. The damping force is then $-dD/d\dot{x} = -c\dot{x}$, and the work done on a small incremental change δx in x is $-c\dot{x}\,\delta x$.

In general two-dimensional motion we resolve the velocity into components $\dot{x}$ and $\dot{y}$, and define the dissipation function by the expression $D = \frac{1}{2}(c_1\dot{x}^2 + c_2\dot{y}^2)$, where c_1 and c_2 are the damping coefficients in the x- and y-directions respectively. In the general case of a system of particles, the dissipation function is obtained by a summation over all the particles of the system

$$D = \tfrac{1}{2}\sum (c_{1i}\dot{x}_i^2 + c_{2i}\dot{y}_i^2) \tag{9.4}$$

This expression may be directly expanded to include the three-dimensional case.

If the generalized coordinates of the system are $(q_1, \ldots, q_n)$, the *generalized friction force* R_i corresponding to the coordinate q_i is determined by giving an increment δq_i to q_i, with all other coordinates kept constant. The work done by the viscous friction forces is then

$$\delta(\text{W.D.}) = [R_i]\,\delta q_i = -\sum (c_{1i}\dot{x}_i\,\delta x_i + c_{2i}\dot{y}_i\,\delta y_i)$$

Substituting $\delta x_i = (\partial x_i/\partial q_i)\,\delta q_i$ and $\delta y_i = (\partial y_i/\partial q_i)\,\delta q_i$ results in

$$R_i\,\delta q_i = -\sum \left(c_{1i}\dot{x}_i \frac{\partial x_i}{\partial q_i} + c_{2i}\dot{y}_i \frac{\partial y_i}{\partial q_i}\right)\delta q_i$$

The function $x_i = F_i(q_1, \ldots, q_n)$ has the differential

$$dx_i = \frac{\partial x_i}{\partial q_1} dq_1 + \cdots + \frac{\partial x_i}{\partial q_n} dq_n$$

from which

$$\frac{dx_i}{dt} = \dot{x}_i = \frac{\partial x_i}{\partial q_1} \dot{q}_1 + \cdots + \frac{\partial x_i}{\partial q_n} \dot{q}_n$$

Differentiating this expression partially with respect to $\dot{q}_i$ gives the result

$$\partial \dot{x}_i / \partial \dot{q}_i = \partial x_i / \partial q_i \tag{9.5}$$

and in a similar way $\partial \dot{y}_i / \partial \dot{q}_i = \partial y_i / \partial q_i$; this operation is sometimes called 'cancellation of the dots'.

Substituting (9.5) in the expression for $R_i \, \delta q_i$ leads to

$$R_i = -\sum \left(c_{1i} \dot{x}_i \frac{\partial \dot{x}_i}{\partial \dot{q}_i} + c_{2i} \dot{y}_i \frac{\partial \dot{y}_i}{\partial \dot{q}_i} \right)$$

If we take the partial derivative of the dissipation function (9.4), we find that

$$\frac{\partial D}{\partial \dot{q}_i} = \sum \left(c_{1i} \dot{x}_i \frac{\partial \dot{x}_i}{\partial \dot{q}_i} + c_{2i} \dot{y}_i \frac{\partial \dot{y}_i}{\partial \dot{q}_i} \right)$$

so that

$$R_i = -\partial D / \partial \dot{q}_i \tag{9.6}$$

This result may be extended directly to three-dimensional motion. The generalized viscous damping force R_i may then be obtained by partial differentiation of the dissipation function D, after this function has been expressed in generalized coordinates.

Example 9.1. Determine the generalized viscous friction force R_θ for the system shown in Fig. 9.4. The angle θ has been taken as the generalized coordinate, and $\theta = \alpha$ at static equilibrium of the system.

Solution. Measuring the position of the mass A by the coordinate x from the static equilibrium position, we have the dissipation function $D = \frac{1}{2} c \dot{x}^2$. When $x = 0$, $\theta = \alpha$; from the geometry of the figure we have $x=2l(\cos\alpha-\cos\theta)$ and $\dot{x}=2l(\sin\theta)\dot{\theta}$. The dissipation function, in the generalized coordinate θ, is then

$$D = \tfrac{1}{2}c[(2l\sin\theta)\dot{\theta}]^2 = (2cl^2\sin^2\theta)\dot{\theta}^2$$

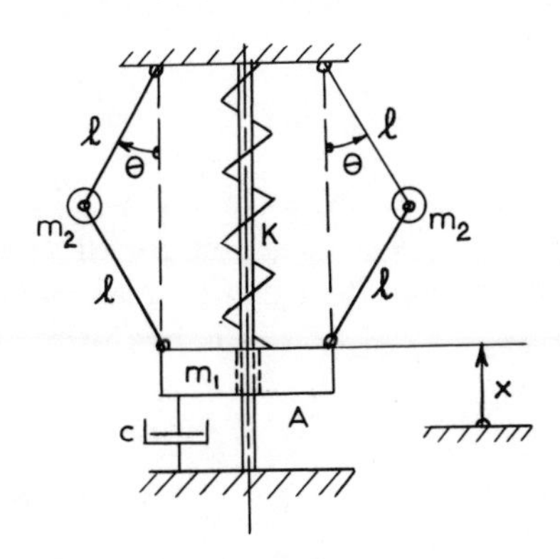

Fig. 9.4

The generalized viscous friction force is

$$R_\theta = -\partial D/\partial\dot{\theta} = -(4cl^2\sin^2\theta)\dot{\theta}$$

So far we have worked with viscous friction forces proportional to the *absolute* velocities of the particles. There are many cases, however, in which the viscous damping force is proportional to the *relative* velocity between two masses; an example of this is shown in Fig. 9.5, where the viscous friction force between the two masses is proportional to the *relative* velocity $\dot{x}_1-\dot{x}_2$. The dissipation function must then be taken as $\frac{1}{2}(\dot{x}_1-\dot{x}_2)^2$, and in the case of the system in Fig. 9.5, we have $D=\frac{1}{2}c_1\dot{x}_1^2+\frac{1}{2}c_2(\dot{x}_1-\dot{x}_2)^2$.

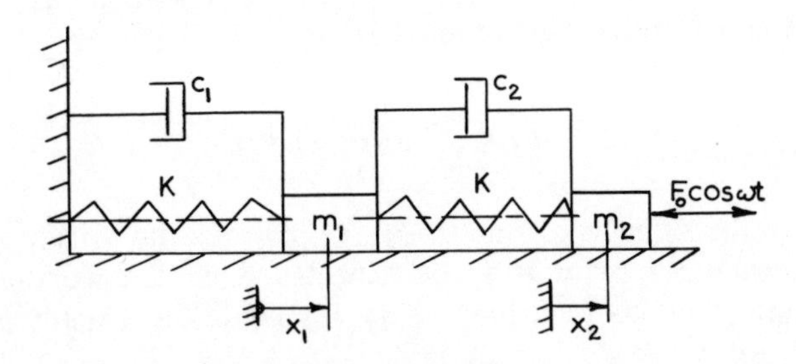

Fig. 9.5

9.4 Lagrange's equations

The D'Alembert equation (2.18) for a particle of mass m_i is $(\mathbf{F}_i - m_i\ddot{\mathbf{r}}_i)\cdot\delta\mathbf{r}_i = 0$, or for a particle in plane motion

$$(F_{ix} - m_i\ddot{x}_i)\,\delta x_i = 0$$
$$(F_{iy} - m_i\ddot{y}_i)\,\delta y_i = 0$$

For a system of particles, including a rigid body as a special case, we write D'Alembert's equation for all particles and find the sum for the total system. If we assume plane motion, this gives

$$\sum\left[(F_{ix} - m_i\ddot{x}_i)\,\delta x_i + (F_{iy} - m_i\ddot{y}_i)\,\delta y_i\right] = 0$$

This equation may be stated in the form

$$\sum m_i(\ddot{x}_i\,\delta x_i + \ddot{y}_i\,\delta y_i) = \sum (F_{ix}\,\delta x_i + F_{iy}\,\delta y_i) \qquad \text{(a)}$$

where the summation is understood to include all particles of the system. From the functions (9.1):

$$x_i = G_i(q_1, \ldots, q_n) \qquad y_i = H_i(q_1, \ldots, q_n)$$

we find the virtual displacements $\delta x_i = (\partial x_i/\partial q_1)\delta q_1$ and $\delta y_i = (\partial y_i/\partial q_1)\delta q_1$, by giving an increment δq_1 to the coordinate q_1 and keeping all other coordinates constant.

Substituting these expressions in eq. (a), and omitting the subscript i for ease of writing, we find

$$\sum m\left(\ddot{x}\frac{\partial x}{\partial q_1} + \ddot{y}\frac{\partial y}{\partial q_1}\right)\delta q_1 = \sum\left(F_x\frac{\partial x}{\partial q_1} + F_y\frac{\partial y}{\partial q_1}\right)\delta q_1$$

Cancelling δq_1, the right-hand side of this expression is the generalized force Q_1 corresponding to the coordinate q_1, so the equation may be stated in the form

$$\sum m\left(\ddot{x}\frac{\partial x}{\partial q_1} + \ddot{y}\frac{\partial y}{\partial q_1}\right) = Q_1 \qquad \text{(b)}$$

Since the right-hand side of eq. (b) involves the work done by all external forces, it is to be expected from the work–energy equation that the left-hand side involves the kinetic energy $T = \frac{1}{2}\sum m(\dot{x}^2 + \dot{y}^2)$.

Equation (b) contains second derivatives, while the kinetic

energy contains first derivatives; we therefore rewrite eq. (b) in the form

$$\frac{d}{dt}\sum m\left(\dot{x}\,\frac{\partial x}{\partial q_1}+\dot{y}\,\frac{\partial y}{\partial q_1}\right)-\sum m\left[\dot{x}\,\frac{d}{dt}\left(\frac{\partial x}{\partial q_1}\right)+\dot{y}\,\frac{d}{dt}\left(\frac{\partial y}{\partial q_1}\right)\right]=Q_1 \quad \text{(c)}$$

Taking partial derivatives of the kinetic energy leads to

$$\begin{aligned}\frac{\partial T}{\partial \dot{q}_1}&=\sum m\left(\dot{x}\,\frac{\partial \dot{x}}{\partial \dot{q}_1}+\dot{y}\,\frac{\partial \dot{y}}{\partial \dot{q}_1}\right)\\ \frac{\partial T}{\partial q_1}&=\sum m\left(\dot{x}\,\frac{\partial \dot{x}}{\partial q_1}+\dot{y}\,\frac{\partial \dot{y}}{\partial q_1}\right)\end{aligned} \quad \text{(d)}$$

These expressions cannot be directly substituted in eq. (c), but applying the relationship (9.5), $\partial\dot{x}/\partial\dot{q}_1=\partial x/\partial q_1$ and $\partial\dot{y}/\partial\dot{q}_1=\partial y/\partial q_1$, to the right-hand side of the first eq. (d), this expression takes the form

$$\frac{\partial T}{\partial \dot{q}_1}=\sum m\left(\dot{x}\,\frac{\partial x}{\partial q_1}+\dot{y}\,\frac{\partial y}{\partial q_1}\right) \quad \text{(e)}$$

The second eq. (d) may be transformed as follows: the function $x=G(q_1,\ldots,q_n)$ has the differential

$$dx=\frac{\partial x}{\partial q_1}\,dq_1+\cdots+\frac{\partial x}{\partial q_n}\,dq_n$$

from which

$$\frac{dx}{dt}=\dot{x}=\frac{\partial x}{\partial q_1}\,\dot{q}_1+\cdots+\frac{\partial x}{\partial q_n}\,\dot{q}_n$$

Differentiating partially with respect to q_1 leads to

$$\frac{\partial \dot{x}}{\partial q_1}=\frac{\partial^2 x}{\partial q_1^2}\,\dot{q}_1+\cdots+\frac{\partial^2 x}{\partial q_1\,\partial q_n}\,\dot{q}_n \quad \text{(f)}$$

From $x=G(q_1,\ldots,q_n)$, we have $\partial x/\partial q_1=F(q_1,\ldots,q_n)$, and

$$\begin{aligned}d\left(\frac{\partial x}{\partial q_1}\right)&=\frac{\partial(\partial x/\partial q_1)}{\partial q_1}\,dq_1+\cdots+\frac{\partial(\partial x/\partial q_n)}{\partial q_n}\,dq_n\\ &=\frac{\partial^2 x}{\partial q_1^2}\,dq_1+\cdots+\frac{\partial^2 x}{\partial q_1\,\partial q_n}\,dq_n\end{aligned}$$

from which

$$\frac{d}{dt}\left(\frac{\partial x}{\partial q_1}\right) = \frac{\partial^2 x}{\partial q_1^2}\dot{q}_1 + \cdots + \frac{\partial^2 x}{\partial q_1 \partial q_n}\dot{q}_n \tag{g}$$

Comparing eqs. (f) and (g) shows that

$$\frac{\partial \dot{x}}{\partial q_1} = \frac{d}{dt}\left(\frac{\partial x}{\partial q_1}\right) \tag{9.7}$$

Equation (9.7) shows that the order in which the differentiations are performed does not matter, that is

$$\frac{\partial}{\partial q_1}\left(\frac{dx}{dt}\right) = \frac{d}{dt}\left(\frac{\partial x}{\partial q_1}\right)$$

Substituting (9.7), and a similar expression for $\partial \dot{y}/\partial q_1$, in the second eq. (d) leads to

$$\frac{\partial T}{\partial q_1} = \sum m\left[\dot{x}\frac{d}{dt}\left(\frac{\partial x}{\partial q_1}\right) + \dot{y}\frac{d}{dt}\left(\frac{\partial y}{\partial q_1}\right)\right] \tag{h}$$

The expressions (e) and (h) may now be substituted in eq. (c) to give the result

$$\frac{d}{dt}\left(\frac{\partial T}{\partial \dot{q}_1}\right) - \frac{\partial T}{\partial q_1} = Q_1 \tag{i}$$

This is Lagrange's equation for the coordinate q_1, and a similar equation may be derived for the coordinate q_i by using an increment δq_i. The process then leads to n Lagrangian equations of motion corresponding to the n degrees of freedom of the system.

Equation (i) may now be stated in the general form:

$$\frac{d}{dt}\left(\frac{\partial T}{\partial \dot{q}_i}\right) - \frac{\partial T}{\partial q_i} = Q_i \qquad i = 1, 2, \ldots, n \tag{9.8}$$

The eqs. (9.8) hold for both conservative and non-conservative systems, and were first stated by the great French mathematician J. L. Lagrange in his famous book *Analytical Mechanics*, which was published in Paris in 1788. Lagrange's equations are of great importance and widely used in engineering and physics.

The equations may obviously be expanded directly to three-dimensional motion.

Example 9.2. Figure 9.6 shows a symmetrical rotor rotating about a fixed axis through O. An external driving torque $M_0 \sin \omega t$ acts on the rotor, and the moment of inertia about the axis of rotation is I_0.

Determine the equation of motion by using Lagrange's equation (9.8).

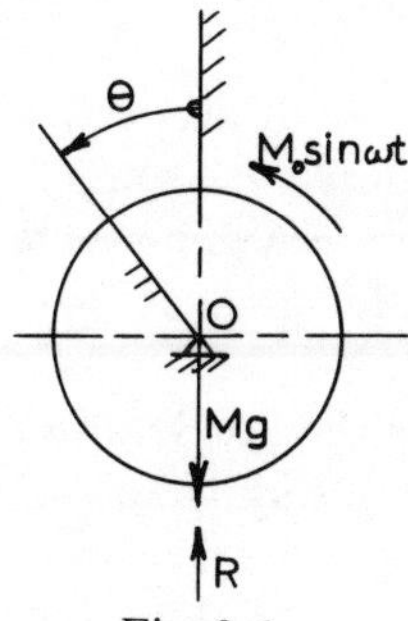

Fig. 9.6

Solution. Taking the generalized coordinate as the angle θ as shown, Lagrange's equation has the form:

$$\frac{d}{dt}\frac{\partial T}{\partial \dot{\theta}} - \frac{\partial T}{\partial \theta} = Q_\theta$$

$$T = \tfrac{1}{2}I_0\omega^2 = \tfrac{1}{2}I_0\dot{\theta}^2 \qquad \frac{\partial T}{\partial \dot{\theta}} = I_0\dot{\theta} \qquad \frac{d}{dt}\frac{\partial T}{\partial \dot{\theta}} = I_0\ddot{\theta}$$

$$\text{and} \qquad \frac{\partial T}{\partial \theta} = 0$$

Giving an increment $\delta\theta$ to θ, we find $\delta(\text{W.D.}) = (M_0 \sin \omega t)\, \delta\theta = Q_\theta \delta\theta$. The result is $Q_\theta = M_0 \sin \omega t$, and the equation of motion is

$$I_0\ddot{\theta} = M_0 \sin \omega t$$

Example 9.3. Determine the equation of motion of the system in Fig. 9.3, by Lagrange's equation.

Solution. If we take the displacement x from static equilibrium as the generalized coordinate, the equation of motion is

$$\frac{d}{dt}\left(\frac{\partial T}{\partial \dot{x}}\right) - \frac{\partial T}{\partial x} = Q_x$$

$$T = \tfrac{1}{2}m\dot{x}^2 \qquad \frac{d}{dt}\frac{\partial T}{\partial \dot{x}} = m\ddot{x} \qquad \frac{\partial T}{\partial x} = 0$$

The generalized force, from Section 9.3.4, is

$$Q_x = F_0 \sin \omega t - kx - c\dot{x}$$

The equation of motion is thus

$$m\ddot{x} + c\dot{x} + kx = F_0 \sin \omega t$$

Lagrange's equations are of no particular advantage in the two simple problems in Example 9.2 and 9.3; we shall find, however, that as the complexity of the systems increases, Lagrange's method becomes of increasing advantage compared to other methods.

Example 9.4. Determine the acceleration of the body of mass m_1 in Fig. 9.7. The body is sliding on an inclined plane and the coefficient of friction between the body and the plane is μ. The pulleys may be treated as solid flat discs of radius r and mass M; there is no slip of the string on the pulleys.

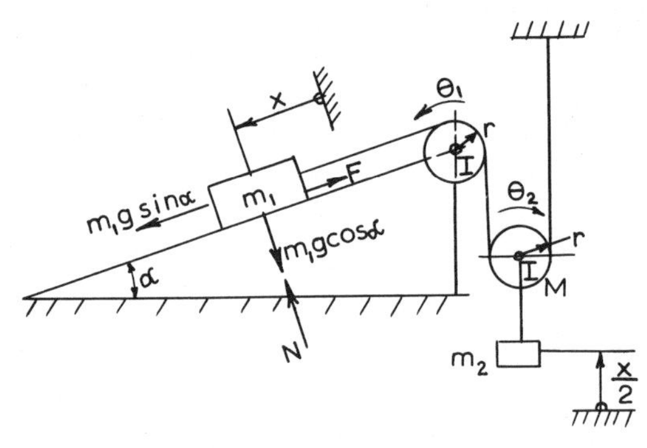

Fig. 9.7

Solution. Taking the distance x as coordinate, the equation of motion is

$$\frac{d}{dt}\frac{\partial T}{\partial \dot{x}} - \frac{\partial T}{\partial x} = Q_x$$

The kinetic energy of the system is

$$T = \tfrac{1}{2}m_1\dot{x}^2 + \tfrac{1}{2}I\dot{\theta}_1^2 + \left(\tfrac{1}{2}I\dot{\theta}_2^2 + \tfrac{1}{2}M\frac{\dot{x}^2}{4}\right) + \tfrac{1}{2}m_2\frac{\dot{x}^2}{4}$$

We have $I = \frac{1}{2}Mr^2$, and for no slip on the pulleys $r\theta_1 = x$, $r\theta_2 = x/2$, from which $\dot{\theta}_1 = \dot{x}/r$ and $\dot{\theta}_2 = \dot{x}/2r$; substituting this, we find

$$T = \tfrac{1}{2}[m_1 + \tfrac{7}{8}M + \tfrac{1}{4}m_2]\dot{x}^2 = \tfrac{1}{2}A\dot{x}^2$$

$$\frac{d}{dt}\frac{\partial T}{\partial \dot{x}} = A\ddot{x} \qquad \text{and} \qquad \frac{\partial T}{\partial x} = 0$$

Giving an increment δx to x, the work done is

$$Q_x \delta x = m_1 g(\sin\alpha)\,\delta x - m_1 g\mu(\cos\alpha)\,\delta x - Mg(\delta x/2) - m_2 g(\delta x/2)$$

and

$$Q_x = m_1 g(\sin\alpha - \mu\cos\alpha) - \tfrac{1}{2}(M + m_2)g = B$$

The equation of motion is $A\ddot{x} = B$, and the acceleration of the mass is $\ddot{x} = B/A$, with the above expressions for the constants A and B.

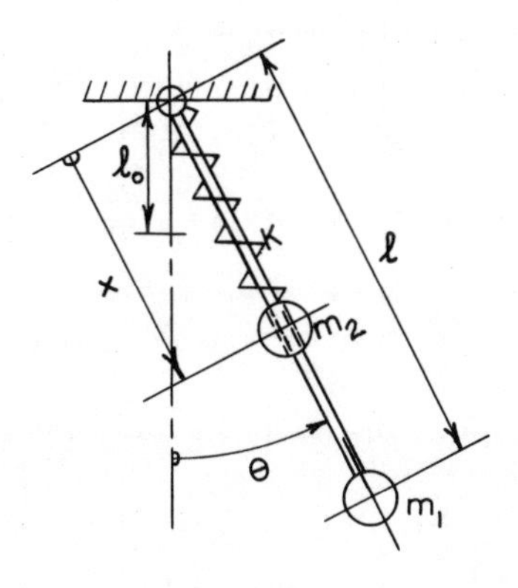

Fig. 9.8

Example 9.5. Figure 9.8 shows a slender, rigid rod pinned at O, with a bob of mass m_1 attached at the lower end. The rod may be considered massless, and is free to rotate about O in the vertical plane. A second bob of mass m_2 is free to slide along the smooth rod under the action of gravity and of the spring of constant K. The unstretched length of the spring is l_0.

Determine the equations of motion of the system, and consider the special case where m_2 is fixed to the rod at a distance l_1 from O, and the case where the rod is fixed in a position where $\theta = \beta$.

Solution. The generalized coordinates may be taken as x and θ as shown on the figure. The velocity of the bob of mass m_1 is $|\mathbf{V}_1| = l\dot{\theta}$. The velocity of the other mass is $\mathbf{V}_2 = x\dot{\theta}$ at angle θ, $+\dot{x}$ at $270° + \theta$, so that $V_2^2 = x^2\dot{\theta}^2 + \dot{x}^2$, and

$$T = \tfrac{1}{2}m_1 l^2\dot{\theta}^2 + \tfrac{1}{2}m_2(\dot{x}^2 + x^2\dot{\theta}^2)$$

Hence

$$\frac{d}{dt}\frac{\partial T}{\partial \dot{x}} = m_2\ddot{x} \qquad \frac{\partial T}{\partial x} = m_2\dot{\theta}^2 x$$

$$\frac{\partial T}{\partial \dot{\theta}} = m_1 l^2\dot{\theta} + m_2 x^2\dot{\theta} \qquad \frac{d}{dt}\frac{\partial T}{\partial \dot{\theta}} = m_1 l^2\ddot{\theta} + m_2(x^2\ddot{\theta} + 2x\dot{x}\dot{\theta})$$

$$\frac{\partial T}{\partial \theta} = 0$$

To find Q_x, we keep θ constant and find the work done by all external forces on a small increment δx. The spring force is dealt with by cutting the spring and considering the spring force as an external force. At position x, the spring force is $K(x - l_0)$ and $\delta(\text{W.D.})_S = -K(x - l_0)\,\delta x$.

The work done by gravity is $\delta(\text{W.D.})_g = m_2 g \cos\theta\,\delta x$. In total then

$$\delta(\text{W.D.}) = -K(x - l_0)\,\delta x + m_2 g \cos\theta\,\delta x = Q_x\,\delta x$$

and $Q_x = m_2 g \cos\theta - K(x - l_0)$.

Keeping x constant and giving θ an increment $\delta\theta$, we find $\delta(\text{W.D.}) = -m_2 g x \sin\theta\,\delta\theta - m_1 g l \sin\theta\,\delta\theta = Q_\theta\,\delta\theta$, or $Q_\theta = -m_2 g x \sin\theta - m_1 g l \sin\theta$. Substituting in Lagrange's equations gives the equations of motion

$$m_2\ddot{x} - m_2\dot{\theta}^2 x + K(x - l_0) - m_2 g \cos\theta = 0 \qquad \text{(a)}$$

$$(m_1 l^2 + m_2 x^2)\ddot{\theta} + 2m_2 x\dot{x}\dot{\theta} + g \sin\theta(m_2 x + m_1 l) = 0 \qquad \text{(b)}$$

If m_2 is fixed to the bar a distance l_1 from O, we have $x = l_1$, $\dot{x} = \ddot{x} = 0$. Equation (b), for the coordinate θ, then gives the equation of motion

$$(m_1 l^2 + m_2 l_1^2)\ddot{\theta} + (m_1 l + m_2 l_1)g \sin\theta = 0$$

This is the usual equation for rotation about a fixed principal axis of a rigid body with moment of inertia $m_1 l^2 + m_2 l_1^2$, and restoring torque $(m_1 l + m_2 l_1)g \sin\theta$.

If $\theta=\beta$, $\dot{\theta}=\ddot{\theta}=0$, eq. (a), for the coordinate x, gives the equation of motion

$$m_2\ddot{x}+K(x-l_0)-m_2g\cos\beta=0$$

the usual type of equation for motion along a fixed x-axis.

9.5 Lagrange's equations for conservative systems

For conservative systems, the generalized forces are usually found by using the potential-energy function V in the expression (9.2): $Q_i=-\partial V/\partial q_i$.

If we are dealing with a conservative system, this expression may be substituted in the Lagrange equation (9.8), which then takes the form

$$\frac{d}{dt}\frac{\partial T}{\partial \dot{q}_i}-\frac{\partial T}{\partial q_i}+\frac{\partial V}{\partial q_i}=0 \qquad i=1,2,\ldots,n \tag{9.9}$$

This gives the form of Lagrange's equations valid *for conservative systems only*. If we define a certain function $L=T-V$, we have

$$\frac{\partial L}{\partial \dot{q}_i}=\frac{\partial T}{\partial \dot{q}_i}-\frac{\partial V}{\partial \dot{q}_i}=\frac{\partial T}{\partial \dot{q}_i}$$

since the potential energy cannot contain time derivatives, or $\partial V/\partial \dot{q}_i=0$. We also find

$$\frac{\partial L}{\partial q_i}=\frac{\partial T}{\partial q_i}-\frac{\partial V}{\partial q_i}$$

Substituting these expressions in eq. (9.9) gives an alternative form of Lagrange's equations:

$$\frac{d}{dt}\frac{\partial L}{\partial \dot{q}_i}-\frac{\partial L}{\partial q_i}=0 \qquad i=1,2,\ldots,n \tag{9.10}$$

This is the shortest possible form of Lagrange's equation, and it is valid for *conservative systems* only. The function $L=T-V$ is called the *Lagrangian function*.

Example 9.6. Figure 9.9 shows a simple pendulum. Neglecting air resistance, determine the equation of motion by eq. (9.10).

Solution. Taking the angle θ as generalized coordinate, the kinetic energy is $T=\frac{1}{2}mV^2=\frac{1}{2}ml^2\dot{\theta}^2$.

Taking the vertical position of the pendulum as datum position, the potential energy is $V=mgl(1-\cos\theta)$.

The Lagrangian function $L=T-V=\frac{1}{2}ml^2\dot{\theta}^2-mgl(1-\cos\theta)$. Hence

$$\frac{d}{dt}\frac{\partial L}{\partial \dot{\theta}} = ml^2\ddot{\theta} \qquad \frac{\partial L}{\partial \theta} = -mgl\sin\theta$$

The equation of motion from eq. (9.10) is then $ml^2\ddot{\theta}+mgl\sin\theta=0$, or $\ddot{\theta}+(g/l)\sin\theta=0$, as found before for this case.

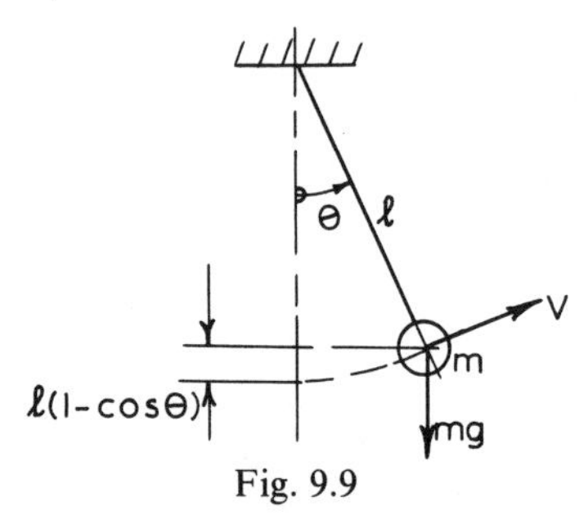

Fig. 9.9

Example 9.7. Solve the previous example (9.5), by using the potential energy function V and the Lagrangian L.

Solution. Taking the potential energy $V=0$ when the bar is vertical and the spring unstretched, we find

$$V = m_1gl(1-\cos\theta)+\tfrac{1}{2}K(x-l_0)^2-m_2g(x\cos\theta-l_0)$$

$$L = T-V = \tfrac{1}{2}m_1l^2\dot{\theta}^2+\tfrac{1}{2}m_2(\dot{x}^2+x^2\dot{\theta}^2)$$
$$-m_1gl(1-\cos\theta)-\tfrac{1}{2}K(x-l_0)^2+m_2g(x\cos\theta-l_0)$$

We find from this that

$$\frac{\partial L}{\partial \dot{x}} = m_2\dot{x} \qquad \frac{d}{dt}\frac{\partial L}{\partial \dot{x}} = m_2\ddot{x}$$

$$\frac{\partial L}{\partial x} = m_2x\dot{\theta}^2-K(x-l_0)+m_2g\cos\theta$$

$$\frac{\partial L}{\partial \dot{\theta}} = m_1l^2\dot{\theta}+m_2x^2\dot{\theta} \qquad \frac{d}{dt}\frac{\partial L}{\partial \dot{\theta}} = m_1l^2\ddot{\theta}+m_2x^2\ddot{\theta}+2m_2x\dot{x}\dot{\theta}$$

$$\frac{\partial L}{\partial \theta} = -m_1gl\sin\theta-m_2gx\sin\theta$$

Substituting in eqs. (9.10) gives directly the two equations of motion that were found in Example 9.5.

Example 9.8. Figure 9.10 shows a horizontal uniform bar of mass $2m$ and centre of mass C. The bar rolls without sliding on two identical cylindrical rotors, each of radius r and mass $2m$.

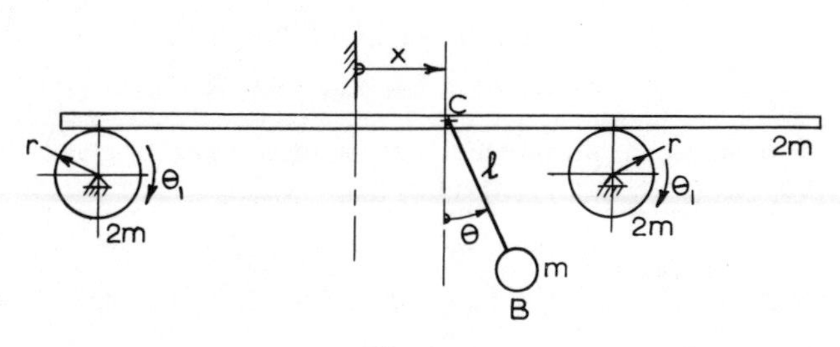

Fig. 9.10

A pendulum of length l and with a bob of mass m is attached at C and swings in the vertical plane.

(a) Determine the equations of motion of the system.

(b) Linearize the equations for small displacements and velocities.

(c) Determine the natural frequency of the system for small motions.

Solution. (a) The moment of inertia of a roller is $I=\frac{1}{2}(2m)r^2=mr^2$. Since there is no slip $x=r\theta_1$, and $\theta_1=x/r$, $\dot{\theta}_1=\dot{x}/r$. The velocity of the bob is $\mathbf{V}_B=\mathbf{V}_C+\mathbf{V}_{BC}=\dot{x}\rightarrow\ +l\dot{\theta}\nearrow$, so that

$$V_B^2 = \dot{x}^2+l^2\dot{\theta}^2+2l\dot{x}\dot{\theta}\cos\theta$$

The kinetic energy is

$$T = \tfrac{1}{2}(2m)\dot{x}^2+\tfrac{1}{2}m(\dot{x}^2+l^2\dot{\theta}^2+2l\dot{x}\dot{\theta}\cos\theta)$$
$$+2(\tfrac{1}{2}mr^2)\dot{\theta}_1^2 = \tfrac{5}{2}m\dot{x}^2+\tfrac{1}{2}ml^2\dot{\theta}^2+ml\dot{x}\dot{\theta}\cos\theta$$

The potential energy is $V=mgl(1-\cos\theta)$, and

$$L = T-V = \tfrac{5}{2}m\dot{x}^2+\tfrac{1}{2}ml^2\dot{\theta}^2+ml\dot{x}\dot{\theta}\cos\theta-mgl(1-\cos\theta)$$

$$\frac{\partial L}{\partial \dot{x}} = 5m\dot{x}+ml\dot{\theta}\cos\theta \qquad \frac{\partial L}{\partial x} = 0$$

$$\frac{d}{dt}\frac{\partial L}{\partial \dot{x}} = 5m\ddot{x}+ml\ddot{\theta}\cos\theta-ml\dot{\theta}^2\sin\theta$$

$$\frac{\partial L}{\partial \dot{\theta}} = ml^2\dot{\theta} + ml\dot{x}\cos\theta \qquad \frac{\partial L}{\partial \theta} = -ml\dot{x}\dot{\theta}\sin\theta - mgl\sin\theta$$

$$\frac{d}{dt}\frac{\partial L}{\partial \dot{\theta}} = ml^2\ddot{\theta} + ml\ddot{x}\cos\theta - ml\dot{x}\dot{\theta}\sin\theta$$

Substituting in Lagrange's equations, we obtain

$$5\ddot{x} + l(\cos\theta)\ddot{\theta} - l(\sin\theta)\dot{\theta}^2 = 0 \qquad \text{(a)}$$

$$(\cos\theta)\ddot{x} + l\ddot{\theta} + g\sin\theta = 0 \qquad \text{(b)}$$

(b) For small displacements and velocities, we may substitute $\sin\theta \to \theta$, $\cos\theta \to 1$ and $\dot{\theta}^2 \to 0$. Hence

$$5\ddot{x} + l\ddot{\theta} = 0 \qquad \text{(a}'\text{)}$$

$$\ddot{x} + l\ddot{\theta} + g\theta = 0 \qquad \text{(b}'\text{)}$$

(c) Substituting $\ddot{x} = -(l/5)\ddot{\theta}$ from (a′) in (b′), we get

$$\ddot{\theta} + (5g/4l)\theta = 0$$

this is simple harmonic motion with frequency $\omega = \frac{1}{2}\sqrt{(5g/l)}$ rad/s.

9.6 Lagrange's equations for systems with viscous damping

It was shown in Section 9.34 that viscous damping forces may be generalized by application of the dissipation function D.

We may now, for convenience, divide the external forces acting on a system into three groups: conservative forces, viscous damping forces and other non-conservative forces. The generalized force Q_i, corresponding to the coordinate q_i, for the first group, may be expressed by $Q_i = -\partial V/\partial q_i$, by introducing the potential energy function V.

The generalized force for the second group is expressed by $Q_i = -\partial D/\partial \dot{q}_i$, by the dissipation function D.

The generalized force, for the third group of non-conservative forces, is determined by the incremental work done by this group alone on a change ∂q_i, so that $Q_i \partial q_i = \partial(\text{W.D.})$, or by the force function if this is known.

Lagrange's equation (9.8)

$$\frac{d}{dt}\frac{\partial T}{\partial \dot{q}_i} - \frac{\partial T}{\partial q_i} = Q_i$$

may now be stated in the form

$$\frac{d}{dt}\frac{\partial T}{\partial \dot{q}_i}-\frac{\partial T}{\partial q_i}+\frac{\partial V}{\partial q_i}+\frac{\partial D}{\partial \dot{q}_i} = Q_i \qquad i = 1, 2, \ldots, n \qquad (9.11)$$

In (9.11) Q_i is the generalized force corresponding to the coordinate q_i for *non-conservative forces* (other than viscous friction forces) *only*.

By introducing the Lagrangian function $L=T-V$, eq. (9.11) may be given the shorter form:

$$\frac{d}{dt}\frac{\partial L}{\partial \dot{q}_i}-\frac{\partial L}{\partial q_i}+\frac{\partial D}{\partial \dot{q}_i} = Q_i \qquad i = 1, 2, \ldots, n \qquad (9.12)$$

All the various forms of Lagrange's equation are contained in eq. (9.12).

Example 9.9. Develop the equations of motion for the system in Fig. 9.5, by using eq. (9.12).

Solution. Taking the generalized coordinates x_1 and x_2 from the unstrained positions of the springs, we have from eq. (9.12) the equations of motion

$$\frac{d}{dt}\frac{\partial L}{\partial \dot{x}_1}-\frac{\partial L}{\partial x_1}+\frac{\partial D}{\partial \dot{x}_1} = Q_{x_1} \qquad (a)$$

$$\frac{d}{dt}\frac{\partial L}{\partial \dot{x}_2}-\frac{\partial L}{\partial x_2}+\frac{\partial D}{\partial \dot{x}_2} = Q_{x_2}$$

The kinetic and potential energies are $T=\frac{1}{2}m_1\dot{x}_1^2+\frac{1}{2}m_2\dot{x}_2^2$ and $V=\frac{1}{2}K_1x_1^2+\frac{1}{2}K_2(x_1-x_2)^2$, so that

$$L = \tfrac{1}{2}m_1\dot{x}_1^2+\tfrac{1}{2}m_2\dot{x}_2^2-\tfrac{1}{2}K_1x_1^2-\tfrac{1}{2}K_2(x_1-x_2)^2$$

$$\frac{d}{dt}\frac{\partial L}{\dot{x}_1} = m_1\ddot{x}_1 \qquad \frac{\partial L}{\partial x_1} = -K_1x_1-K_2(x_1-x_2)$$

$$\frac{d}{dt}\frac{\partial L}{\partial \dot{x}_2} = m_2\ddot{x}_2 \qquad \frac{\partial L}{\partial x_2} = K_2(x_1-x_2)$$

The dissipation function is

$$D = \tfrac{1}{2}c_1\dot{x}_1^2+\tfrac{1}{2}c_2(\dot{x}_1-\dot{x}_2)^2$$

Hence

$$\frac{\partial D}{\partial \dot{x}_1} = c_1\dot{x}_1 + c_2(\dot{x}_1 - \dot{x}_2) \qquad \frac{\partial D}{\partial \dot{x}_2} = -c_2(\dot{x}_1 - \dot{x}_2)$$

Giving an increment δx_1 to x_1, we find the work done by non-conservative forces other than viscous forces: $\delta(\text{W.D.}) = Q_{x_1}\,\delta x_1 = 0$, and $Q_{x_1} = 0$. In the same manner for an increment δx_2: $\delta(\text{W.D.}) = Q_{x_2}\,\delta x_2 = F_0(\cos\omega t)\,\delta x_2$, and $Q_{x_2} = F_0\cos\omega t$.

The equations of motion are then

$$m_1\ddot{x}_1 + (c_1 + c_2)\dot{x}_1 - c_2\dot{x}_2 + (K_1 + K_2)x_1 - K_2x_2 = 0 \qquad \text{(a)}$$

$$m_2\ddot{x}_2 + c_2(\dot{x}_2 - \dot{x}_1) + K_2(x_2 - x_1) = F_0\cos\omega t \qquad \text{(b)}$$

Example 9.10. In Example 9.1, Fig. 9.4, determine (a) the kinetic and potential energy of the system as a function of θ and the given system constants, (b) the equation of motion of the system, and (c) the equation of linear motion of the body A of mass m_1, if $c = 0$ and $m_2 = 0$.

Solution. (a) The kinetic energy is

$$T = \tfrac{1}{2}m_1\dot{x}^2 + 2[\tfrac{1}{2}m_2(l\dot{\theta})^2]$$

from Example 9.1. We have

$$x = 2l(\cos\alpha - \cos\theta) \qquad \text{and} \qquad \dot{x} = 2l(\sin\theta)\dot{\theta}$$

Substituting this gives $T = l^2(2m_1\sin^2\theta + m_2)\dot{\theta}^2$.

Static considerations at the equilibrium position shows that the spring force is $F = (m_1 + m_2)g$. From the spring-force diagram in Fig. 9.11, the potential energy of the spring is

$$V_S = -\{[(m_1 + m_2)g - Kx]x + \tfrac{1}{2}Kx^2\} = -(m_1 + m_2)gx + \tfrac{1}{2}Kx^2$$

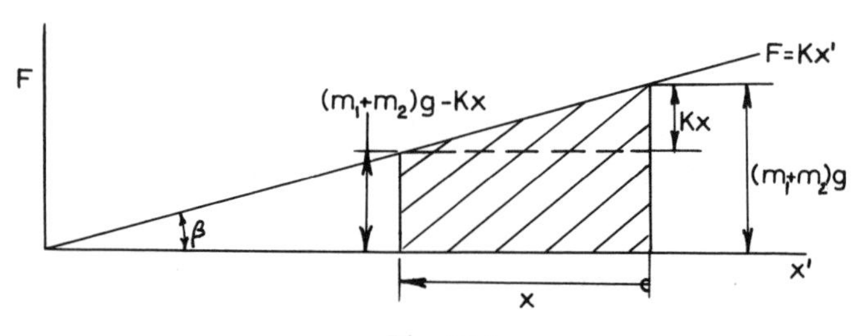

Fig. 9.11

The potential energy due to gravity is

$$V_{m_1} = m_1 g x \qquad V_{m_2} = 2m_2 g l(\cos\alpha - \cos\theta) = m_2 g x$$

The total potential energy is thus $V = \frac{1}{2}Kx^2 = 2Kl^2(\cos\alpha - \cos\theta)^2$.

(b) Lagrange's equation of motion in the coordinate θ is

$$\frac{d}{dt}\frac{\partial L}{\partial \dot{\theta}} - \frac{\partial L}{\partial \theta} + \frac{\partial D}{\partial \dot{\theta}} = Q_\theta$$

$$L = T - V = l^2(2m_1 \sin^2\theta + m_2)\dot{\theta}^2 - 2Kl^2(\cos\alpha - \cos\theta)^2$$

$$\frac{\partial L}{\partial \dot{\theta}} = 2l^2(2m_1 \sin^2\theta + m_2)\dot{\theta}$$

$$\frac{d}{dt}\frac{\partial L}{\partial \dot{\theta}} = 2l^2(2m_1 \sin^2\theta + m_2)\ddot{\theta} + 8l^2 m_1 \dot{\theta}^2 \sin\theta\cos\theta$$

$$\frac{\partial L}{\partial \theta} = 4m_1 l^2 \dot{\theta}^2 \sin\theta\cos\theta - 4Kl^2(\cos\alpha - \cos\theta)\sin\theta$$

The generalized viscous damping force R_θ was determined in Example 9.1 and was $R_\theta = -\partial D/\partial\dot{\theta} = -4cl^2\dot{\theta}\sin^2\theta$.

The generalized force Q_θ for non-conservative forces is $Q_\theta = 0$. Substituting in Lagrange's equation gives the equation of motion

$$(m_1 \sin^2\theta + m_2/2)\ddot{\theta} + \sin\theta(m_1\cos\theta + c\sin\theta)\dot{\theta}^2 + K\sin\theta(\cos\alpha - \cos\theta) = 0$$

If $c = 0$ and $m_2 = 0$, the equation of motion takes the form

$$(m_1 \sin\theta)\ddot{\theta} + (m_1\cos\theta)\dot{\theta}^2 + K(\cos\alpha - \cos\theta) = 0$$

Multiplying by $2l/m_1$ leads to

$$[(2l\sin\theta)\ddot{\theta} + (2l\cos\theta)\dot{\theta}^2] + 2l(K/m_1)(\cos\alpha - \cos\theta) = 0$$

Substituting

$$x = 2l(\cos\alpha - \cos\theta),$$

$$\dot{x} = (2l\sin\theta)\dot{\theta}$$

and

$$\ddot{x} = (2l\sin\theta)\ddot{\theta} + (2l\cos\theta)\dot{\theta}^2$$

gives the equation of motion involving the coordinate x:

$$\ddot{x} + (K/m_1)x = 0$$

This is the usual equation for a mass on a spring. The motion is simple harmonic with frequency $f=(1/2\pi)\sqrt{(K/m_1)}$.

9.7 Some general remarks on Lagrange's equations

The following are some of the advantages of Lagrange's equations:

1. The most suitable coordinates may be used to describe the motion; these may be absolute or a mixture of absolute and relative coordinates.
2. The equations of motion are established by a series of simple differentiations of the kinetic energy expressed in the chosen coordinates. The method is the same for all types of systems.
3. The necessary kinematic analysis involves only the velocities and not the accelerations.
4. Fixed reactions, internal forces and normal forces do not enter the equations.

When a set of generalized coordinates has been established, it should be assumed that the system is moving in the positive direction of *all* coordinates when the velocities are determined, otherwise mistakes may be made in the direction of the velocity components.

When a set of equations of motion of a system has been determined, we often want the equations of motion of a simplified system where some of the coordinates of the original system are kept constant. Suppose that the equations of motion are established in coordinates x and θ for a system with two degrees of freedom, and we want the equation of motion for a simpler system in which $x=$constant; the equation of motion for *this* system is then the original equation of motion in the coordinate θ. This equation may of course be simplified where appropriate by substituting $x=$constant, $\dot{x}=\ddot{x}=0$.

The equations for systems with more than two degrees of freedom may be treated in a similar way, if some of the coordinates are fixed, to form a simplified system.

Problems

9.1 A simple pendulum in plane motion is suspended on an elastic string as shown in Fig. 9.12. Under the action of gravity in the middle position the length is x_0. Taking coordinates as shown, find the equations of motion. Simplify the equations by assuming small vibrations with small $\dot{x}$ and $\dot{\theta}$.

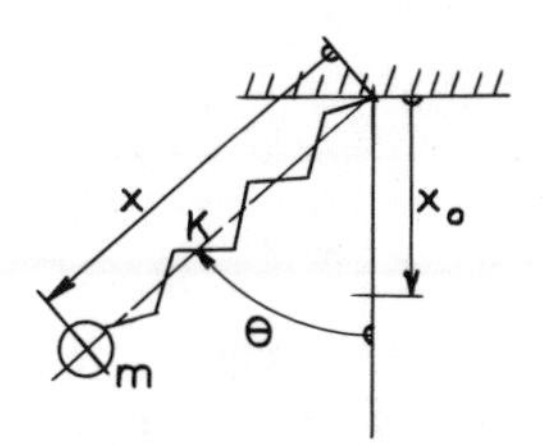

Fig. 9.12

9.2 A spherical pendulum (Fig. 9.13) consists of a particle of mass m supported by a massless string of length l. Using θ and φ as generalized coordinates, derive the two differential equations of motion. Show that these equations reduce to previously known results when θ and φ are successively held constant.

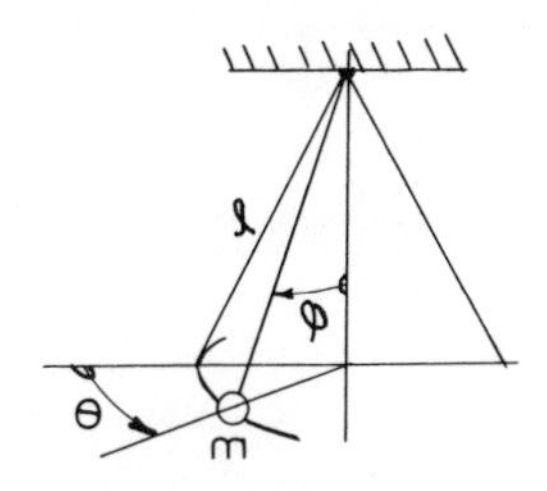

Fig. 9.13

9.3 Find the equations of motion for the particle in Fig. 9.14. The string has cross-section A and modulus E. The particle is performing vibrations along the horizontal x-axis. Assume that there is no initial tension in the string and that the tension produced by the weight of the particle is negligible.

Use $\sqrt{(l^2+x^2)} \simeq l + x^2/2l$. Find the equation of motion if the initial tension S is large and assumed constant for small vibrations. The spring constant is $K = AE/2l$.

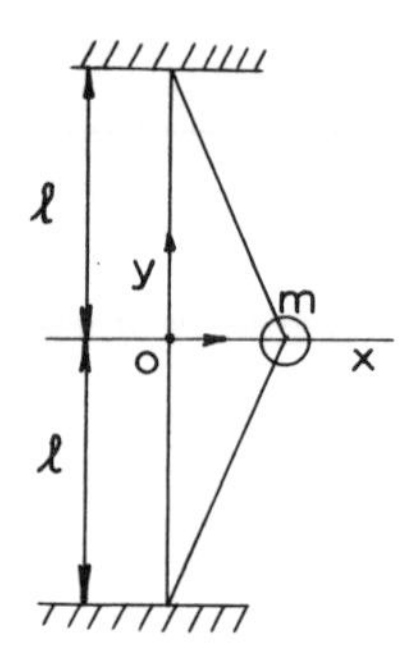

Fig. 9.14

9.4 Figure 9.15 shows a double pendulum. The string of each pendulum is of length l and the bob of mass m; the pendulum swings in a vertical plane due to the action of gravity.

(a) Without assuming small displacements, determine the equations of motion of the system.

(b) For small displacements and velocities, linearize the equations under (a).

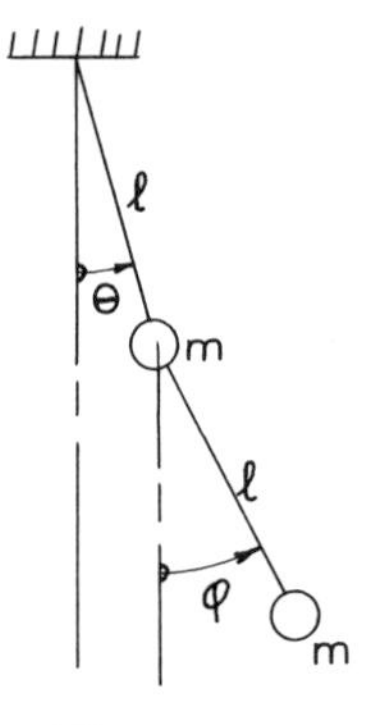

Fig. 9.15

9.5 Figure 9.16 shows a pendulum consisting of a rigid slender bar OA of length l and a bob A of mass M_1. The pendulum swings in a vertical plane about a fixed point O as shown. A second bob B of mass M_2 is guided by a circular seat of radius r and centre O_1, and slides freely along the rod OA. The two masses may be assumed to be concentrated, and the mass of the bar OA and all friction may be neglected.

(a) Determine the kinetic and potential energy functions of the system in terms of the system constants.

(b) Determine the equation of motion of the system.

(c) Show from the results of (b) that if $M_2=0$ the equation of motion takes the usual form for a simple pendulum of length l. Show also that if $M_1=0$ the equation takes the usual form for a simple pendulum of length r.

(d) Simplify the equation of motion by assuming small displacements and determine the frequency of the system if $l=20$ cm, $r=5$ cm, and $M_1=M_2=M$.

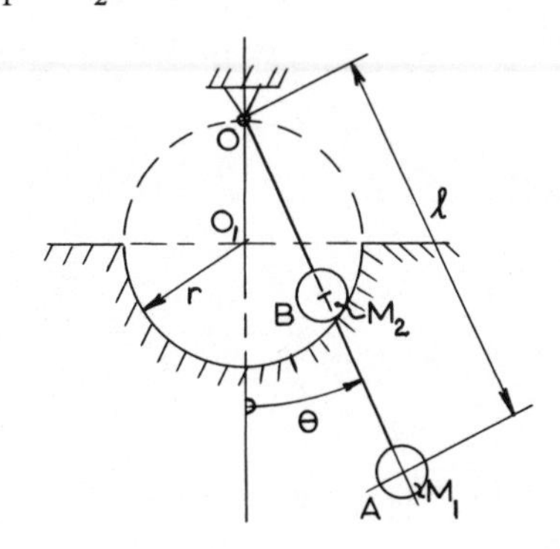

Fig. 9.16

9.6 A homogeneous right circular cylinder of mass m and radius r is rolling without slipping on a triangular prism also of mass m (Fig. 9.17). The prism of angle α is sliding without friction on a horizontal plane. Find the vertical and horizontal acceleration of the centre of gravity of the cylinder, and the acceleration of the prism, in terms of the given quantities.

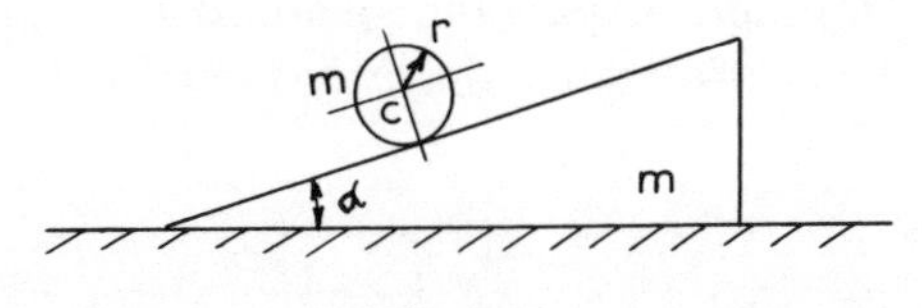

Fig. 9.17

9.7 A right circular homogeneous cylinder (Fig. 9.18) of radius r and mass m rolls without slipping in a semi-circular groove of radius R cut in a block of mass M which is constrained to move without friction in a vertical guide. The block is supported on a spring with spring constant K. Assume that the cylinder and block are always in contact.

(a) Find the equations of motion of the system without assuming small displacements.

(b) Assuming small displacements and velocities, find the simplified equations of motion.

(c) Find the natural frequency of the rolling cylinder if the block is fixed.

(d) Find the natural frequency of the system if the cylinder is fixed to the block.

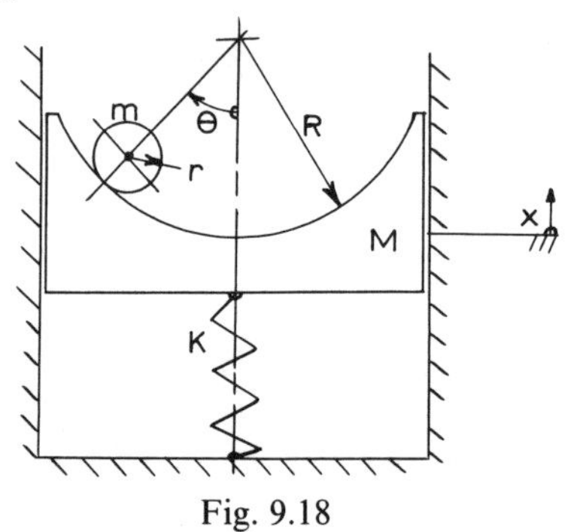

Fig. 9.18

9.8 Figure 9.19 shows a vibrating system consisting of a mass M which is sliding without friction on a horizontal plane. The mass M is connected by a light spring of spring constant K to a vertical wall. A simple pendulum consisting of a light string of length l and a concentrated mass m is attached to M as shown. A force $P(t)$ acts upon the mass M in the positive x-direction.

Establish the equations of motion of the system without assuming that θ is small. Simplify the equations, if θ is assumed small and terms of higher order than the second in θ and $\dot{\theta}$ are neglected.

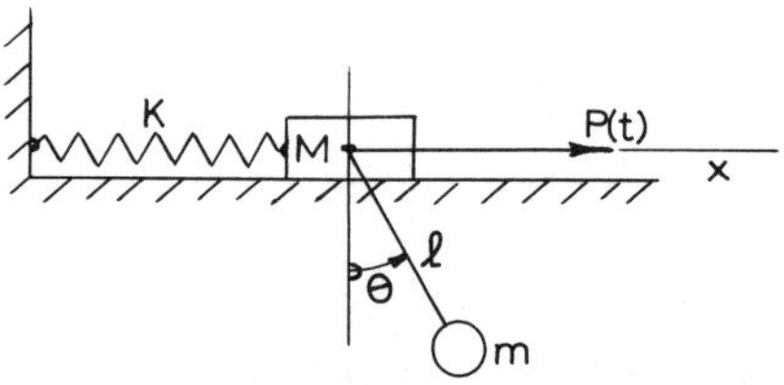

Fig. 9.19

Establish the equation of rotational motion of the pendulum for small angular displacements for the particular case of $M=0$ and $P(t)=0$ and discuss the motion in detail.

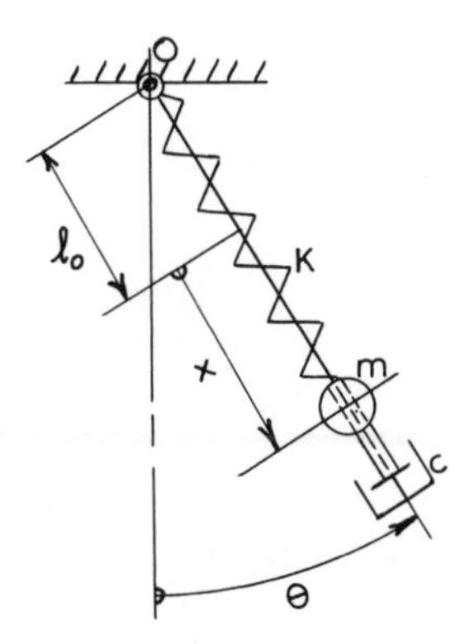

Fig. 9.20

9.9 Figure 9.20 shows a pendulum which is able to swing in the vertical plane about point O. The pendulum consists of a light rigid rod on which a bob of mass m is sliding without friction. The bob is connected to point O through a spring of spring constant K.

The motion of the bob along the bar is damped by a light dashpot with coefficient of damping c. The length of the spring is l_0 in the vertical statical equilibrium position. Determine:

(a) The kinetic and potential energy functions of the system in terms of the system constants and the coordinates x and θ as shown on the figure.

(b) The equations of motion of the system.

(c) The equation of motion if $\theta=0$ and constant. The equation of motion if x is constant and $x=b$.

9.10 Find the equation of motion for the system in Fig. 9.21.

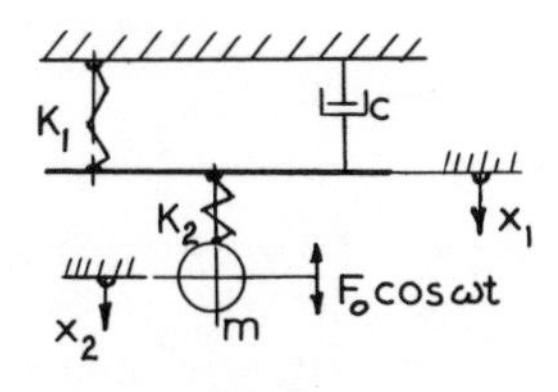

Fig. 9.21

CHAPTER TEN

Vibrations with One Degree of Freedom

The case of a mass in linear motion under the action of a force proportional to the displacement was discussed in Section 2.4.4. This was the case of a mass on a spring and was shown to be a simple harmonic vibratory motion.

To set up a more realistic mathematical model, consider the system shown in Fig. 10.1. The spring and the dashpot are assumed light, and all the mass of the system is assumed concentrated in the body of mass m as shown. All damping in the system is assumed to be of the viscous friction type and concentrated in the dashpot with damping coeffiicient c. The damping force is then proportional to the velocity and in the opposite direction, so that it may be expressed as $-c\dot{x}$.

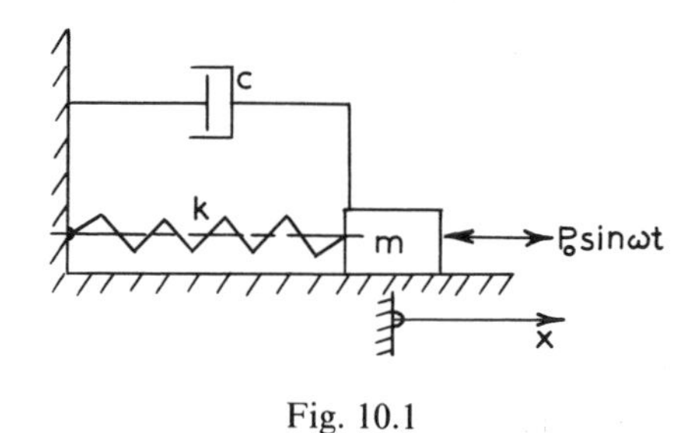

Fig. 10.1

An external exciting force $P = P_0 \sin \omega t$, of force amplitude P_0 and frequency ω, is acting on the body as shown. The displacement x is measured from the static equilibrium position and forces and motion are taken positive to the right.

Newton's second law gives directly the equation of motion:

$$m\ddot{x} = P_0 \sin \omega t - c\dot{x} - kx \qquad \text{or} \qquad \ddot{x} + \frac{c}{m}\dot{x} + \frac{k}{m}x = P_0 \sin \omega t$$

Substituting the constants $\omega_0^2 = k/m$ and $n = c/2m$ leads to

$$\ddot{x} + 2n\dot{x} + \omega_0^2 x = \frac{P_0}{m} \sin \omega t \tag{10.1}$$

The investigation of the motion is divided into four cases for convenience.

10.1 Free vibrations without damping

If $P_0 = 0$ and $c = 0$, we have free vibrations without damping, and the equation of motion (10.1) takes the form

$$\ddot{x} + \omega_0^2 x = 0 \tag{10.2}$$

The general solution of this equation is

$$x = A \cos \omega_0 t + B \sin \omega_0 t$$

With starting conditions $x = x_0$ and $\dot{x} = \dot{x}_0$ at $t = 0$, the solution is

$$x = \sqrt{[x_0^2 + (\dot{x}_0/\omega_0)^2]} \cos(\omega_0 t - \varphi) \qquad \tan \varphi = \dot{x}_0/x_0\omega_0$$

as shown in Chapter 2. The motion is simple harmonic motion. The amplitude is $A_0 = \sqrt{[x_0^2 + (\dot{x}_0/\omega_0)^2]}$ and φ is the phase angle. The constant $\omega_0 = \sqrt{(k/m)}$ is the *natural circular frequency*, the *frequency* is $f = \omega_0/2\pi$ and the period is $\tau = 1/f$. If the static deflection of the spring is Δ_S under a load mg, we have $\Delta_S = mg/k$, and

$$\omega_0 = \sqrt{(kg/mg)} = \sqrt{(g/\Delta_S)}$$

If the spring is vertical, the frequency of the system may be determined by measuring the static deflection Δ_S occurring when the mass is attached to the spring.

The following Examples were simple harmonic motion: 2.8, 2.19, 8.9 and 8.10.

Example 10.1. The vibratory system shown in Fig. 10.2 is in static equilibrium when the rigid light bar OB is horizontal. For small rotational motions of the bar about O, the spring may be assumed to stay vertical and the body of mass m to move in vertical rectilinear motion. Determine the frequency for small displacements of the system.

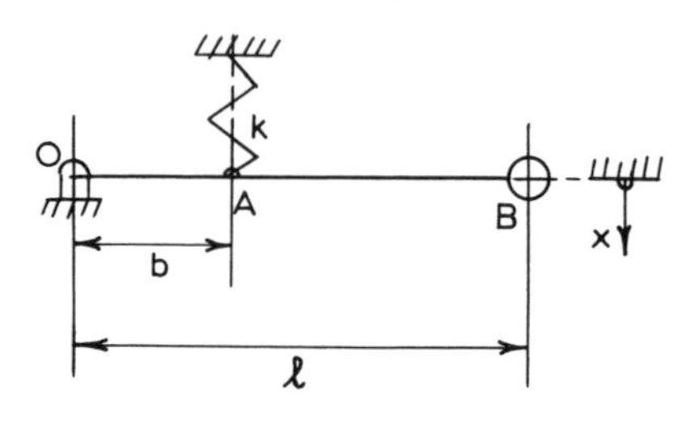

Fig. 10.2

Solution. In the equilibrium position, the force in the spring is P. Taking moments about O gives the result $Pb = lmg$, or $P = (l/b)mg$.

If the mass is displaced a distance x, the point A is displaced $\Delta = xb/l$, and the total spring force is $P + k\Delta = (l/b)mg + kxb/l$. The force F acting on the mass from the bar is then found by taking moments about O:

$$Fl = \left(\frac{l}{b}mg + kx\frac{b}{l}\right)b \qquad \text{or} \qquad F = mg + kx(b/l)^2$$

The total restoring force on the mass is then

$$-(F - mg) = -k\left(\frac{b}{l}\right)^2 x$$

Newton's second law gives the equation of motion

$$m\ddot{x} = -k\left(\frac{b}{l}\right)^2 x \qquad \text{or} \qquad \ddot{x} + \frac{k}{m}\left(\frac{b}{l}\right)^2 x = 0$$

This is simple harmonic motion with circular frequency $\omega_0 = (b/l)\sqrt{(k/m)}$. If $b \to 0$, the frequency becomes very low; the maximum frequency is for $b = l$ when $\omega_0 = \sqrt{(k/m)}$.

The effective spring constant at the mass is evidently $k(b/l)^2$.

Example 10.2. In the system in Fig. 10.3 the body of mass m performs small vertical vibrations. The bar AB is rigid and massless, and C is the middle point of the bar. Determine the frequency of the vibrations.

Solution. In static equilibrium, the force in the middle spring is mg and that in the other two springs $mg/2$. If the mass is moved down a distance x from static equilibrium, the increase in the force in the middle spring is a force P, and the increase in the forces in the other two springs is $P/2$. If the displacement of points A and B are Δ_1 and Δ_2, the displacement of point C is $(\Delta_1+\Delta_2)/2$. We have $\Delta_1 = P/2k_1$ and $\Delta_2 = P/2k_2$, so that

$$(\Delta_1+\Delta_2)\tfrac{1}{2} = \tfrac{1}{4}P\left(\frac{1}{k_1}+\frac{1}{k_2}\right)$$

The extension of the middle spring is $x-\frac{1}{2}(\Delta_1+\Delta_2)$, and the force

$$P = k_3[(x-\tfrac{1}{2}(\Delta_1+\Delta_2)] = k_3x-\tfrac{1}{4}Pk_3\left(\frac{1}{k_1}+\frac{1}{k_2}\right)$$

or

$$P = \frac{k_3x}{1+\frac{1}{4}k_3(1/k_1+1/k_2)}$$

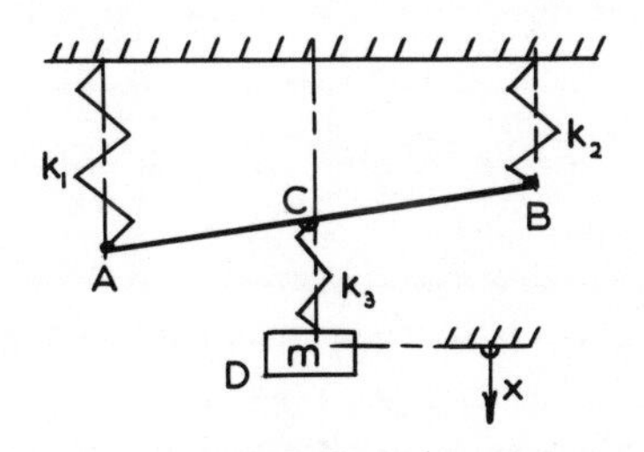

Fig. 10.3

When $x=1$, P is the total spring constant K, so that

$$K = \frac{4k_1k_2k_3}{4k_1k_2+k_3(k_1+k_2)}$$

The circular frequency is $\omega_0 = \sqrt{(K/m)}$.

10.2 Free vibrations with viscous damping

In this case the equation of motion is obtained from (10.1), with $P_0=0$:

$$\ddot{x}+2n\dot{x}+\omega_0^2x = 0 \tag{10.3}$$

In solving this equation, we assume a solution of the form $x = e^{st}$, where e is the base of the natural logarithms, $e = 2{\cdot}71828\ldots$, t is the time and s is a constant that must be adjusted so that $x = e^{st}$ satisfies eq. (10.3).

Differentiating x gives $\dot{x} = se^{st}$ and $\ddot{x} = s^2 e^{st}$; substituting in (10.3), we find that s must satisfy the equation $s^2 + 2ns + \omega_0^2 = 0$. This equation has the roots $s = -n \pm \sqrt{(n^2 - \omega_0^2)}$. Three possible cases occur for the roots s:

(a) $n^2 = \omega_0^2$. This means that the roots are $s_1 = s_2 = -n = -\omega_0$, and we have the solution $x = e^{-nt}$. It may be seen by direct substitution in (10.3), that a second solution is $x = te^{-nt}$. The solutions may be multiplied by an arbitrary constant, so a set of solutions is given by $x = C_1 e^{-nt}$ and $x = C_2 te^{-nt}$. The sum of two solutions is also a solution, so the complete general solution is:

$$x = e^{-nt}(C_1 + C_2 t)$$

The constants C_1 and C_2 may be determined in each particular case from the starting conditions $x = x_0$ and $\dot{x} = \dot{x}_0$ at $t = 0$; the result is

$$x = e^{-nt}\,[x_0 + (x_0\omega_0 + \dot{x}_0)t]$$

It is easy to show that for $t \to \infty$, $x \to 0$; the mass then creeps back to the equilibrium position after release.

If $\dot{x}_0$ is sufficiently large and directed towards equilibrium, it is possible for the mass to move through the equilibrium position; it will then creep back to equilibrium from the opposite side of the starting displacement. In no case will the curve $x = f(t)$ for the motion intercept the t-axis more than once.

The motion is thus an *aperiodic* motion and *not* a vibratory motion. The damping coefficient in this case is found from $n = c/2m = \omega_0$; therefore $c = 2m\omega_0$. This value of the damping coefficient is called the *critical damping* $c_C = 2m\omega_0 = 2\sqrt{(km)}$, since this value of c is the *minimum value* for aperiodic motion. Introducing a *damping ratio* $d = c/c_C$, we have in this case $d = 1$, and the system is called a *critically damped system*.

(b) $n^2 > \omega_0^2$. The roots of the characteristic equation are here

$$\left.\begin{matrix} s_1 \\ s_2 \end{matrix}\right\} = -n \pm \sqrt{(n^2 - \omega_0^2)}$$

These are both *real* and *negative*. A pair of solutions is $x = e^{s_1 t}$ and

$x=e^{s_2 t}$. The general solution to eq. (10.3) in this case is $x=C_1e^{s_1t}+C_2e^{s_2t}$, where C_1 and C_2 are arbitrary constants for the general solution; for any particular case these constants may be found from the starting conditions $x=x_0$, $\dot{x}=\dot{x}_0$ at $t=0$.

The motion is *aperiodic*, and the remarks concerning the motion for case (a) hold also in this case. The damping ratio is $d=c/c_C=c/2m\omega_0=n/\omega_0$; since $n^2>\omega_0^2$, we have in this case $d>1$. The system is called an *overdamped system.*

This type of motion is encountered in hydraulic door stops.

(c) $n^2<\omega_0^2$. The roots of the characteristic equation are

$$\left.\begin{matrix} s_1 \\ s_2 \end{matrix}\right\} = -n\pm\sqrt{(n^2-\omega_0^2)} = -n\pm i\sqrt{(\omega_0^2-n^2)}$$

These are *conjugate complex roots*. Introducing the real constant $p_1=\sqrt{(\omega_0^2-n^2)}$, we have $s_1=-n+p_1i$ and $s_2=-n-p_1i$.

The general solution of eq. (10.3) in this case is

$$x = C_1e^{s_1t}+C_2e^{s_2t} = e^{-nt}(C_1e^{ip_1t}+C_2e^{-ip_1t})$$

Introducing Euler's formula, $e^{\pm i\theta}=\cos\theta\pm i\sin\theta$, we find

$$\begin{aligned} x &= e^{-nt}[C_1(\cos p_1t+i\sin p_1t)+C_2(\cos p_1t-i\sin p_1t)] \\ &= e^{-nt}[(C_1+C_2)\cos p_1t+(C_1-C_2)i\sin p_1t] \end{aligned}$$

Introducing new arbitrary constants $A_1=C_1+C_2$ and $A_2=C_1-C_2$, we may write this as

$$x = e^{-nt}(A_1\cos p_1t+A_2i\sin p_1t)$$

The solution is of the form $x=u+iv$; differentiating we find $\dot{x}=\dot{u}+i\dot{v}$ and $\ddot{x}=\ddot{u}+i\ddot{v}$. Substituting in eq. (10.3), the result is

$$(\ddot{u}+2n\dot{u}+\omega_0^2u)+i(\ddot{v}+2n\dot{v}+\omega_0^2v) = 0$$

This expression is satisfied only if $\ddot{u}+2n\dot{u}+\omega_0^2u=0$ and $\ddot{v}+2n\dot{v}+\omega_0^2v=0$, which means that $x=u$ and $x=v$ satisfies eq. (10.3), and a set of solutions is $x=e^{-nt}A_1\cos p_1t$ and $x=e^{-nt}A_2\sin p_1t$. The general solution of eq. (10.3) is then in this case:

$$x = e^{-nt}(A_1\cos p_1t+A_2\sin p_1t)$$

The constants A_1 and A_2 must be found from the starting conditions $x=x_0$, $\dot{x}=\dot{x}_0$ at $t=0$; we find in this way the solution

$$x = e^{-nt}\left(x_0\cos p_1t+\frac{\dot{x}_0+nx_0}{p_1}\sin p_1t\right)$$

where $\omega_0 = \sqrt{(k/m)}$; $d = c/c_C = c/2m\omega_0 = n/\omega_0$, so that $d < 1$ and

$$p_1 = \sqrt{(\omega_0^2 - n^2)} = \omega_0\sqrt{(1-d^2)}$$

This is called an *underdamped* case of motion. The solution may be simplified by writing

$$A \cos p_1 t + B \sin p_1 t = C \cos (p_1 t - \varphi)$$
$$= C \cos p_1 t \cos \varphi + C \sin p_1 t \sin \varphi$$

so that $C \cos \varphi = A$, $C \sin \varphi = B$, and $C = \sqrt{(A^2 + B^2)}$ with $\tan \varphi = B/A$. We find

$$C = \sqrt{\left[x_0^2 + \left(\frac{\dot{x}_0 + nx_0}{p_1}\right)^2\right]} \qquad \tan \varphi = \frac{\dot{x}_0 + nx_0}{p_1 x_0}$$

$$x = e^{-nt} C \cos (p_1 t - \varphi) \quad (10.4)$$

Figure 10.4 shows a graph of this function. The curve is called a *damped sine curve*, and shows the motion of damped free vibrations. When $\cos (p_1 t - \varphi) = \pm 1$, the curve touches the curves $x = \pm Ce^{-nt}$, which are exponentially decreasing as shown. The maximum displacements occur slightly before the tangential points. The points of intersection with the t-axis are found for $x = 0$, and at these points $\cos (p_1 t - \varphi) = 0$.

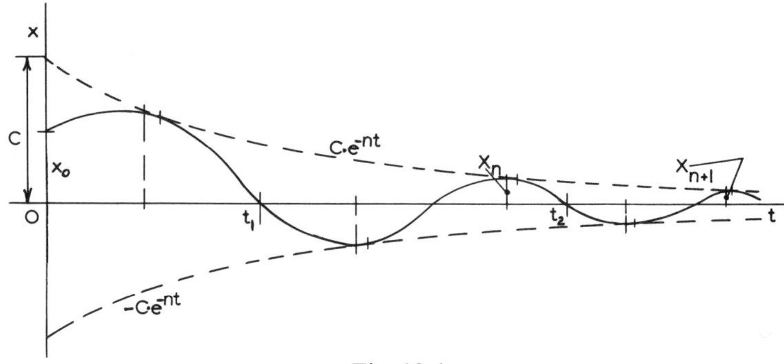

Fig. 10.4

The curve is not periodic in the usual sense of the word, since the displacements are diminishing; however, it is customary to talk about a 'period' as the time from $t = t_1$ to $t = t_2$ as shown. We find

$$p_1 t_1 - \varphi = \frac{\pi}{2} \text{ and } p_1 t_2 - \varphi = \frac{5\pi}{2}$$

the period is then $\tau = t_2 - t_1 = 2\pi/p_1$. The constant $p_1 = \omega_0\sqrt{(1-d^2)}$ is called the damped natural circular frequency. The *damped natural frequency* is

$$f = \frac{1}{\tau} = \frac{p_1}{2\pi} = \frac{\omega_0}{2\pi}\sqrt{(1-d^2)}$$

It may be seen that damping slows down the motion – the frequency has a smaller value than the undampened case, where $d=0$.

The rate of decay of the motion may be defined as the ratio x_n/x_{n+1}. The maximum amplitudes occur approximately at those points where $\cos(p_1 t - \varphi) = 1$, so that

$$x_n = Ce^{-nt_n} \qquad x_{n+1} = Ce^{-n(t_n+\tau)} \qquad x_n/x_{n+1} = e^{n\tau} = e^{\delta}$$

The constant δ defining the rate of decay is determined by

$$\delta = \ln\left(\frac{x_n}{x_{n+1}}\right) = n\tau = d\omega_0\tau$$

δ is called the *logarithmic decrement*. Since

$$\tau = \frac{2\pi}{p_1} = \frac{2\pi}{\omega_0\sqrt{(1-d^2)}}$$

we find

$$\delta = d\omega_0\tau = \frac{2\pi d}{\sqrt{(1-d^2)}} \qquad (10.5)$$

If the damping is negligible, $d = c/c_C \to 0$ and $\delta \to 0$; for increasing damping δ increases. For *small damping* $d^2 \ll 1$, and we have approximately $\delta = 2\pi d$.

Example 10.3. The system in Fig. 10.1 is executing free damped vibrations ($P_0 = 0$) and $m = 7{\cdot}66$ kg, $k = 52{\cdot}5$ N/cm and $c = 0{\cdot}438$ Ns/cm. Determine c_C, δ, x_n/x_{n+1} and the damped natural frequency.

Given that the system is released from rest with $x_0 = 1$ cm, determine the approximate displacement after 10 cycles and the time elapsed.

Solution

$$\omega_0 = \sqrt{(k/m)} = \sqrt{(52{\cdot}5\times 100/7{\cdot}66)} = 26{\cdot}2 \text{ rad/s}$$

$$c_C = 2m\omega_0 = 2\times 7{\cdot}66\times 26{\cdot}2 = 401 \text{ Ns/m} = 4{\cdot}01 \text{ Ns/cm}$$

$$d = c/c_C = 0{\cdot}438/4{\cdot}01 = 0{\cdot}109.$$

The damping is about 11% of critical damping

$$\delta = \frac{2\pi d}{\sqrt{(1-d^2)}} = \frac{2\pi\times 0{\cdot}109}{\sqrt{(1-0{\cdot}109^2)}} = 0{\cdot}690$$

$x_n/x_{n+1}=e^{\delta}=1{\cdot}994$; we have approximately $x_{n+1}=\frac{1}{2}x_n$.
The damped natural frequency is

$$f_1 = \frac{p_1}{2\pi} = \frac{\omega_0}{2\pi}\sqrt{(1-d^2)} = \frac{26{\cdot}2}{2\pi}\sqrt{(1-0{\cdot}109^2)} = 4{\cdot}14 \text{ cycles/s}$$

The undamped natural frequency is

$$f_0 = \frac{\omega_0}{2\pi} = 4{\cdot}17 \text{ cycles/s}$$

Damping has then lowered the frequency by less than 1% in this case.

If the system is released with $\dot{x}_0=0$ and $x_0=1$ cm at $t=0$, we find that the maximum displacement after 10 cycles is approximately $(\frac{1}{2})^{10}=\frac{1}{1024}<\frac{1}{1000}$ cm. The time elapsed is $10/4{\cdot}14=2{\cdot}41$ s. The vibrations have clearly been damped out, for all practical purposes, after say 5 s.

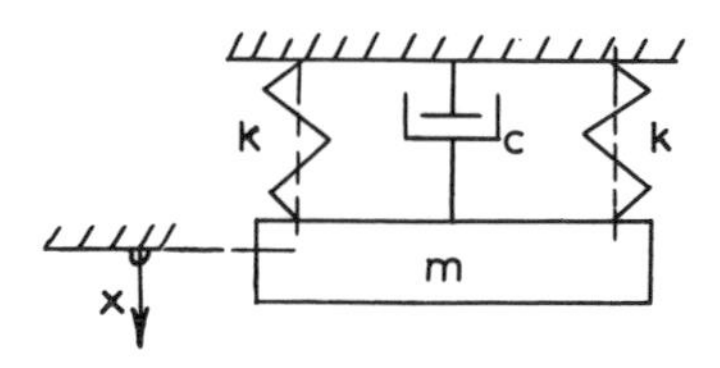

Fig. 10.5

Example 10.4. Figure 10·5 shows a damped vibratory system. The system constants are: $k=87{\cdot}5$ N/cm, $m=22{\cdot}7$ kg and $c=3{\cdot}50$ Ns/cm. The system is initially at rest, when the body of mass m is given an impact which starts it moving in the positive x-direction with initial velocity $\dot{x}_0=12{\cdot}7$ cm/s. Determine the frequency of the ensuing damped vibrations, the logarithmic decrement and the maximum displacement from the equilibrium position.

Solution

$$\omega_0 = \sqrt{(k/m)} = \sqrt{\left(\frac{2\times 87{\cdot}5\times 100}{22{\cdot}7}\right)} = 27{\cdot}8 \text{ rad/s}$$

$$c_C = 2m\omega_0 = 2\times 22{\cdot}7\times 27{\cdot}8 = 1261 \text{ Ns/m} = 12{\cdot}61 \text{ Ns/cm}$$

$$d = \frac{c}{c_C} = \frac{3{\cdot}50}{12{\cdot}61} = 0{\cdot}277$$

$$f_1 = \frac{27{\cdot}8}{2\pi}\sqrt{(1-0{\cdot}277^2)} = 4{\cdot}25 \text{ cycles/s}$$

$$\delta = \frac{2\pi d}{\sqrt{(1-d^2)}} = 1{\cdot}81$$

From the solution (10.4), $x = e^{-nt}C\cos(p_1t-\varphi)$, the maximum displacement occurs approximately when $\cos(p_1t-\varphi)=1$ the first time; this happens at a time $t=t_1$, and $p_1t_1-\varphi=0$, or $t_1=\varphi/p_1$ and $x_{max}=e^{-nt_1}C$.

$$p_1 = \omega_0\sqrt{(1-d^2)} = 26{\cdot}7 \text{ rad/s}$$

From (10.4),

$$\tan\varphi = (\dot{x}_0+nx_0)/p_1x_0 \to \infty \qquad \varphi = \pi/2 \qquad n = c/2m = 350/(2\times 22{\cdot}7) = 7{\cdot}71 \text{ s}^{-1}$$

$$C = \dot{x}_0/p_1 = 12{\cdot}70/26{\cdot}7 = 0{\cdot}476 \text{ cm} \qquad nt_1 = n\varphi/p_1 = 0{\cdot}453$$

We have now $x_{max} = e^{-0{\cdot}453}\times 0{\cdot}476 = 0{\cdot}303$ cm.

10.3 Forced vibrations without damping

If we assume that there is no damping in the system of Fig. 10.1, we have $c=0$, and the equation of motion is, from (10.1),

$$\ddot{x}+\omega_0^2x = \frac{P_0}{m}\sin\omega t \tag{10.6}$$

The general solution to this non-homogeneous equation is known to be the sum of the general solution to the homogeneous equation $\ddot{x}+\omega_0^2x=0$, and a particular solution to the total equation (10.6).

The general solution to the homogeneous equation is $x = A\cos\omega_0 t + B\sin\omega_0 t$, with $\omega_0 = \sqrt{(k/m)}$. The constants A and B are arbitrary. This solution has been discussed before under free undamped vibrations.

To find a particular solution to (10.6), we substitute a function of the same form as the right-hand function $(P_0/m)\sin\omega t$. Taking $x = x_0 \sin\omega t$, we have $\ddot{x} = -x_0\omega^2\sin\omega t$; substituting in (10.6) and cancelling $\sin\omega t$, we find that $x = x_0\sin\omega t$ is a solution to (10.6), if $x_0(\omega_0^2 - \omega^2) = P_0/m$, or

$$x_0 = \frac{P_0}{m(\omega_0^2 - \omega^2)} = \frac{P_0}{k - m\omega^2} = \frac{P_0}{k}\,\frac{1}{1-(\omega/\omega_0)^2}$$

Introducing $x_S = P_0/k$, where x_S is the static deflection for a force P_0, we have

$$x = x_0\sin\omega t = x_S\frac{1}{1-(\omega/\omega_0)^2}\sin\omega t$$

The general solution to (10.6) is thus

$$x = (A\cos\omega_0 t + B\sin\omega_0 t) + \frac{x_S}{1-(\omega/\omega_0)^2}\sin\omega t \qquad (10.7)$$

The constants A and B are found from the starting conditions $x = x_0$, $\dot{x} = \dot{x}_0$ at $t = 0$.

The first two terms in (10.7) are the *free* vibrations of the system. In any practical case some damping due to air resistance, dry friction and internal material damping is always present; the free vibration part of the solution is therefore soon damped out. This part of the solution is called the *transient state*. We are mainly interested in the last term of the solution (10.7), which is a vibratory motion with frequency ω equal to the forcing frequency, and a constant amplitude

$$x_0 = x_S\frac{1}{1-(\omega/\omega_0)^2}$$

This solution is called the *steady-state* solution. The factor

$$\left|\frac{x_0}{x_S}\right| = \frac{1}{|1-(\omega/\omega_0)^2|}$$

is called *the magnification factor*, since it is the factor by which the static deflection x_S must be multiplied to give the maximum deflection x_0. We are usually only interested in the magnitude of the vibrations, any phase difference being of minor importance; only the numerical value of the magnification factor is therefore of interest.

The variation of the magnification factor with the ratio (ω/ω_0) is shown in Fig. 10.6 for $d=0$. It will be seen that for $\omega/\omega_0 \rightarrow 0$, $x_0/x_S \rightarrow 1$. For $\omega/\omega_0 \rightarrow 1$, $|x_0/x_S| \rightarrow \infty$, and this condition is called a *resonance condition*. We say that the force is in resonance with the natural frequency of the system. For large values of (ω/ω_0), we find that $x_0/x_S \rightarrow 0$, which means that for large values of the forcing frequency compared to the natural frequency, the body is practically motionless.

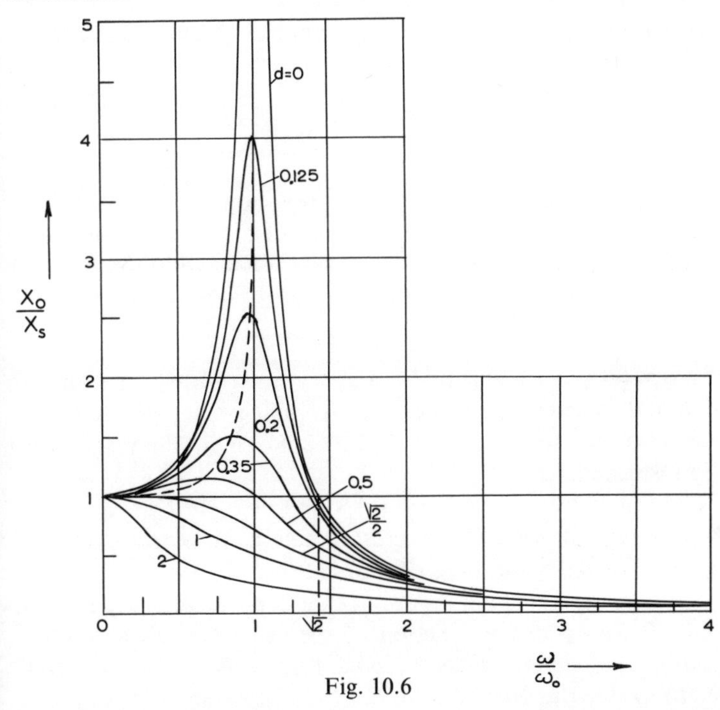

Fig. 10.6

Example 10.5. For the system in Fig. 10.5, the system constants are $k=87{\cdot}5$ N/cm, $m=22{\cdot}7$ kg and $c=0$. The system is vibrated by an external force in the x-direction: $P=201 \cdot \sin(13 \cdot t)$ newton.

Determine the magnification factor and the steady state motion of the mass.

Solution

$$\omega_0 = \sqrt{(k/m)} = \sqrt{(2\times 87{\cdot}5\times 100/22{\cdot}7)} = 27{\cdot}8 \text{ rad/s}$$

The magnification factor is $1/[1-(13/27{\cdot}8)^2] = 1{\cdot}28$.

$$x_S = \frac{P_0}{k} = \frac{201}{175} = 1{\cdot}15 \text{ cm} \qquad x_0 = 1{\cdot}15\times 1{\cdot}28 = 1{\cdot}47 \text{ cm}$$

The steady-state motion is $x = x_0 \sin \omega t = 1{\cdot}47 \sin 13t$ cm.

10.4 Forced vibrations with viscous damping

Turning finally to the complete system in Fig. 10.1, the equation of motion is, from (10.1),

$$\ddot{x} + 2n\dot{x} + \omega_0^2 x = \frac{P_0}{m} \sin \omega t$$

with $\omega_0^2 = k/m$ and $n = c/2m$.

The equation is a second-order non-homogeneous ordinary differential equation with constant coefficients. The corresponding homogeneous equation is $\ddot{x} + 2n\dot{x} + \omega_0^2 x = 0$, with the solution obtained before:

$$x = e^{-nt}(A_1 \cos p_1 t + A_2 \sin p_1 t)$$

where $p_1 = \omega_0 \sqrt{(1-d^2)}$ and A_1 and A_2 are arbitrary constants, to be found in each particular case from the starting conditions. This solution is for $c < c_C$; if $c \geqslant c_C$, the solution for these cases must be substituted instead.

It is known that the general solution to eq. (10.1) is the sum of the general solution to the corresponding homogeneous equation and a *particular* solution to eq. (10.1).

The forcing function is a sine function; to find a particular solution we assume a function of the same type as a solution. Taking $x = A \sin \omega t + B \cos \omega t$, where A and B are constants to be determined so that this is a solution to (10.1), we find by substitution in (10.1):

$$[A(\omega_0^2-\omega^2) - 2n\omega B - P_0/m]\sin \omega t + [2n\omega A + (\omega_0^2-\omega^2)B] \cos \omega t = 0$$

To satisfy this equation for all values of time t, the square brackets must vanish. This gives the equations

$$A(\omega_0^2-\omega^2)-2n\omega B = P_0/m$$
$$2n\omega A+(\omega_0^2-\omega^2)B = 0$$

Substituting $\omega_0^2=k/m$ and $n=c/2m$, we obtain the solution

$$A = \frac{P_0(k-m\omega^2)}{(k-m\omega^2)^2+c^2\omega^2} \qquad B = \frac{-P_0c\omega}{(k-m\omega^2)^2+c^2\omega^2}$$

The general solution to (10.1) is thus

$$x = e^{-nt}(A_1 \cos p_1t+A_2 \sin p_1t)+(A \sin \omega t+B \cos \omega t)$$

The constants A_1 and A_2 must be determined by the starting conditions $x=x_0$, $\dot{x}=\dot{x}_0$ at $t=0$, while A and B have the above values.

The solution thus consists of two parts, the first part being that of the free damped vibrations, the *transient state*, superimposed on the forced vibrations with frequency ω equal to the forcing frequency, the *steady state*.

As was shown before, the transient state soon vanishes because of the damping, and we are mainly interested in the steady-state vibrations:

$$x = A \sin \omega t+B \cos \omega t$$

Writing this as

$$x = x_0 \sin (\omega t-\varphi) = (x_0 \cos \varphi) \sin \omega t-(x_0 \sin \varphi) \cos \omega t$$

we find $x_0 \cos \varphi=A$ and $x_0 \sin \varphi=-B$, so that $x_0=\sqrt{(A^2+B^2)}$ and $\tan \varphi=-B/A$. Substituting the expressions for A and B found before, we obtain

$$x_0 = \frac{P_0}{\sqrt{[(k-m\omega^2)^2+c^2\omega^2]}} \qquad \text{and} \qquad \tan \varphi = \frac{c\omega}{k-m\omega^2}$$

The steady-state motion is thus $x=x_0 \sin (\omega t-\varphi)$, which is simple harmonic motion with frequency ω equal to the forcing frequency. The amplitude is x_0 and the phase angle φ determined by the above expressions.

To get the formulae in a dimensionless form more convenient for analysis, we substitute $\omega_0^2=k/m$, $d=c/c_C$, $c_C=2m\omega_0$ and $x_S=P_0/k$; the result is

$$\frac{x_0}{x_S}=\frac{1}{\sqrt{\{[1-(\omega/\omega_0)^2]^2+(2d\omega/\omega_0)^2\}}} \tag{10.8}$$

x_0/x_S is thus a function of (ω/ω_0) and d. Figure 10.6 shows this relationship for various values of the damping ratio d. It may be seen that for small values of (ω/ω_0), $x_0 \simeq x_S$. When $(\omega/\omega_0)\rightarrow 1$, x_0/x_S increases to a value close to the maximum value, and this is approximately the *resonance condition*. When ω is large compared to ω_0, the mass is practically stationary, this is the case for ω greater than about 3.

So far we have considered two circular frequencies, the natural undamped frequency $\omega_0=\sqrt{(k/m)}$, and the natural damped frequency $p_1=\omega_0\sqrt{(1-d^2)}$, where the damping ratio $d<1$.

In measurements of vibrations, the maximum amplitude, velocity or acceleration may be measured. The frequency for *maximum amplitude* is usually called the resonance frequency; it may be determined from (10.8) for the condition $d(x_0/x_S)/d(\omega/\omega_0)=0$, and is $p_2=\omega_0\sqrt{(1-2d^2)}$. This maximum exists only if $d\leqslant\sqrt{2}/2=0{\cdot}707$; for $d>0{\cdot}707$ there is no maximum amplitude, the amplitude x_0 during vibrations is always less than x_S in this case, as may be seen in Fig. 10.6.

The velocity is $\dot{x}=x_0\omega \cos(\omega t-\varphi)$, and $\dot{x}_{max}=x_0\omega$; the frequency for *maximum velocity* is then found for

$$d\dot{x}_{max}/d\omega=x_0+\omega\, dx_0/d\omega=0$$

which gives the result $p_3=\omega_0$.

The acceleration is $\ddot{x}=-x_0\omega^2 \sin(\omega t-\varphi)$, with $\ddot{x}_{max}=x_0\omega^2$. The frequency for *maximum acceleration* is found from

$$d\ddot{x}_{max}/d\omega = 2x_0\omega+\omega^2\, dx_0/d\omega = 0$$

The result is $p_4 = \omega_0/\sqrt{(1-2d^2)}$. There is a maximum acceleration only for $d<\sqrt{2}/2=0{\cdot}707$.

For a damping of 10% of the critical damping, we have $d=0{\cdot}10$. The five circular frequencies are in this case the undamped natural frequency ω_0, the damped natural frequency $p_1=\omega_0\sqrt{(1-d^2)}=0{\cdot}995\omega_0$, the frequency for maximum displacement $p_2=\omega_0\sqrt{(1-2d^2)}=0{\cdot}990\omega_0$, the frequency for maximum velocity $p_3=\omega_0$, and the frequency for maximum acceleration $p_1=1{\cdot}010\omega_0$.

The highest value of these frequencies is about 2% higher than the lowest. For small values of the damping coefficient c all these frequencies are evidently very close together.

The phase angle was determined by the expression $\tan \varphi = c\omega/(k - m\omega^2)$; introducing $\omega_0^2 = k/m$, $d = c/c_C$ and $c_C = 2m\omega_0$, this may be given the dimensionless form:

$$\tan \varphi = \frac{2d\omega/\omega_0}{1-(\omega/\omega_0)^2} \tag{10.9}$$

Figure 10.7 shows the phase angle φ as a function of ω/ω_0 for various values of d. For $d=0$, the phase angle is $\varphi = 0$ for $(\omega/\omega_0) < 1$ and $\varphi = 180°$ for $\omega/\omega_0 > 1$. There is a discontinuous jump at $\omega/\omega_0 = 1$. For all values of d the phase angle is $\varphi = 90°$ at the condition $\omega/\omega_0 = 1$, which is the resonance condition for *small* damping.

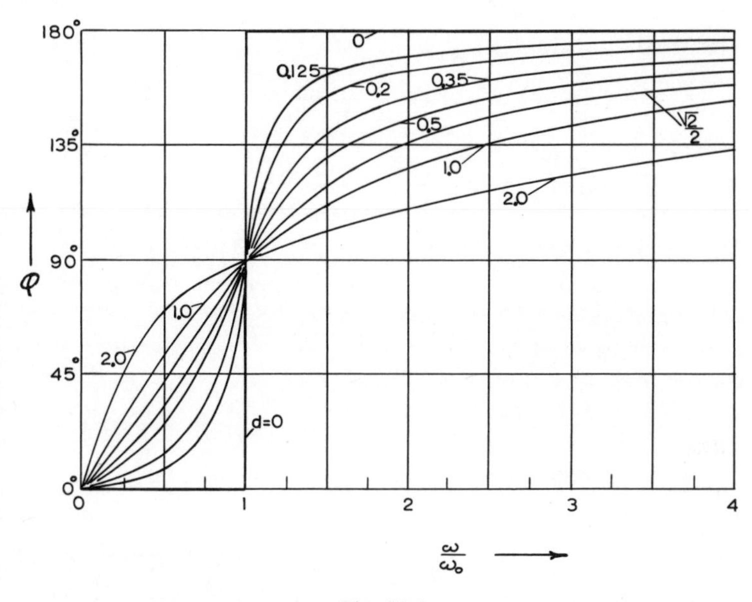

Fig. 10.7

Example 10.6. The system in Fig. 10.1 has the system constants $k = 43{\cdot}8$ N/cm, $m = 18{\cdot}2$ kg, $c = 1{\cdot}49$ Ns/cm, $P_0 = 44{\cdot}5$ N and $\omega = 15$ rad/s. Determine the steady-state motion of the mass.

Solution

$$x_S = \frac{P_0}{k} = \frac{44{\cdot}5}{43{\cdot}8} = 1{\cdot}016 \text{ cm} \qquad \omega_0 = \sqrt{\frac{4380}{18{\cdot}2}} = 15{\cdot}5 \text{ rad/s}$$

$$c_C = 2m\omega_0 = 2\times 18{\cdot}2\times 15{\cdot}5 = 564 \text{ N s/m} = 5{\cdot}64 \text{ N s/cm}$$

$$d = \frac{c}{c_C} = \frac{1{\cdot}49}{5{\cdot}64} = 0{\cdot}265 \cdot (\omega/\omega_0) = \frac{15}{15{\cdot}5} = 0{\cdot}967$$

$$1-(\omega/\omega_0)^2 = 0{\cdot}065\times 2d(\omega/\omega_0) = 0{\cdot}513$$

From (10.8)

$$x_0 = \frac{1{\cdot}016}{\sqrt{(0{\cdot}065^2+0{\cdot}513^2)}} = 1{\cdot}835 \text{ cm}$$

From (10.9)

$$\tan\varphi = \frac{0{\cdot}513}{0{\cdot}065} = 7{\cdot}90 \qquad \varphi = 82{\cdot}7° = 1{\cdot}45 \text{ rad}$$

The steady state motion is

$$x = x_0 \sin(\omega t-\varphi) = 1{\cdot}835 \sin(15t-1{\cdot}45) \text{ cm}$$

10.5 Vibration isolation

The forces transmitted to the ground by a vibrating system are transmitted through the spring and damper, since these are the only connections to the ground.

The steady state motion is $x=x_0 \sin(\omega t-\varphi)$, with $\dot{x}=x_0\omega \cos(\omega t-\varphi)$. The transmitted force P_T is

$$P_T = kx+c\dot{x} = kx_0 \sin(\omega t-\varphi)+cx_0\omega\cos(\omega t-\varphi)$$

The maximum value P_{Tm} of the transmitted force is

$$P_{Tm} = \sqrt{(k^2x_0^2+c^2x_0^2\omega^2)} = x_0\sqrt{(k^2+c^2\omega^2)}$$

$$= kx_0\sqrt{\left[1+\left(\frac{c\omega}{k}\right)^2\right]}$$

which is superimposed on the weight force. Substituting $c=dc_C=2m\omega_0 d$ gives

$$P_{Tm} = kx_0\sqrt{\left[1+\left(\frac{2m\omega_0\omega d}{k}\right)^2\right]} = kx_0\sqrt{\left[1+\left(2d\frac{\omega}{\omega_0}\right)^2\right]}$$

If the springs were completely rigid, the maximum transmitted force would be P_0, so the ratio P_{Tm}/P_0 is a measure of the quality of the vibration isolation of the system. This dimensionless ratio is called the *transmissibility* of the system and is denoted t_r.

Introducing eq. (10.8) for x_0, we have

$$t_r = \frac{P_{Tm}}{P_0} = \sqrt{\left(\frac{1+(2d\omega/\omega_0)^2}{[1-(\omega/\omega_0)^2]^2+(2d\omega/\omega_0)^2}\right)} \qquad (10.10)$$

In an ideal case t_r should be zero; in practical cases it is made as small as possible.

The transmissibility t_r is plotted in Fig. 10.8 as a function of ω/ω_0 for various values of the damping ratio d. The effect of damping in lowering the transmitted force for $\omega/\omega_0 < \sqrt{2}$ is clear from the figure; for $\omega/\omega_0 = \sqrt{2}$ the maximum force is transmitted in full.

The mounting springs should be designed with as small a value of k as possible so that $\omega_0 = \sqrt{k/m}$ is as small as possible, and ω/ω_0 large, in order to minimize the transmitted force.

Actually damping is detrimental for $\omega/\omega_0 > \sqrt{2}$ as may be seen from Fig. 10.8, since a greater damping ratio gives a greater transmitted force in this region; however some damping is necessary to avoid dangerous force amplitudes at the approximate resonance condition $\omega/\omega_0 = 1$, so that a compromise is necessary.

Example 10.7. A refrigerator unit weighs 290 N and is to be supported on three springs of spring constant k each.

Given that the unit operates at 580 rev/min, determine the value of the spring constant k if only 10% of the shaking force of the unit is to be transmitted to the support. Determine also the static deflection due to the weight of the unit when the unit is placed on the springs. Damping may be neglected.

Solution. From eq. (10.10) we find, for $d=0$,

$$t_r = \frac{1}{-[1-(\omega/\omega_0)^2]} = 0{\cdot}10$$

from which $(\omega/\omega_0)^2 = 11$, and $\omega/\omega_0 = 3{\cdot}32$. Hence $\omega = \pi \times 580/30 = 60{\cdot}7$ rad/s, so that $\omega_0 = \omega/3{\cdot}32 = 18{\cdot}3$ rad/s. Thus $K = m\omega_0^2 = (290/9{\cdot}81) \times 18{\cdot}3^2 = 9880$ N/m $= 98{\cdot}8$ N/cm. Since $K = 3k$, we have $k = 98{\cdot}8/3 = 32{\cdot}9$ N/cm. The static deflection is

$$\Delta_S = W/K = 290/98{\cdot}8 = 2{\cdot}93 \text{ cm}$$

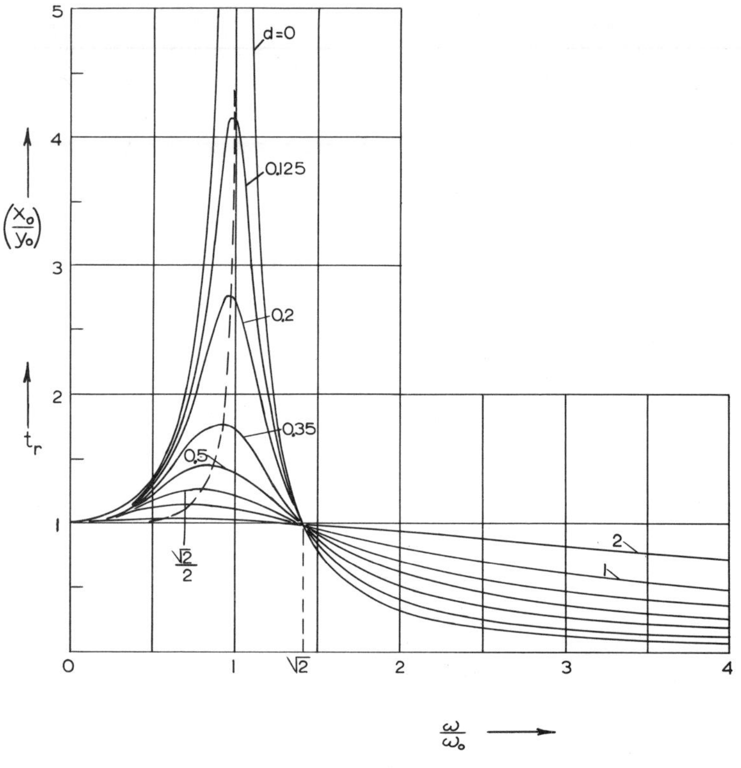

Fig. 10.8

10.6 Vibrating support

In many practical cases instruments have to be mounted on supports that are vibrating, as in the case of an aircraft instrument panel. The vibrations may cause malfunction of the instruments, and it becomes necessary to isolate the instruments as far as possible from the supports.

To investigate this situation, consider the system shown in Fig. 10.9. The displacement of the vibrating support is given by the function $y = y_0 \sin \omega t$, and the velocity is $\dot{y} = y_0 \omega \cos \omega t$. The absolute motion of the body of mass m is given by the coordinate x as shown. The velocity of the damper piston relative to the damper cylinder is $\dot{x} - \dot{y}$, while the extension of the spring is $x - y$.

The equation of motion of the body of mass m is thus

$$m\ddot{x}+c(\dot{x}-\dot{y})+k(x-y) = 0$$

Substituting the expressions for y and $\dot{y}$, we obtain

$$\begin{aligned} m\ddot{x}+c\dot{x}+kx &= cy_0\omega \cos \omega t+ky \sin \omega t \\ &= \sqrt{(c^2y_0^2\omega^2+k^2y_0^2)} \sin (\omega t-\beta) \\ &= y_0\sqrt{(k^2+c^2\omega^2)} \sin (\omega t-\beta) = P_0 \sin (\omega t-\beta) \end{aligned}$$

where $P_0=y_0\sqrt{(k^2+c^2\omega^2)}$.

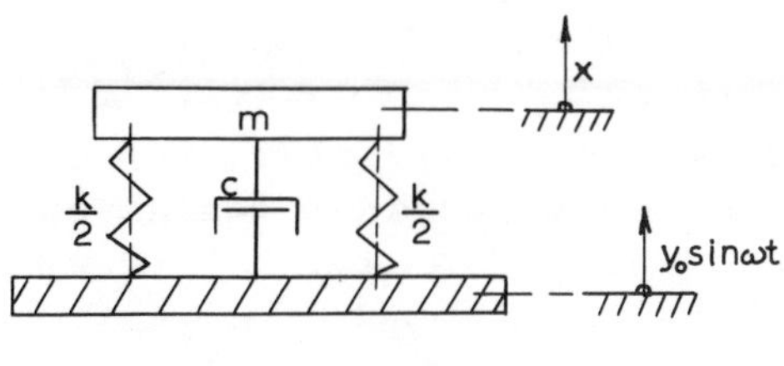

Fig. 10.9

For a given frequency ω, the vibrating support then acts as a sinusoidal exciting force of frequency ω and force amplitude P_0. The phase angle β is of no particular interest in this case.

The equation of motion is of the same type as eq. (10.1), and the same solution therefore applies. We find

$$\begin{aligned} P_0 &= y_0\sqrt{(k^2+c^2\omega^2)} = y_0k\sqrt{\left[1+\left(\frac{c\omega}{k}\right)^2\right]} \\ &= y_0k\sqrt{\left[1+\left(\frac{2m\omega_0\omega d}{k}\right)^2\right]} = y_0k\sqrt{\left[1+\left(\frac{2d\omega}{\omega_0}\right)^2\right]} \end{aligned}$$

From the expression (10.8) for the steady-state amplitude x_0, we find the amplitude ratio:

$$\left(\frac{x_0}{y_0}\right) = \sqrt{\left(\frac{1+(2d\omega/\omega_0)^2}{[1-(\omega/\omega_0)^2]^2+(2d\omega/\omega_0)^2}\right)} \qquad (10.11)$$

This is exactly the same as expression (10.10) for the transmissibility t_r, and the curves in Fig. 10.8 therefore also show the amplitude ratio x_0/y_0, in the present case, as a function of ω/ω_0 for various values of the damping ratio d.

The ratio x_0/y_0 shows the effectiveness of the vibration mounting in reducing the amplitude of the body of mass m. For any value of

$\omega/\omega_0 > \sqrt{2}$, the amplitude of the body will be less than that of the vibrating support.

Damping is again detrimental in the region $\omega/\omega_0 > \sqrt{2}$, but a compromise must be made, since damping is essential to prevent large amplitudes in passing through the approximate resonance condition $\omega/\omega_0 = 1$.

Example 10.8. An aircraft instrument board, including the instruments, weighs 178 N, and is supported on four rubber mounts with spring constant 224 N/cm each. The total damping ratio is $d = 0{\cdot}10$. The rubber supports are mounted on a vibrating surface which has a frequency equal to the engine revolutions.

Determine the percentage of the vibrating motion of the supporting surface that is transmitted to the instrument panel at an engine revolution of (a) 2200 rev/min, (b) 1500 rev/min. (c) Determine the percentage transmissibility for (a) and (b) if $d = 0$.

Solution. The percentage of the motion transmitted is, from eq. (10.11),

$$\frac{x_0}{y_0} \times 100\% = 100\sqrt{\left(\frac{1+(2d\omega/\omega_0)^2}{[1-(\omega/\omega_0)^2]^2+(2d\omega/\omega_0)^2}\right)}$$

(a) $d = 0{\cdot}10$, $\omega = (\pi/30) \times 2200 = 230$ rad/s.

$$\omega_0 = \sqrt{(K/m)},\ K = 4k = 4 \times 224 = 896 \text{ N/cm} = 89\,600 \text{ N/m}.$$

$$m = 178/9{\cdot}81 \text{ kg},\ \omega_0 = \sqrt{(89600 \times 9{\cdot}81/178)} = 70{\cdot}2 \text{ rad/s}.$$

$$\omega/\omega_0 = 230/70{\cdot}2 = 3{\cdot}28;\ 1-(\omega/\omega_0)^2 = -9{\cdot}74;$$

$$2d\omega/\omega_0 = 0{\cdot}656.$$

$$(x_0/y_0) \times 100\% = 100\sqrt{[1{\cdot}430/(9{\cdot}74^2+0{\cdot}430)]} = 12{\cdot}3\%$$

(b) $\omega = \pi/30 \times 1500 = 157$ rad/s; $\omega_0 = 70{\cdot}2$ rad/s.

$$\omega/\omega_0 = 2{\cdot}24,\ 2d\omega/\omega_0 = 0{\cdot}448.$$

$$(x_0/y_0) \times 100\% = 100\sqrt{[1{\cdot}20/(4{\cdot}01^2+0{\cdot}20)]} = 27{\cdot}3\%$$

(c) $d = 0$; for part (a),

$$(x_0/y_0) \times 100\% = 100/|1-(\omega/\omega_0)^2| = 100/9{\cdot}74 = 10{\cdot}3\%$$

For question (b):

$$(x_0/y_0)\times 100\% = 100/4{\cdot}01 = 24{\cdot}9\%.$$

In both cases the transmitted motion is somewhat smaller for $d=0$.

10.7 Forced vibrations due to a rotating unbalance

Figure 10.10 shows a body of total mass m which is supported on springs with total spring constant k. The motion is damped by a dashpot of damping coefficient c. The body is guided to move in the vertical direction only. An unbalance of magnitude $m'r$ is rotating at a constant angular velocity ω as shown.

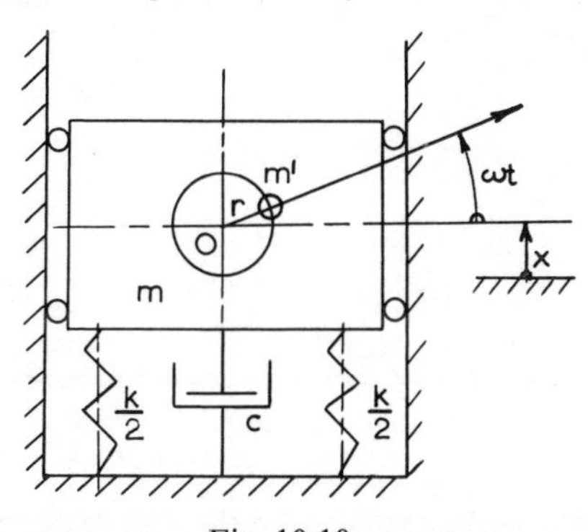

Fig. 10.10

Assuming that the vertical motion is small, the changes in the unbalanced force due to the motion of the centre O may be ignored. The force on the unbalance is $m'r\omega^2$ directed towards O and it acts therefore as a rotating exciting force of constant amplitude $m'r\omega^2$ on the body as shown. Resolving the force in a vertical and horizontal component, the vertical exciting force is $m'r\omega^2 \sin \omega t$.

The equation of motion of the body is thus

$$m\ddot{x}+c\dot{x}+kx = (m'r\omega^2)\sin \omega t$$

This is of the same form as eq. (10.1) if we take $P_0=m'r\omega^2$, and the solutions to eq. (10.1) may be applied directly to the case of a rotating unbalance.

Considering the steady-state vibrations only, the solution is

$$x = x_0 \sin(\omega t - \varphi)$$

with

$$x_0 = \frac{P_0}{\sqrt{[(k - m\omega^2)^2 + c^2\omega^2]}}$$

as found before.

Substituting $P_0 = m'r^2$, $\omega_0^2 = k/m$, $d = c/c_C$ and $c_C = 2m\omega_0$, we obtain the solution in dimensionless form:

$$\left(\frac{mx_0}{m'r}\right) = \frac{(\omega/\omega_0)^2}{\sqrt{[1 - (\omega/\omega_0)^2]^2 + (2d\omega/\omega_0)^2}} \tag{10.12}$$

Curves for $(mx_0/m'r)$ as a function of (ω/ω_0), for various values of the damping ratio d, are plotted in Fig. 10.11.

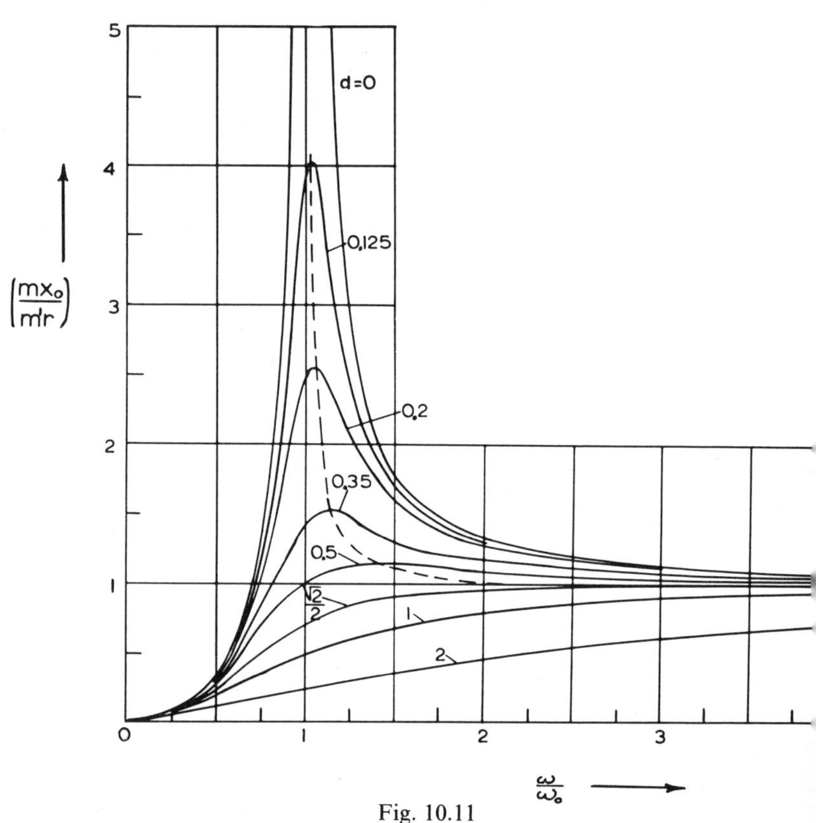

Fig. 10.11

For small damping a resonance condition with maximum amplitude of the body is found approximately at $\omega/\omega_0 = 1$. The effect of damping in lowering the maximum amplitude when the system goes through resonance is obvious from the curves. For larger values of ω/ω_0, say $\omega/\omega_0 > 5$, the effect of damping is negligible; however, damping is essential for passing through resonance safely.

The rotational angular velocity ω to give maximum amplitude of vibration is found at $\omega = \omega_0/\sqrt{(1-2d^2)}$; a maximum amplitude is found only for $d < \sqrt{2}/2 = 0{\cdot}707$.

The expression for the phase angle φ is the same as for forced vibration:

$$\tan\varphi = \frac{c\omega}{k - m\omega^2} = \frac{2d\omega/\omega_0}{1-(\omega/\omega_0)^2}$$

This relationship is plotted in Fig. 10.7.

For a rotating unbalance system, the transmissibility

$$t_r = \frac{P_{\mathrm{Tm}}}{P_0} = \frac{kx_{\mathrm{C}}\sqrt{[1+(2d\omega/\omega_0)^2]}}{m'r\omega^2}$$

Substituting x_0 from eq. (10.12) gives the result:

$$t_r = \sqrt{\left(\frac{1+(2d\omega/\omega_0)^2}{[1-(\omega/\omega_0)^2]^2+(2d\omega/\omega_0)^2}\right)}$$

This expression is exactly the same as (10.10), and the relationship is plotted in Fig. 10.8.

Example 10.9. Four equal vibration mounts are used to support a machine which has a total weight of 1985 N. The machine has a rotating unbalance of 3·46 kg cm, and runs at 600 rev/min. The total damping is 30% of the critical damping.

Given that 20% of the exciting force is transmitted to the foundation, determine (a) the spring constant of each mount, (b) the deflection of the springs when the machine is placed on them statically, (c) the maximum transmitted vibratory force, (d) the maximum amplitude of vibration.

The problem may be treated as a system with one degree of freedom, and only vertical translational motion considered.

Solution. (a) The transmissibility is

$$t_r = \sqrt{\left(\frac{1+(2d\omega/\omega_0)^2}{[1-(\omega/\omega_0)^2]^2+(2d\omega/\omega_0)^2}\right)} = 0{\cdot}20$$

Introducing $x=(\omega/\omega_0)^2$ and $d=c/c_C=0{\cdot}30$, we obtain

$$\frac{1+0{\cdot}36\,x}{1-2x+x^2+0{\cdot}36x} = 0{\cdot}04 \qquad \text{or} \qquad x^2-10{\cdot}65x-24 = 0$$

Hence

$$x = 5{\cdot}32+\sqrt{(28{\cdot}3+24)} = 12{\cdot}56$$
$$\omega = (\pi/30)\times 600 = 62{\cdot}8 \text{ rad/s} \qquad \omega_0^2 = \omega^2/12{\cdot}56 = 314$$
$$K = m\omega^2 = 1985/9{\cdot}81\times 314 = 63\ 500 \text{ N/m} = 635 \text{ N/cm}$$

For each mount $k=K/4=158{\cdot}6$ N/cm.

(b) The static deflection $\Delta_S=W/K=1985/635=3{\cdot}12$ cm.

(c) The maximum transmitted vibratory force is found from $t_r=P_{Tm}/P_0=0{\cdot}20$.

$$P_0 = m'r\omega^2 = \frac{3{\cdot}46\times 1}{100}\times 62{\cdot}8^2 = 136{\cdot}7 \text{ N}$$

$P_{Tm}=0{\cdot}20\ P_0=27{\cdot}3$ N, which is superimposed on the weight force 1985 N.

(d) The maximum vibration amplitude x_0 is determined from $P_{Tm}=Kx_0\sqrt{[1+(2d\omega/\omega_0)^2]}$.

$$(2d\omega/\omega_0)^2 = (2\times 0{\cdot}30\times 62{\cdot}8/17{\cdot}7)^2 = 4{\cdot}52$$
$$x_0 = 27{\cdot}3/635\sqrt{5{\cdot}52} = 0{\cdot}018 \text{ cm}$$

10.8 Electrical analogy

Figure 10.12 shows an electric circuit with an alternating-current generator in series with a condenser C, an inductance L and a resistance R. The differential equation for the charge Q on the condenser is

$$L\ddot{Q}+R\dot{Q}+(1/C)Q = E_0 \sin \omega t$$

comparing this equation with eq. (10.1),

$$m\ddot{x}+c\dot{x}+kx = P_0 \sin \omega t$$

we see that the two equations are mathematically of the same form.

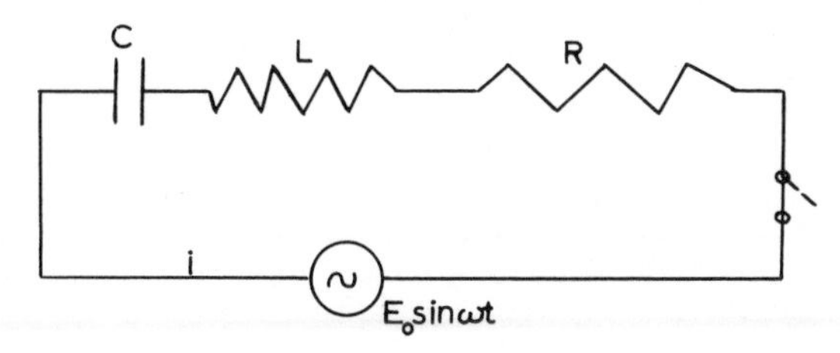

Fig. 10.12

All the solutions and results from the mechanical system equation may thus be applied directly to the electrical system, which is an electrical analogue of the mechanical system. The translation from the electrical system to the mechanical system may be obtained from Table 10.1.

Table 10.1

Linear mechanical system	Electrical system
Mass m	Inductance L
Stiffness k	$\dfrac{1}{\text{capacitance}} = \dfrac{1}{C}$
Damping c	Resistance R
Impressed force $P_0 \sin \omega t$	Impressed voltage $E_0 \sin \omega t$
Displacement x	Condenser charge Q
Velocity $\dot{x}=v$	Current $\dot{Q}=i$
Acceleration $\ddot{x}=\dot{v}$	$\ddot{Q}=di/dt$

The behaviour of the mechanical system may then be studied by taking the necessary electrical measurements from the electrical analogue. Effects of changes in the system may be studied much more simply and cheaply in this way. Electrical analogue computers work on these principles.

Problems

10.1 A spherical ball of radius R floats half submerged in water. Supposing that the ball is depressed slightly and released, write the differential equation of motion and determine the period of vibration. If $R = 0{\cdot}61$ m, what is the numerical value of the period?

10.2 Determine the equation for the period of small oscillations of the system shown in Fig. 10.13.

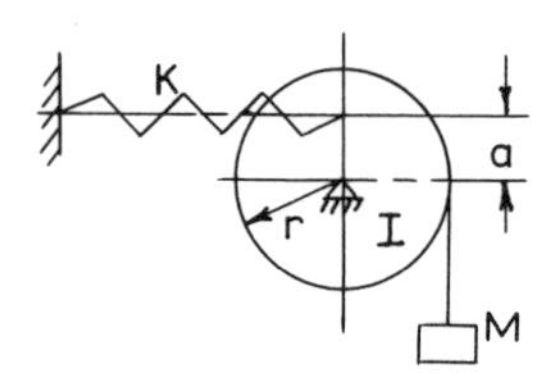

Fig. 10.13

10.3 Find the frequency of small vibrations of the inverted pendulum in Fig. 10.14. Assume that K is sufficiently large, so that the pendulum is stable.

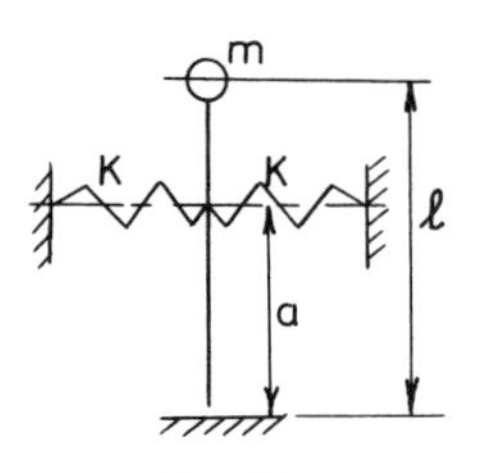

Fig. 10.14

10.4 Determine the damped natural frequency and the critical damping coefficient for the system shown in Fig. 10.15.

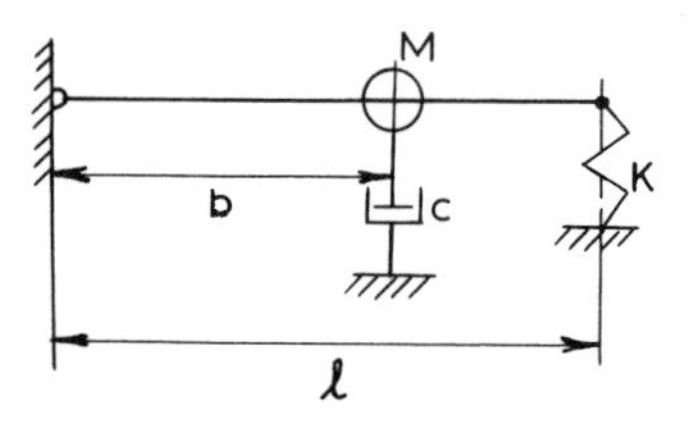

Fig. 10.15

10.5 An undamped spring–mass system, which under gravity has a static deflection of 2·54 cm, is acted upon by a sinusoidal exciting force which has a frequency of 4 cycles/s. Find the damping ratio if the amplitude of the steady-state forced vibrations is reduced to half of the amplitude of the undamped forced vibrations.

10.6 A mass of 0·908 kg is attached to the end of a spring with a stiffness of 7 N/cm. Determine the critical damping coefficient.

Given that the damping coefficient $c = 0{\cdot}252$ Ns/cm, determine the damped natural frequency and compare it with the natural frequency of the undamped system.

10.7 An instrument of mass 9·07 kg is to be spring mounted on a vibrating surface which has a sinusoidal motion of amplitude 0·397 mm and frequency 60 cycles/s. If the instrument is mounted rigidly on the surface, what is the maximum force to which it is subjected? Find the spring constant for the support which will limit the maximum acceleration of the instrument to $g/2$. Assume that negligible damping forces have caused the transient vibrations to die out.

10.8 A machine having a total weight of 89 000 N has an unbalance such that it is subjected to a force of amplitude 22 200 N at a frequency of 600 cycles/min. Find the spring constant for the supporting springs if the maximum force transmitted to the foundation is 2220 N. Assume that damping may be neglected.

CHAPTER ELEVEN

Some Topics in the Theory of Machines

11.1 Critical speeds of rotating shafts

Figure 11.1(a) shows a thin shaft in two bearings. A disc of mass M and centre of mass C is attached to the shaft. In any practical case, it is impossible to locate C exactly on the centreline of the shaft, and the disc is, therefore, shown with a small eccentricity e. Because of this eccentricity, the centre of mass is in circular motion when the shaft is rotating, and a force must act on the disc towards the centre of rotation O; the disc then acts on the shaft with an equal and opposite force—the so called centrifugal force, directed radially outwards. This force bends the shaft and it rotates in a bent condition. We will here, for simplicity, neglect the mass of the shaft. The gravity force on the disc introduces some secondary effects which we shall also ignore.

The rotation of the shaft in a bent position is called whirling. To clarify the position, Fig. 11.1(b) shows a cross-section of the shaft during one rotation. The point A on the shaft is always on the inside and the fibres at A are always in compression.

Figure 11.1(c) shows the disc in dynamic equilibrium for a constant angular velocity ω of the shaft. The centre C moves in a circle with radius $x+e$, and the inertia force, as shown, is $M(x+e)\omega^2$. The inertia force is balanced by the elastic force P in the shaft. If the spring constant of the shaft is K for the given position of the disc, $P=Kx$. For a simply supported uniform shaft with the disc at the middle point, $K=48\ EI/l^3$, where EI is the flexural rigidity of the shaft; for other positions of the disc, K has different values.

From dynamic equilibrium we have $M(x+e)\omega^2=Kx$, from

which

$$x = \frac{Me\omega^2}{K - M\omega^2} = \frac{e}{K/M\omega^2 - 1}$$

The natural circular frequency for vibrations of the disc on the shaft as a spring is $\omega_0^2 = K/M$; substituting this leads to

$$x = \frac{e}{(\omega_0/\omega)^2 - 1} \qquad (11.1)$$

We see from (11.1), that as $\omega \rightarrow \omega_0$, but $\omega < \omega_0$, x becomes very large. In practical cases, damping and bearing stiffness will limit x, but in general we shall get large deflections of the shaft whirling in a bent condition as the angular velocity approaches ω_0.

The value of the angular velocity $\omega_c = \omega_0 = \sqrt{(K/M)}$ is called the *critical* angular velocity of the shaft. The critical speed of the shaft is then $N_c = (30/\pi)\omega_0$ rev/min, if ω_0 is in rad/s.

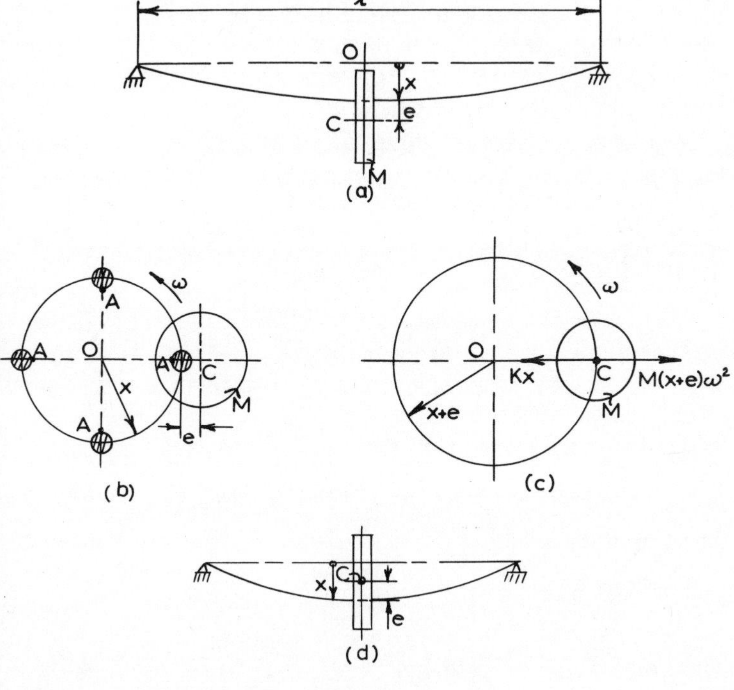

Fig. 11.1

The fact that the frequency of the so-called 'standing' vibrations of the shaft is the same as the critical speed is of great practical importance. The same relationship may be shown to exist for shafts with several discs and distributed mass.

When the angular velocity is increased past the critical value, the shaft will again run quietly. The centre of mass is now on the inside as shown in Fig. 11.1(d). This is known from experiments; no satisfactory explanation seems to exist for this.

A force balance gives in this case $M(x-e)\omega^2 = Kx$, from which

$$x = \frac{e}{1-K/M\omega^2} = \frac{e}{1-(\omega_0/\omega)^2}$$

with $\omega > \omega_0$. In this case, as ω increases ($\omega > \omega_0$), x decreases, and at high speeds $x \simeq e$. The centre of mass then approaches the centre-line joining the supports, and the deflected shaft rotates about the centre of mass.

Example 11.1. The rotor of a machine weighs 222 N and is keyed to the centre of a steel shaft located in two bearings with 50·8 cm between bearing centres. The diameter of the shaft is 5·1 cm and the modulus is $E = 21 \times 10^7$ kN/m². The rotor is rotating at 4000 rev/min and the eccentricity is $e = 0·0508$ mm.

Neglecting the mass of the shaft, determine the critical speed, the amplitude of vibration and the dynamic bearing force.

Solution. $I = (\pi/64)d^4 = 33·2/10^8 m^4$. From strength of materials, the spring constant is

$$K = \frac{48\,EI}{l^3} = \frac{48 \times (21 \times 10^{10}) \times 33·2}{0·508^3 \times 10^8} = 25·6 \times 10^6 \text{ N/m}$$

$$\omega_0 = \sqrt{K/M)} = \sqrt{(25·6 \times 10^6 \times 9·81/222)} = 1062 \text{ rad/s}$$

The critical speed is $N_c = (30/\pi)\omega_0 = 10\ 150$ rev/min.

$$\omega = (2\pi/60) \times 4000 = 419 \text{ rad/s} \qquad (\omega_0/\omega)^2 = 6·43$$

The amplitude is

$$x = \frac{e}{(\omega_0/\omega)^2 - 1} = \frac{0·0508}{5·43} = 0·0094 \text{ mm}$$

The total force is $M(x+e)\omega^2 = Kx = 25·6 \times 10^6 \times 0·0094/10^3 =$ 240 N. The dynamic bearing force $= 240/2 = 120$ N.

11.2 Flexural vibrations of beams

11.2.1 Rayleigh's method

Figure 11.2 shows a mass M on a simply supported shaft for which we will neglect the mass. The static deflection under the weight force of the mass is Δ and the spring constant of the shaft is K, so that $\Delta = Mg/K$.

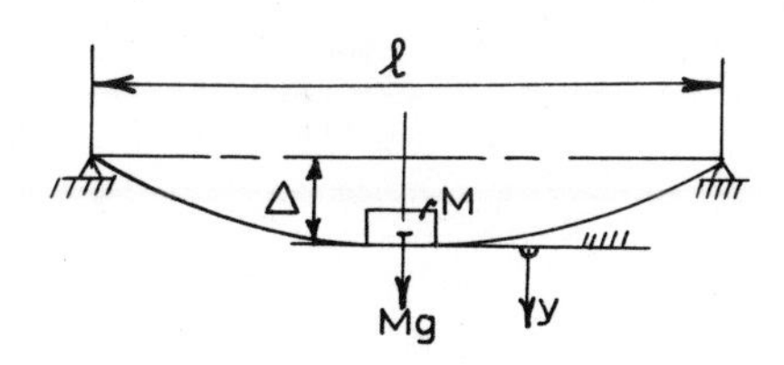

Fig. 11.2

Considering the system as a conservative system, we take the datum position for the potential energy V at the static equilibrium position, and measure the deflection of the mass from this position by the coordinate y.

The potential energy is now $V = (\frac{1}{2}Ky^2 + Mgy) - Mgy = \frac{1}{2}Ky^2$. The kinetic energy is $T = \frac{1}{2}M\dot{y}^2$. Applying the principle of conservation of mechanical energy, we have $\frac{1}{2}M\dot{y}^2 + \frac{1}{2}Ky^2 = \text{constant}$. Differentiating leads to $M\dot{y}\ddot{y} + Ky\dot{y} = 0$, or $\ddot{y} + (K/M)y = 0$ as equation of motion for the mass. This is simple harmonic motion with frequency $\omega_0^2 = K/M$. The frequency of vibration is then

$$f = \omega_0/2\pi = (1/2\pi)\sqrt{(K/M)} = (1/2\pi)\sqrt{(g/\Delta)}$$

If the mass is at the middle point of the beam, $K = 48EI/l^3$; otherwise K may be determined from the principles of strength of materials.

This energy method can be extended to include the distributed mass of the shaft. This general energy method is called Rayleigh's method after Lord Rayleigh who first developed the method in his book *Theory of Sound* (1877). Rayleigh's method consists essentially in assuming a reasonable deflection curve during vibrations. The kinetic and potential energies of the system are worked out and the sum is taken as constant, since the system is assumed to be conservative.

To illustrate Rayleigh's method, consider the beam in Fig. 11.3. The beam is vibrating in the vertical plane and the assumed deflection curve during vibrations is the curve $y=f(x)$. The mass of the beam for a unit length is μ, each element dx of the beam has the mass $\mu\, dx$ and is assumed to be in simple harmonic motion given by $z=y \sin \omega_0 t$. The velocity of the element is $\dot{z}=y\omega_0 \cos \omega_0 t$. When $t=0$ the element is passing the neutral position and the velocity is a maximum $y\omega_0$. The total kinetic energy of the beam as it passes the neutral position is then

$$T_{\max} = \int_0^l \tfrac{1}{2}(\mu\, dx)y^2\omega_0^2 = \tfrac{1}{2}\mu\omega_0^2 \int_0^l y^2\, dx$$

In this position there is no potential energy.

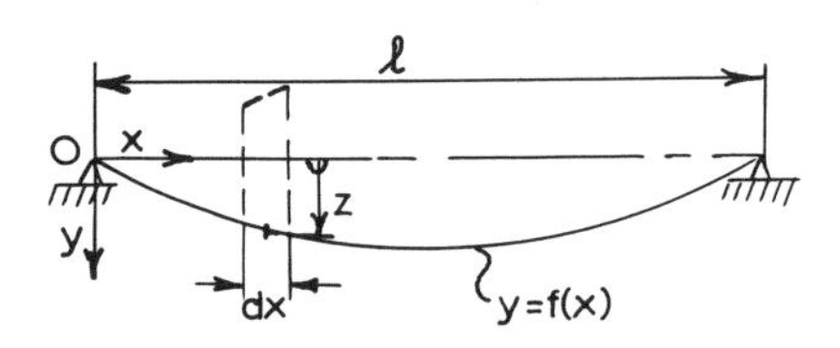

Fig. 11.3

At the position of maximum deflection, there is no kinetic energy, but the potential energy due to bending only is

$$V_{\max} = \int_0^l \frac{M_x^2\, dx}{2EI}$$

this is a formula from the theory of strength of materials. Introducing $M_x=EI\, d^2y/dx^2$, we find

$$V_{\max} = \frac{EI}{2}\int_0^l \left(\frac{d^2y}{dx^2}\right)^2 dx = \frac{EI}{2}\int_0^l (y'')^2\, dx$$

we have now

$$T_{\max}+0 = 0+V_{\max} \qquad \text{or} \qquad \tfrac{1}{2}\mu\omega_0^2 \int_0^l y^2\, dx = \frac{EI}{2}\int_0^l (y'')^2\, dx$$

from which

$$\omega_0^2 = \frac{EI}{\mu}\frac{\int_0^l (y'')^2\, dx}{\int_0^l y^2\, dx} \tag{11.2}$$

Example 11.2. For the uniform simply supported beam in Fig. 11.3, the deflection curve during vibrations is assumed to be $y = y_0 \sin(\pi x/l)$. Determine the fundamental frequency of vibration by Rayleigh's method.

Solution. We have $y'' = -y_0(\pi^2/l^2)\sin(\pi x/l)$, so that the integrals cancel in (11.2) in this particular case. Substituting in (11.2) gives the result

$$\omega_0^2 = \frac{EI}{\mu}\frac{\pi^4}{l^4}$$

and the fundamental frequency is

$$f = (\pi/2l^2)\sqrt{(EI/\mu)}$$

The value of the frequency determined in Example 11.2 is the same as the known exact value. The reason for this is that we have used the known exact curve for deflections during vibrations as our assumed curve.

In general Rayleigh's method gives too high a value for the frequency, usually 2–10% too high. The reason is that assuming a non-exact curve means introducing constraints to force the beam to vibrate in the assumed shape, and these additional constraints can only increase the frequency. If several curves are assumed, the one giving the lowest value for ω_0 is the best choice.

Rayleigh's method works in general only for the fundamental mode of vibration. The deflection curve for a higher mode cannot be determined with sufficient accuracy, and the higher frequencies are critically dependent on the assumed curve, which is not the case for the fundamental frequency.

Example 11.3. If the beam in Example 11.2 has a body of mass M attached at mid-span, determine the fundamental frequency by Rayleigh's method. Use the same deflection curve as in Example 11.2.

Solution. The kinetic energy of the beam alone is

$$T_B = \tfrac{1}{2}\mu\omega_0^2 \int_0^l y^2\,dx = \tfrac{1}{4}\mu l\omega_0^2 y_0^2$$

Introducing the total beam mass $m = \mu l$, we have

$$T_B = \tfrac{1}{4}m\omega_0^2 y_0^2$$

The kinetic energy of the mass is $T_M = \frac{1}{2}MV_{max}^2 = \frac{1}{2}My_0^2\omega_0^2$. The potential energy of bending is the same as before for the same deflection curve,

$$V_{max} = \tfrac{1}{2}EI \int_0^l (y'')^2\, dx = \frac{\pi^4 EI}{4l^3}\, y_0^2$$

We have then

$$\tfrac{1}{4}m\omega_0^2 y_0^2 + \tfrac{1}{2}My_0^2\omega_0^2 = \frac{\pi^4}{4}\frac{EIy_0^2}{l^3}$$

from which

$$\omega_0^2 = \frac{48{\cdot}7\, EI}{(M + \frac{1}{2}m)l^3}$$

For a light beam we have $\omega_0^2 = K/M = 48EI/Ml^3$. It may be seen that to use the simple formula $\omega_0^2 = K/M$ we get a more accurate result by adding half of the total beam mass to the mass in the centre.

Example 11.4. Figure 11.4 shows a body of mass M, supported by a uniform helical spring of total mass m and spring constant K. The frequency of vibrations is found from $\omega_0^2 = K/M$, if the mass of the spring is neglected. Determine, by Rayleigh's method, the part of the spring *mass* that must be added to the end mass to give greater accuracy when the formula $\omega_0^2 = K/M$ is used.

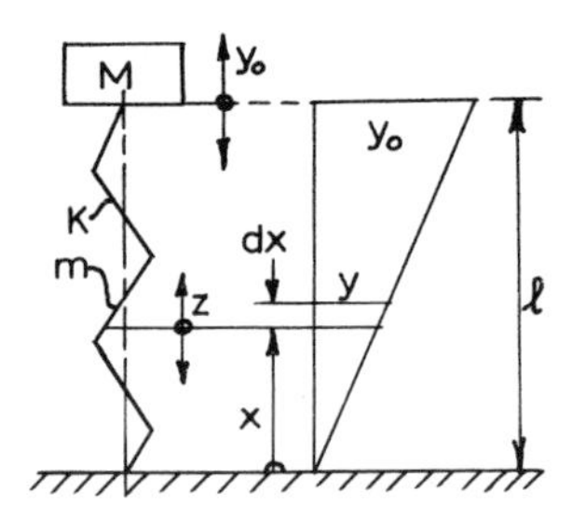

Fig. 11.4

Solution. Taking linear deflections as shown from the static equilibrium position, we take the motion of an element dx of the spring as determined by $z = y \sin \omega_0 t$. The velocity is $\dot{z} = y\omega_0 \cos \omega_0 t$, and $\dot{z}_{max} = y\omega_0 = (x/l)y_0\omega_0$. The maximum kinetic energy of the spring is

$$T_S = \int_0^l \frac{1}{2}\left(\frac{m}{l}\,dx\right)\dot{z}^2_{max} = \frac{1}{2}\frac{m}{l^3}\,y_0^2\omega_0^2 \int_0^l x^2\,dx = \tfrac{1}{6}my_0^2\omega_0^2$$

The maximum kinetic energy of the end mass is $T_M = \frac{1}{2}My_0^2\omega_0^2$. The maximum potential energy of the spring is $V = \frac{1}{2}Ky_0^2$, from which

$$\tfrac{1}{2}My_0^2\omega_0^2 + \tfrac{1}{6}my_0^2\omega_0^2 = \tfrac{1}{2}Ky_0^2 \qquad \text{or} \qquad \omega_0^2 = \frac{K}{M+m/3}$$

To get a more realistic result in calculating the fundamental frequency of the system from the formula $\omega_0^2 = K/M$, we must add $m/3$ to the end mass M, where m is the total mass of the spring.

Example 11.5. Repeat Example 11.4 for the case of a torsional pendulum as shown in Fig. 11.5. The torsional spring constant of the shaft is K.

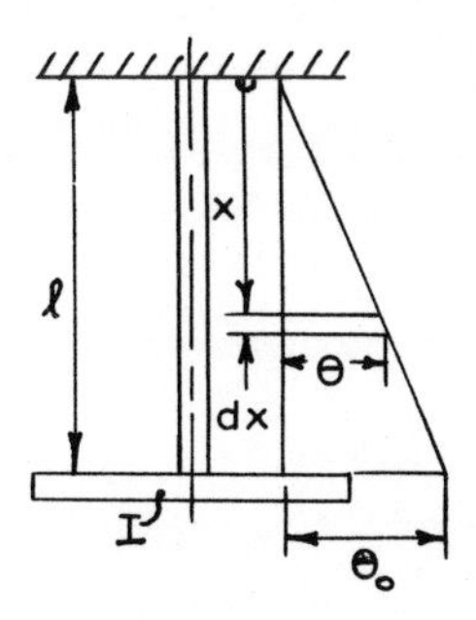

Fig. 11.5

Solution. Taking the rotational motion of an element dx of the shaft as determined by $\beta = \theta \sin \omega_0 t$, the angular velocity is $\dot{\beta} = \theta\omega_0 \cos \omega_0 t$ and $\dot{\beta}_{max} = \theta\omega_0 = (x/l)\theta_0\omega_0$.

If the total shaft inertia is I_S, the maximum kinetic energy of the shaft is

$$T_S = \int_0^l \frac{1}{2}\left(\frac{I_S}{l}\,dx\right)\dot{\beta}^2_{max} = \frac{1}{2}\frac{I_S}{l^3}\,\theta_0^2\omega_0^2 \int_0^l x^2\,dx = \tfrac{1}{6}I_S\theta_0^2\omega_0^2$$

The maximum kinetic energy of the end inertia is $\frac{1}{2}I\theta_0^2\omega_0^2$, and the maximum potential energy of twist of the shaft is $V_{\max}=\frac{1}{2}K\theta_0^2$. We have then

$$\tfrac{1}{2}I\theta_0^2\omega_0^2+\tfrac{1}{6}I_S\theta_0^2\omega_0^2 = \tfrac{1}{2}K\theta_0^2 \qquad \text{or} \qquad \omega_0^2 = \frac{K}{I+\frac{1}{3}I_S}$$

11.2.2 Rayleigh's method for multimass systems

Figure 11.6 shows a shaft with a series of n masses. The deflection curve is the static deflection due to the weight of the discs on the

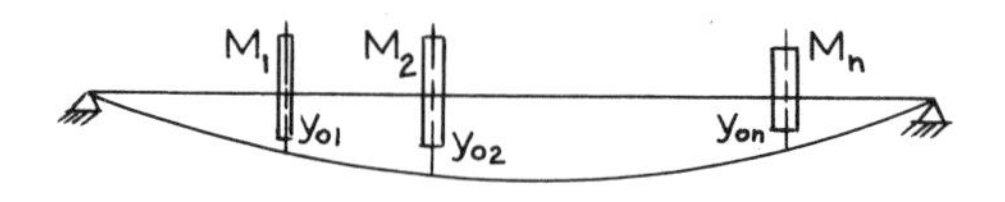

Fig. 11.6

shaft. Assuming simple harmonic motion of each mass we have $y_n=y_{0n}\sin\omega_0 t$ and velocity $\dot{y}_n=y_{0n}\omega_0\cos\omega_0 t$. When $t=0$ all the discs are passing through the neutral position with maximum velocity $\dot{y}_{n\max}=y_{0n}\omega_0$. The total kinetic energy in the neutral position is then

$$T_1 = \tfrac{1}{2}M_1(y_{01}^2\omega_0^2)+\cdots = \tfrac{1}{2}\omega_0^2\sum_1^n M_n y_{0n}^2$$

while the potential energy $V_1=0$. At the instant of maximum deflection, the kinetic energy is $T_2=0$, and the potential energy is the strain energy due to bending. This is the work done by the weight forces; since both forces and displacements vary from zero to a maximum value with a linear relationship between them, we have

$$V_2 = \tfrac{1}{2}M_1 g y_{0n}+\cdots = \tfrac{1}{2}g\sum_1^n M_n y_{0n}$$

Taking $V_1+T_1=V_2+T_2$, we find that

$$\tfrac{1}{2}\omega_0^2\sum_1^n M_n y_{0n}^2 = \tfrac{1}{2}g\sum_1^n M_n y_{0n} \qquad \text{or} \qquad \omega_0 = \sqrt{\left(\frac{g\sum_1^n M_n y_{0n}}{\sum_1^n M_n y_{0n}^2}\right)}$$

The fundamental critical speed may now be found from the expression

$$f_c = \frac{30}{\pi}\sqrt{\left(\frac{g\sum My}{\sum My^2}\right)} \tag{11.3}$$

f_c is in rev/min if M is in kg, y in m and $g = 9{\cdot}81$ m/s^2.

Example 11.6. Figure 11.7(a) shows a steel shaft in two bearings. The modulus of elasticity of the shaft is $E = 21 \times 10^7$ kN/m^2, and the cross-sectional second moment of area is $I = 5$ cm^4. The shaft carries two equal discs as shown, each of weight $W = 267$ N. Calculate the fundamental critical speed of the shaft by the formula (11.3), if $l = 1{\cdot}017$ m and $a = 0{\cdot}305$ m. The mass of the shaft may be neglected.

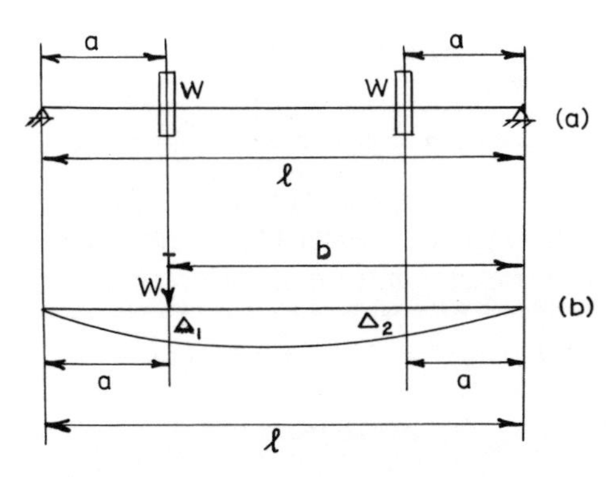

Fig. 11.7

Solution. The deflections due to the left-hand disc alone are shown in Fig. 11.7(b). Formulae from strength of materials give:

$$EI\Delta_1 = W\frac{ba}{6l}(l^2 - b^2 - a^2) = W\frac{a^2b^2}{3l}$$

$$EI\Delta_2 = \frac{W}{6}\left[\frac{b^2}{l}(l^2 - 2b^2) + (b-a)^3\right]$$

With the given figures we find $EI\Delta_1 = 4{\cdot}12$ Nm^3 and $EI\Delta_2 = 3{\cdot}43$ Nm^3.

The total static deflection of each disc in the system in Fig. 11.7(a) is now y, where $EI(\Delta_1+\Delta_2)=EIy=7{\cdot}55\ \text{Nm}^3$,

$$EI = 21\times10^{10}\times5/10^8 = 1{\cdot}05\times10^4\ \text{Nm}^2$$

Hence

$$y = \frac{7{\cdot}55}{1{\cdot}05\times10^4} = 7{\cdot}19\times10^{-4}\ \text{m}$$

Substituting in the formula (11.3) gives the result

$$f_c = \frac{30}{\pi}\sqrt{\left(\frac{g2My}{2My^2}\right)} = \frac{30}{\pi}\sqrt{\frac{g}{y}} = \frac{30}{\pi}\sqrt{\left(\frac{9{\cdot}81}{7{\cdot}19}\times10^4\right)}$$
$$= 1114\ \text{rev/min}$$

11.2.3 Determination of critical speeds by influence numbers

For the system shown in Fig. 11.8, the fundamental critical speed may be determined by Rayleigh's method; however, since the system has two degrees of freedom, neglecting the distributed mass of the shaft, it has two critical speeds. To determine both critical speeds the following method, using influence numbers, may be employed.

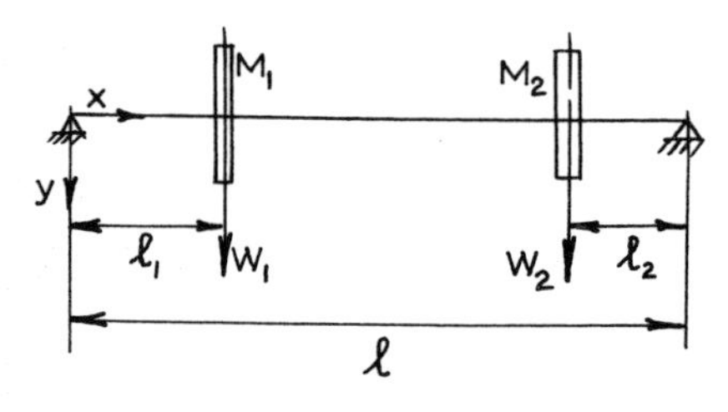

Fig. 11.8

The static deflections due to the weight forces W_1 and W_2 may be expressed by

$$\Delta_1 = a_{11}W_1+a_{12}W_2$$
$$\Delta_2 = a_{21}W_1+a_{22}W_2$$

where $a_{11}, \dots, a_{22}$ are influence numbers for deflections. Solving for W_1 and W_2 results in

$$W_1 = \frac{\Delta_1 a_{22} - \Delta_2 a_{12}}{a_{11}a_{22} - a_{21}a_{12}} \qquad \text{and} \qquad W_2 = \frac{\Delta_2 a_{11} - \Delta_1 a_{21}}{a_{11}a_{22} - a_{21}a_{12}}$$

If we introduce $c = a_{11}a_{22} - a_{21}a_{12}$, the forces P_1 and P_2 necessary to give deflections y_1 and y_2 are

$$P_1 = \frac{1}{c}(a_{22}y_1 - a_{12}y_2) \qquad P_2 = \frac{1}{c}(a_{11}y_2 - a_{21}y_1)$$

The equations of motion of the two masses during vibrations are

$$M_1\ddot{y}_1 + P_1 = 0 \qquad \text{and} \qquad M_2\ddot{y}_2 + P_2 = 0$$

Assuming that the shaft vibrates in one of the two normal configurations, we take $y_1 = \lambda_1 \sin \omega t$ and $y_2 = \lambda_2 \sin \omega t$. Substituting this and the expressions for P_1 and P_2 in the equations of motion leads to

$$(a_{22} - M_1 c\omega^2)\lambda_1 - a_{12}\lambda_2 = 0$$
$$-a_{21}\lambda_1 - (M_2 c\omega^2 - a_{11})\lambda_2 = 0$$

To have solutions $(\lambda_1, \lambda_2) \neq (0, 0)$, the determinant of the coefficients must vanish. Expanding the determinant gives the frequency equation

$$\omega^4 - \frac{1}{c}\left(\frac{a_{11}}{M_2} + \frac{a_{22}}{M_1}\right)\omega^2 + \frac{1}{cM_1M_2} = 0$$

If the system in Fig. 11.8 is symmetrical, we have $M_1 = M_2 = M$ and $l_1 = l_2$, so that $a_{11} = a_{22}$. The reciprocal theorem states that $a_{12} = a_{21}$, and we have $c = a_{11}^2 - a_{21}^2$ in this case. The frequency equation then takes the simpler form:

$$\omega^4 - \frac{2a_{11}\omega^2}{cM} + \frac{1}{cM^2} = 0$$

and the solution is

$$\omega^2 = \frac{a_{11}}{cM} \pm \sqrt{\left(\frac{a_{11}{}^2}{c^2M^2} - \frac{1}{cM^2}\right)} = \frac{a_{11}}{cM} \pm \frac{1}{cM}\sqrt{(a_{11}{}^2 - c)}$$
$$= \frac{1}{cM}(a_{11} \pm a_{21}) = \frac{1}{M(a_{11} \pm a_{21})}$$

The critical speeds are then

$$f_c = 30\omega/\pi = \frac{30}{\pi\sqrt{[M(a_{11} \pm a_{21})]}} \text{ rev/min}$$

a_{11} and a_{21} may be determined from the formulae for Δ_1 and Δ_2 in Example 11.6, for $W = 1$.

Example 11.7. Calculate the two critical speeds of the system in Example 11.6, Fig. 11.7.

Solution. From the results in Example 11.6, we find

$$a_{11} = \frac{\Delta_1}{W} = \frac{4{\cdot}12}{267 \times 1{\cdot}05 \times 10^4} = 1{\cdot}47 \times 10^{-6} \text{ m/N}$$

$$a_{21} = \frac{\Delta_2}{W} = \frac{3{\cdot}43}{267 \times 1{\cdot}05 \times 10^4} = 1{\cdot}223 \times 10^{-6} \text{ m/N}$$

$$M = \frac{W}{g} = \frac{267}{9{\cdot}81} = 27{\cdot}2 \text{ kg}$$

The critical speeds are

$$f_c = \frac{30}{\pi\sqrt{[27{\cdot}2(1{\cdot}47 \pm 1{\cdot}223) \times 10^{-6}]}} = 1114 \text{ or } 3710 \text{ rev/min}$$

This result is in agreement with the result found in Example 11.6 for the fundamental critical speed.

11.3 Torsional two-mass systems

11.3.1 Fundamental two-mass system

Figure 11.9 shows a torsional two-mass system. This consists of two inertias I_1 and I_2, which we shall consider as discrete inertias,

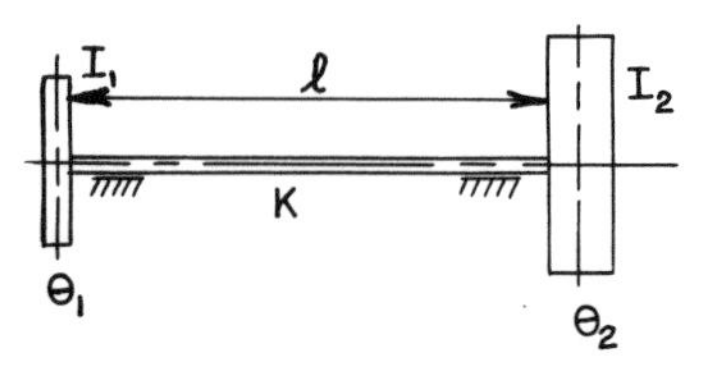

Fig. 11.9

connected by a shaft of length l and torsional spring constant K.

Taking rotational displacements θ_1 and θ_2 from the static condition, the equations of motion may be established from Lagrange's equations with

$$T = \tfrac{1}{2}I_1\dot{\theta}_1^2 + \tfrac{1}{2}I_2\dot{\theta}_2^2 \qquad \text{and} \qquad V = \tfrac{1}{2}K(\theta_1 - \theta_2)^2$$

In this simple case the equations may also be stated directly:

$$I_1\ddot{\theta}_1 + K(\theta_1 - \theta_2) = 0$$
$$I_2\ddot{\theta}_2 + K(\theta_2 - \theta_1) = 0$$

Assuming simple harmonic motion, we substitute $\theta_1 = \beta_1 \sin \omega_0 t$ and $\theta_2 = \beta_2 \sin \omega_0 t$; this leads to the two algebraic equations:

$$(K - I_1\omega_0^2)\beta_1 - K\beta_2 = 0$$
$$-K\beta_1 + (K - I_2\omega_0^2)\beta_2 = 0$$

These two homogeneous equations can only have solutions $(\beta_1, \beta_2) \neq (0, 0)$ if the determinant of the coefficients vanishes. The frequency determinant is then

$$\begin{vmatrix} K - I_1\omega_0^2 & -K \\ -K & K - I_2\omega_0^2 \end{vmatrix} = 0$$

Expanding the determinant gives the frequency equation

$$\omega_0^2[I_1 I_2 \omega_0^2 - (I_1 + I_2)K] = 0$$

The first root is $\omega_0 = 0$, which may be interpreted as the frequency for the system rotating as a rigid body. The second root is

$$\omega_0 = \sqrt{\left(\frac{I_1 + I_2}{I_1 I_2} K\right)} \tag{11.4}$$

The frequency from (11.4) is somewhat too high since the mass of the shaft and damping was neglected. The frequency is evidently independent of the amplitudes β_1 and β_2.

From the equation $(K - I_1\omega_0^2)\beta_1 - K\beta_2 = 0$ above, we find $\beta_1/\beta_2 = K/(K - I_1\omega_0^2)$. Substituting the expression (11.4) for ω_0, we obtain $\beta_1/\beta_2 = -I_2/I_1$; this means that the amplitudes are in a fixed ratio and always in the opposite direction.

Figure 11.10 shows the maximum amplitudes during vibrations. The line is straight because we have neglected the inertia of the shaft; it is called the mode form or normal elastic line; where it intersects the horizontal line is the so-called nodal section, this section of the shaft remains stationary during vibrations.

From the mode form we find the relationship $|\beta_1/\beta_2| = I_2/I_1 = a/b$; since we have $a+b=l$, the solution for a is $a = lI_2/(I_1+I_2)$. This relationship determines the position of the nodal section or node.

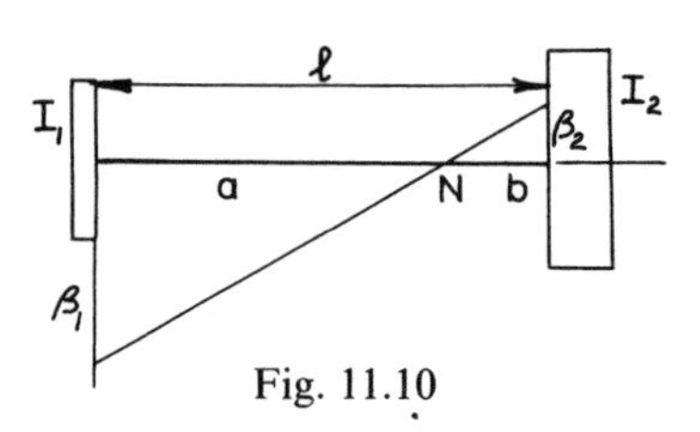

Fig. 11.10

It may be seen from $a/b = I_2/I_1$ that the node is closest to the greatest inertia. The actual displacements cannot be determined, but they retain the same fixed ratio a/b as the initial displacements.

The system behaves like two torsional pendulums of shaft length a and b, vibrating at the same frequency ω_0 and always in the opposite direction.

From Example 7.4 we have $\omega_0^2 = K/I$ for a torsional pendulum, with $K = GI_p/l$.

In the case of the two-mass system, we find

$$\omega_0^2 = \frac{GI_p}{aI_1} = \frac{GI_p}{bI_2} = \frac{GI_p}{l}\frac{(I_1+I_2)}{I_1I_2} = K\frac{(I_1+I_2)}{I_1I_2}$$

as before.

If the total inertia of the shaft is denoted by I_S, we may take it into account by applying the result in Example 11.5. Calling the distances to the nodal section a' and b', we may write

$$\omega_0^2 = \omega_1^2 = \frac{GI_p}{a'}\frac{1}{I_1+\frac{1}{3}I_S a'/l} = \omega_2^2 = \frac{GI_p}{b'}\frac{1}{I_2+\frac{1}{3}I_S b'/l}$$

Substituting $b' = l - a'$, and solving for a', we obtain

$$a' = \frac{l(I_2+\frac{1}{3}I_S)}{I_1+I_2+\frac{2}{3}I_S}$$

This locates the node, and the frequency is

$$\omega_0 = \sqrt{\frac{K'}{I_1+\frac{1}{3}I_S a'/l}}$$

where $K'=GI_p/a'$. If the shaft inertia is neglected, we have $I_S=0$ and $a'=a$ as determined before.

Example 11.8. A marine engine and its propeller have moments of inertia of 3770 kg m^2 and 944 kg m^2 respectively. The shaft is made of steel with $G=8{\cdot}13\times10^7$ kN/m^2. The length of the shaft is 15·24 m and its diameter is 50·8 cm, the density being $0{\cdot}776\times10^4$ kg/m^3.

Determine the torsional frequency and the distance between the propeller and the node (a) if the shaft inertia is neglected, (b) if the shaft inertia is included.

Solution. (a) From the given figures we find $I_p=(\pi/32)0{\cdot}508^4=$ 0·006 53 m^4, and

$$K = \frac{GI_p}{l} = \frac{8{\cdot}13\times10^{10}\times0{\cdot}006\,53}{15{\cdot}24} = 3{\cdot}48\times10^7 \text{ N m/rad}$$

From (11.4), we find

$$f = \frac{1}{2\pi}\sqrt{\left(\frac{4714\times3{\cdot}48\times10^7}{3770\times944}\right)} = 34{\cdot}1 \text{ cycles/s}$$

The distance from the propeller of inertia I_1 to the node is

$$a = \frac{lI_2}{I_1+I_2} = 15{\cdot}24\times\frac{3770}{4714} = 12{\cdot}2 \text{ m}$$

(b) The shaft inertia is $I_S=Md^2/8$.

$$M = (\pi/4)d^2l\rho = (\pi/4)0{\cdot}508^2\times15{\cdot}24\times0{\cdot}776\times10^4$$
$$= 2{\cdot}39\times10^4 \text{ kg}$$

$$I_S = 2{\cdot}39\times10^4\times\frac{0{\cdot}508^2}{8} = 770 \text{ kg m}^2$$

We find

$$a' = \frac{15{\cdot}24(3770+770/3)}{4714+\frac{2}{3}\times770} = 11{\cdot}74 \text{ m}$$

$$K' = \frac{GI_p}{a'} = \frac{Kl}{a'} = 3{\cdot}48\times10^7\times\frac{15{\cdot}24}{11{\cdot}74} = 4{\cdot}51\times10^7 \text{ N m/rad}$$

The frequency is

$$f = \frac{1}{2\pi}\sqrt{\left(\frac{4{\cdot}51\times 10^7}{944+\frac{1}{3}\times 770\times 11{\cdot}74/15{\cdot}24}\right)} = 31{\cdot}6 \text{ cycles/s}$$

11.3.2 Two-mass geared torsional system

Figure 11.11 shows a two-mass torsional system. A set of gear wheels is introduced as shown with gear ratio $r_1/r_2 = n$. The mass of the gears is neglected.

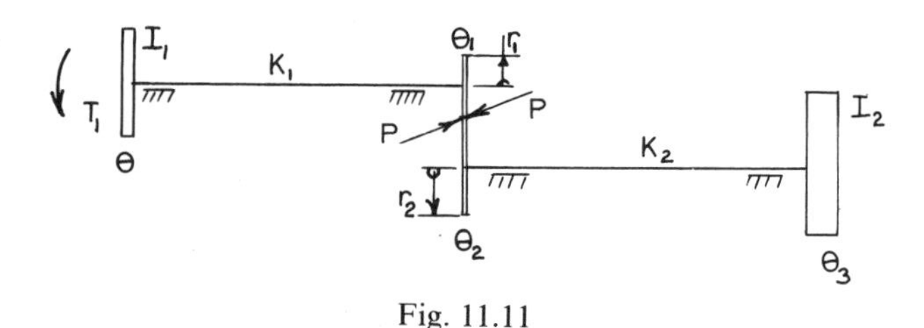

Fig. 11.11

To determine the equivalent torsional spring constant of the system, we clamp the right-hand end of the system and rotate the left-hand end by a torque T_1. Calling the tangential pressure between the teeth of the gear wheels P, we have from static equilibrium that $Pr_1 = T_1$, while the torque T_2 in the second shaft is $T_2 = Pr_2 = T_1 r_2/r_1 = T_1/n$. The rotation θ_2 is now $\theta_2 = T_2/K_2 = T_1/nK_2$.

Since there can be no slip of the wheels, we may state that numerically $r_1\theta_1 = r_2\theta_2$, or $\theta_1 = (r_2/r_1)\theta_2 = \theta_2/n = T_1/n^2K_2$. The total rotation θ of the left-hand end is then

$$\theta = \frac{T_1}{K_1} + \theta_1 = T_1\left(\frac{1}{K_1} + \frac{1}{n^2K_2}\right) = T_1 \times \frac{1}{K}$$

The equivalent spring constant K is then determined from

$$\frac{1}{K} = \frac{1}{K_1} + \frac{1}{n^2K_2}$$

The formula for springs in series is

$$\frac{1}{K} = \sum_1^n \frac{1}{K_n}$$

and the equivalent spring constants may now be taken as shown in Fig. 11.12, with the springs of constants K_1 and n^2K_2 in series. The value of K is $K=K_1K_2n^2/(K_1+n^2K_2)$. The equivalent two-mass system is now the system in Fig. 11.12, where the equivalent inertia I_2' is found from $I_2'=I_2(\omega_2/\omega_1)^2=I_2(r_1/r_2)^2=n^2I_2$.

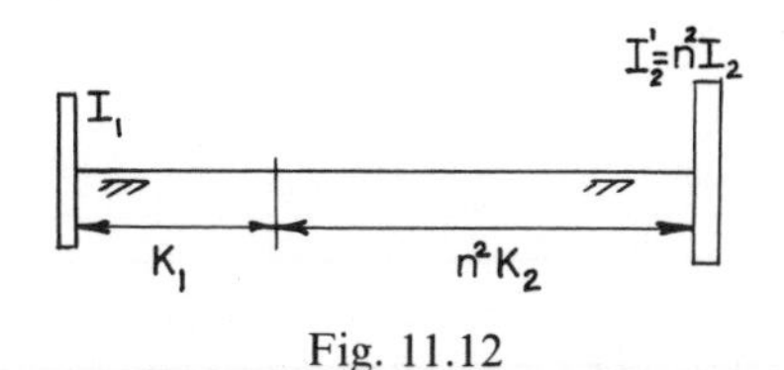

Fig. 11.12

Comparing the system in Fig. 11.12 to the system in Fig. 11.9, we find that the torsional frequency of the geared system is determined by the formula

$$\omega_0^2 = \frac{I_1+n^2I_2}{I_1n^2I_2} \frac{K_1K_2n^2}{K_1+n^2K_2} = \frac{I_1+n^2I_2}{K_1+n^2K_2} \frac{K_1K_2}{I_1I_2}$$

The same result may be found by using the equations of motion of the system, which are:

$$I_1\ddot{\theta}+K_1(\theta-\theta_1) = 0$$
$$I_2\ddot{\theta}_3+K_2(\theta_3+n\theta_1) = 0$$
$$K_1(\theta-\theta_1) = K_2(\theta_3+n\theta_1)n$$

The last equation states that the torques on the gears are balanced, since the gears are assumed light.

Substituting $\theta=\beta \sin \omega_0 t$, $\theta_1=\beta_1 \sin \omega_0 t$ and $\theta_3=\beta_3 \sin \omega_0 t$ in the equations of motion, and equating the determinant of the coefficients to zero, we find the same result for ω_0 as determined before.

11.4 Kinematics of gear trains and epicyclic gears

It is often necessary to combine several gears in various types of machinery. The result of this is called a gear train. Figure 11.13 shows two meshing gear wheels. This combination is called a simple gear train. If several gears are fixed to the same shaft, the combination is called a compound gear train.

In order to discuss the angular velocity relationship in Fig. 11.13, we substitute a pair of friction wheels for the actual gear wheels, and assume rolling without slipping. The friction wheels must have the same velocity relationship as the actual gears. The diameters of the friction wheels are called the pitch circle diameters D_A and D_B.

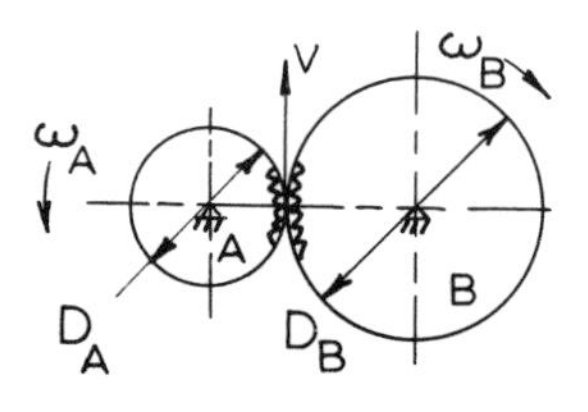

Fig. 11.13

The situation in Fig. 11.13 is called external gearing and the gear wheels evidently rotate in opposite directions. Figure 11.14 shows internal gearing, where the rotation of the gears is in the same direction.

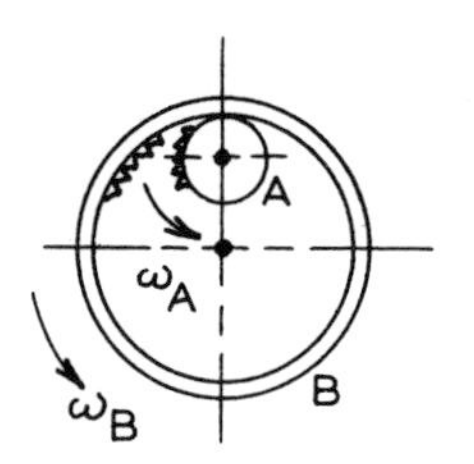

Fig. 11.14

If the number of teeth is called T and the pitch diameter D, we define the diametral pitch P_D as the ratio $P_D = T/D$. The circular pitch P_C is defined as the distance occupied by each tooth along the pitch circle, or $P_C = \pi D/T$, which means that $P_D P_C = \pi$. Meshing or mating gears must have the same pitch.

If we call the number of revolutions per minute N, and the peripheral speed of the gears V, as indicated in Fig. 11.13, we find that

$$V = \omega_A \frac{D_A}{2} = \frac{\pi}{30} N_A \frac{D_A}{2} = \frac{\pi}{30} N_B \frac{D_B}{2}$$

or the gear ratio $= N_A/N_B = D_B/D_A = T_B/T_A = \omega_A/\omega_B$.

Gear trains are sometimes designed so that one gear wheel rolls around the circumference of a fixed wheel, as shown in Fig. 11.15, where the large gear is assumed fixed. Each point on the rolling gear generates a plane curve called an epicycloid, and the gears are called an epicyclic gear train.

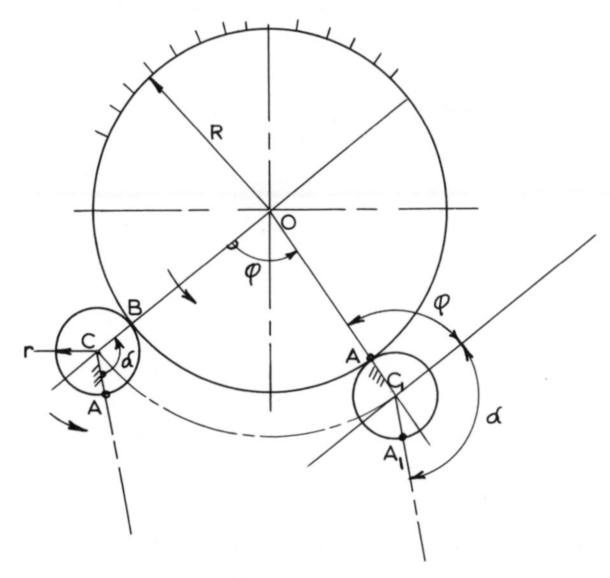

Fig. 11.15

Sometimes the rolling gear is called a planet gear and the fixed gear a sun gear, and the system is called a planetary gear train.

To discuss the angular-velocity relationship in Fig. 11.15 we consider the rolling gear in two positions: in the second position the centre line OC has been rotated through an angle φ as shown. Drawing a line CA on the rolling wheel at an angle α as shown to line OC, we assume that point A is the point of contact between the wheels in the second position. From the figure it may be seen that the total angle of rotation, θ, of the rolling wheel in the plane is given by $\theta=\varphi+\alpha$. Because we assume no slip, we have $r\alpha=R\varphi$, or $\alpha=R\varphi/r$ and $\theta=\varphi+R\varphi/r=(R/r+1)\varphi$. When $\varphi=2\pi$, the line OC has rotated through one revolution, and $\theta=(R/r+1)2\pi$. The number of revolutions of the rolling wheel is then

$$\theta/2\pi = R/r+1 = T_R/T_r+1$$

Before solving any problems, we shall make use of the following simple idea. A gearbox consists of any number of gears all in parallel planes. Suppose now that we fix all gears so that the gearbox becomes a rigid body; if we now rotate the box once about any axis perpendicular to the planes of the gears, each gear wheel will be rotated once about its centre. This result follows in the same way as the moon, during its monthly rotation about the earth, rotates once about its centre, because the same side of the moon always faces the earth.

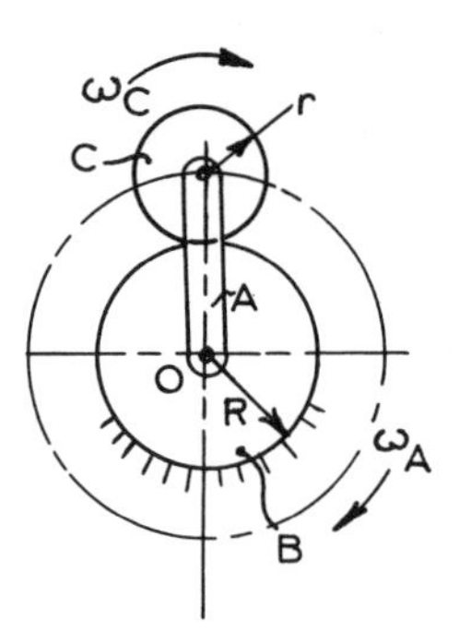

Fig. 11.16

Consider now the epicyclic gear in Fig. 11.16. The wheel C rolls on the fixed wheel B, due to the rotation of the arm A. We want to find the number of rotations of C for one rotation of the arm A. We must take one direction of rotation positive; taking the anti-clockwise direction as positive in this example, the solution may be found by the method shown in Table 11.1. The final result $1+T_B/T_C$ is the same as for the real system, where wheel B is fixed and the arm A is rotated once.

Table 11.1

Procedure	Revolutions of Arm A	B	C
Fix arm A, give one negative rotation to wheel B	0	-1	$+\dfrac{T_B}{T_C}$
Lock system and give one revolution in the positive direction to entire system	$+1$	0	$\dfrac{T_B}{T_C}+1$

The result was, of course, already known from the previous discussion of the system in Fig. 11.15. If, in the system $T_B = 30$, $T_C = 15$ and $N_A = 60$ rev/min ↻, we find $N_C = 60(1 + 30/15) =$ 180 rev/min ↻.

In the case of internal gearing, the situation is as shown in Fig. 11.17. By considerations similar to those in the discussion of external gearing, we find $\theta = \alpha - \varphi$, $r\alpha = R\varphi$, so that $\theta = (R/r - 1)\varphi$.

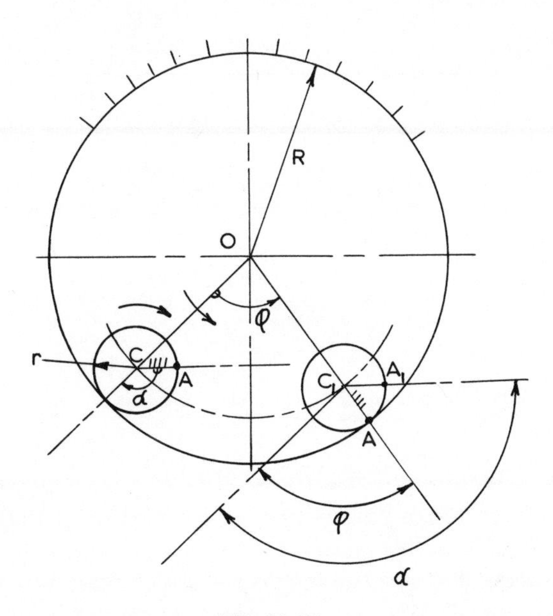

Fig. 11.17

For one revolution of the arm A, we have $\varphi = 2\pi$, and $\theta =$ $(R/r - 1)2\pi$. The number of revolutions of the rolling wheel are thus

$$\frac{\theta}{2\pi} = \frac{R}{r} - 1 = \frac{T_R}{T_r} - 1$$

Consider now the system in Fig. 11.18; a tabular solution is shown in Table 11.2. The result is $1 - T_B/T_C$ revolutions of C for one revolution of the arm A. This result is the same as found in the discussion of Fig. 11.17. Since $T_B > T_C$, $1 - T_B/T_C < 0$; this means that the direction of rotation of C is opposite to that of A.

Table 11.2

Procedure	Revolutions of Arm A	B	C
Fix A, give B one negative revolution	0	-1	$-\frac{T_B}{T_C}$
Lock, give one positive revolution to system	$+1$	0	$1-\frac{T_B}{T_C}$

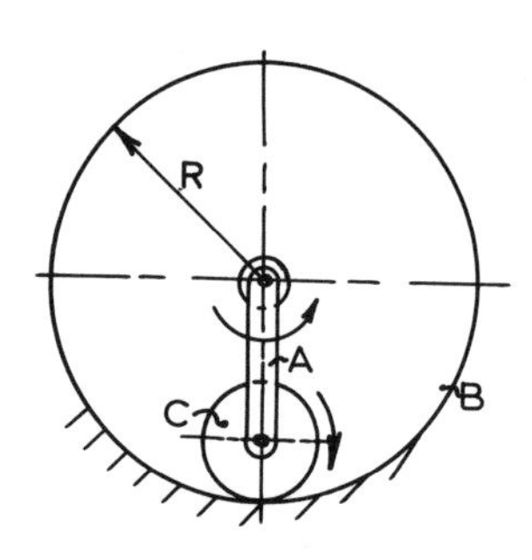

Fig. 11.18

Example 11.9. For the epicyclic system in Fig. 11.19, determine the direction of rotation and number of rotations per minute of the driven shaft D if the speed of the driving shaft is N_A rev/min. The gears B and C form an integral set which rotates freely on the tip of the arm A. All the gears have the same pitch.

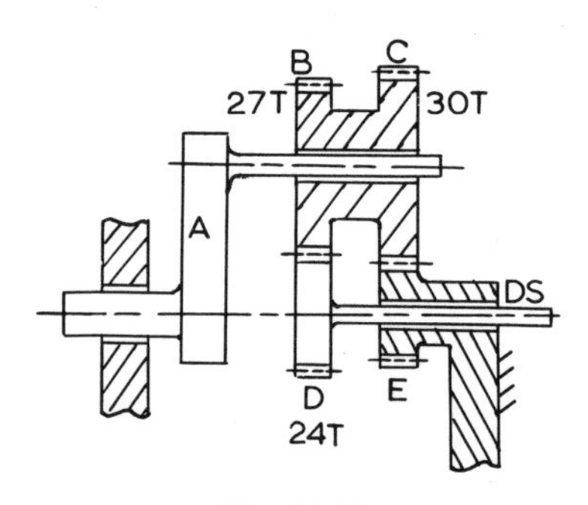

Fig. 11.19

Solution. The number of teeth of wheel E may be determined as follows:

The centre distances are $\frac{1}{2}(D_E+D_C)=\frac{1}{2}(D_D+D_B)$. Since the pitch is the same for all gears, we have $P_C=\pi D/T$, or $D=(P_C/\pi)T$. Substituting this gives

$$T_E+T_C = T_D+T_B \qquad \text{or} \qquad T_E = T_D+T_B-T_C$$
$$= 24+27-30 = 21$$

A tabular solution is shown in Table 11.3. The result is

$$1-\frac{21}{30}\times\frac{27}{24} = 1-\frac{7}{10}\times\frac{9}{8} = 1-\frac{63}{80} = +\frac{17}{80}$$

revolutions of D for one revolution of A. The revolutions are in the same direction.

Table 11.3

	Revolutions of				
	A	E	C	$B=C$	D
Fix A, give E one negative revolution	0	-1	$+\frac{21}{30}$	$+\frac{21}{30}$	$-(\frac{21}{30})\times(\frac{27}{24})$
Lock, give one positive revolution to system	$+1$	0	$1+\frac{21}{30}$	$1+\frac{21}{30}$	$1-(\frac{21}{30})\times(\frac{27}{24})$

We find $N_D=+(17/80)N_A$ rev/min$=+(1/4{\cdot}71)N_A$. The speed reduction is 1/4·71.

11.5 Flywheels and speed fluctuation

Flywheels are used essentially to keep the speed fluctuations within desired limits.

The torque from an internal combustion engine varies greatly during one revolution and may become negative during part of a period. To keep the angular velocity variations small during a period, a flywheel is introduced between the engine and the driven machinery, as shown in Fig. 11.20.

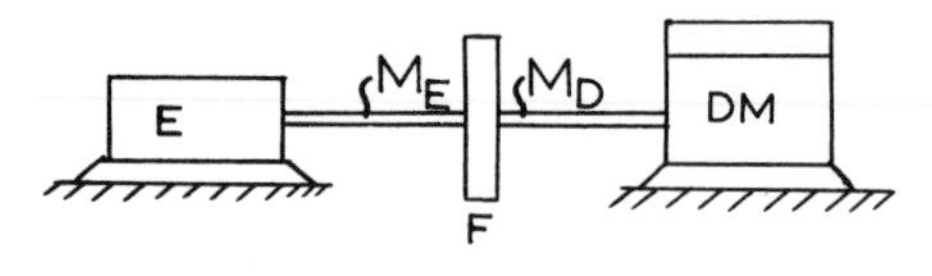

Fig. 11.20

If the torque from the engine is M_E and the resisting torque from the driven machinery is M_D, the equation of motion of the flywheel is $M_E - M_D = I\dot{\omega}$. Figure 11.21 shows the variations in M_E and M_D as a function of the angle of rotation θ.

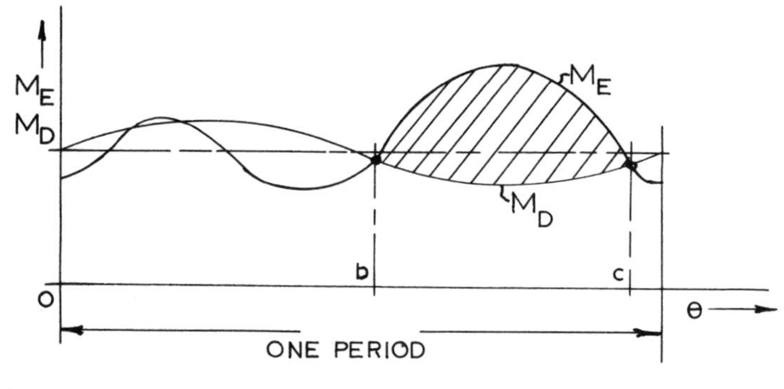

Fig. 11.21

The curves are periodic. A period may be one revolution for a two-stroke engine and two revolutions for a four-stroke engine. These curves are often very complicated.

It may be seen that at the position indicated by the letter b we have minimum angular velocity, since the M_E is smaller than M_D for some time before position b is reached. From b to c there is a surplus of energy, and the velocity of c is a maximum.

The cross-hatched area between the curves is the largest loop area in one period; in this case it is surplus energy, but sometimes it may be an energy deficit. For more complicated cases part of a period may have to be used to determine the maximum surplus or deficit by summing up positive and negative areas.

If we consider motion between positions b and c and denote the cross-hatched area by ΔE, we have $\Delta E = \int_b^c (M_E - M_D)\, d\theta$. This is the work done on the flywheel and equal to the increase in kinetic energy of the flywheel, or $\Delta E = \frac{1}{2}I(\omega_c^2 - \omega_b^2) = \frac{1}{2}I(\omega_c - \omega_b)(\omega_c + \omega_b)$.

Taking the mean speed ω as $\omega \simeq \frac{1}{2}(\omega_c + \omega_b)$ and introducing the speed fluctuation as $\Delta\omega = \omega_c - \omega_b$, we find

$$\Delta E = I\omega(\Delta\omega) = I\omega^2(\Delta\omega/\omega) = I\omega^2\delta \tag{11.5}$$

where $\delta = (\Delta\omega)/\omega$ is called the degree of irregularity. The percentage of speed fluctuation from the mean speed is $(\Delta\omega/\omega) \times 100\% = (\Delta E/I\omega^2) \times 100\%$.

For higher angular velocities ω, the speed fluctuation will be smaller for the same value of the moment of inertia I. The speed fluctuation may be lowered by increasing I or ω or both. The flywheel thus absorbs excess energy or supplies deficiencies in energy without great speed changes. The flywheel cannot, however, stop the speed from increasing indefinitely if the load comes off. The degree of irregularity may be seen from the following values:
Diesel ship engines $\frac{1}{8}$ to $\frac{1}{20}$
Factory transmission belts $\frac{1}{30}$ to $\frac{1}{50}$
Textile mills $\frac{1}{60}$ to $\frac{1}{100}$
Automobile engines $\frac{1}{80}$ to $\frac{1}{150}$. Dynamoes $\frac{1}{100}$ to $\frac{1}{150}$.
A.C. generators $\frac{1}{250}$ to $\frac{1}{300}$.

Example 11.10. A single-cylinder two-stroke engine is running at 2000 rev/min. From the complete turning moment diagram it is found that the area of the loop above the mean torque line is 27·2 N m. Determine the moment of inertia of the flywheel required, if the angular velocity variation is not to exceed ± 5 rev/min.

The engine with flywheel is directly coupled to machinery requiring 0·0967 kW. The mean velocity is 300 rev/min and the excess energy is 12·4% of the work done per revolution. Determine the degree of irregularity.

Solution. $\omega = (\pi/30) \times 2000 = 209{\cdot}5$ rad/s, $\Delta E = 27{\cdot}2$ N m. 10 rev/min $= 1{\cdot}047$ rad/s, $\Delta\omega = \omega_{max} - \omega_{min} = 1{\cdot}047$ rad/s. From (11.5),

$$I = \frac{\Delta E}{\omega(\Delta\omega)} = \frac{27{\cdot}2}{209{\cdot}5 \times 1{\cdot}047} = 0{\cdot}1242 \text{ kg m}^2$$

300 rev/min $= 5$ rev/s, W.D./rev $= 0{\cdot}0967 \times 1000/5 = 19{\cdot}34$ N m. $\Delta E = (12{\cdot}4/100) \times 19{\cdot}34 = 2{\cdot}40$ N m. $\omega = (\pi/30) \times 300 = 31{\cdot}4$ rad/s. From (11.5)

$$\delta = \frac{\Delta E}{I\omega^2} = \frac{2{\cdot}40}{0{\cdot}1242 \times 31{\cdot}4^2} = \frac{1}{51}$$

11.6 Rotating disc cams

Rotating cams are widely used in machinery: in internal-combustion engines they are used for closing and opening valves and activating fuel pumps, and they also feature in machine tools, computers and instruments.

A great variety of cams are produced, both two and three-dimensional. We shall limit this discussion to rotating disc cams giving a reciprocating motion to a follower which is kept in contact with the cam surface through gravity or more often through the action of a spring.

Two types of follower are shown in Fig. 11.22: (a) is a roller type follower while (b) is a flat-faced follower. In Fig. 11.22(b), the axis of the follower is offset from the axis of the cam.

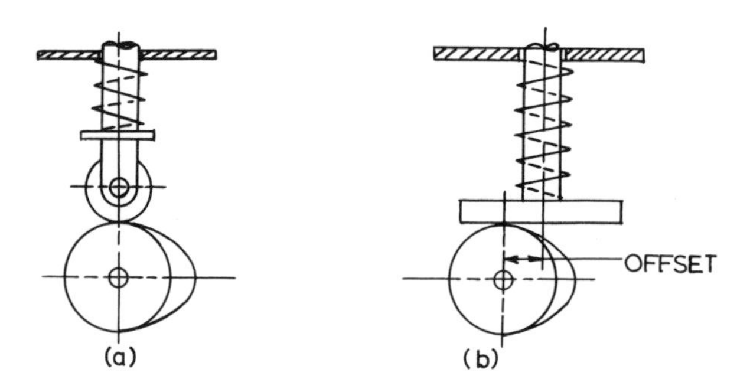

Fig. 11.22

Many types of motion may be imparted to a follower by a cam; we shall consider only two of the most widely used motions. Simple harmonic and parabolic, or motion with constant acceleration.

The cam surface is determined graphically by first plotting a displacement–time curve for the follower. The cam is supposed to rotate with constant angular velocity ω, so that the cam angle of rotation θ is $\theta=\omega t$.

Suppose now that in the time interval $t=0$ to $t=t_0$, the follower has to move with simple harmonic motion through a displacement $s=0$ to $s=s_0$, while the cam rotates through an angle $\theta_0=\omega t_0$. The lift curve may then be determined as shown in Fig. 11.23.

The radius CA of length $s_0/2$ is assumed to be rotating with constant angular velocity π/t_0 from position CO to position CB.

The projection of the point A on the vertical diameter is point D and this point performs a simple harmonic motion along OB.

The vertical scale for displacements s is taken as the same scale used for the cam drawing so that displacements can be directly taken from Fig. 11.23. The horizontal line is the time t or cam angle θ. The time scale is arbitrary. By dividing the half-circle into any convenient number of equal parts and using the same number of equal distances on the horizontal line, the lift curve may be found as shown in the figure.

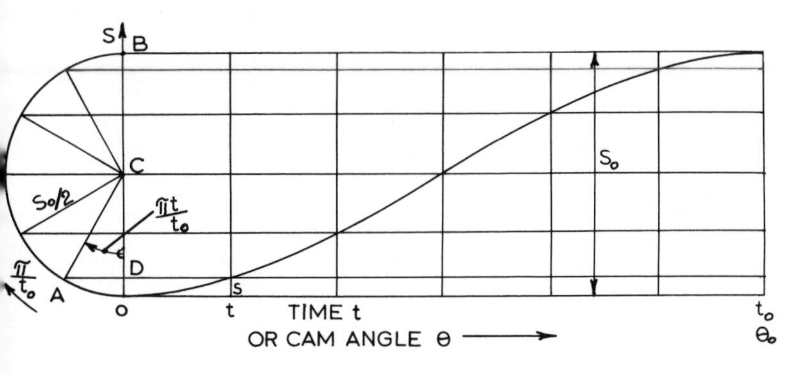

Fig. 11.23

The displacement s is seen to be given by

$$s = \tfrac{1}{2}s_0\left(1-\cos\frac{\pi t}{t_0}\right) = \tfrac{1}{2}s_0\left(1-\cos\frac{\pi\theta}{\theta_0}\right)$$

This is simple harmonic motion with velocity

$$v = \frac{ds}{dt} = \frac{\pi s_0}{2t_0}\sin\frac{\pi t}{t_0} = \frac{\pi s_0 \omega}{2\theta_0}\sin\frac{\pi\theta}{\theta_0}$$

The acceleration is

$$a = \frac{dV}{dt} = \frac{\pi^2 s_0}{2t_0^2}\cos\frac{\pi t}{t_0} = \frac{\pi^2 s_0 \omega^2}{2\theta_0^2}\cos\frac{\pi\theta}{\theta_0}$$

If the follower has to move with constant acceleration, which gives the minimum inertia forces for a given ω and s_0, the lift curve may be taken as two parabolic arcs determined as shown in

Fig. 11.24. The construction follows from the relationship $s=\frac{1}{2}at^2$. The vertical distance is here divided into an *even* number of equal parts.

The displacement is $s=\frac{1}{2}at^2$. When $t=t_0/2$, $s=s_0/2$, from which $s=(2s_0/t_0^2)t^2$, $0\leqslant t\leqq t_0/2$.

The velocity of the follower is $v=ds/dt=(4s_0/t_0^2)t$, and the constant acceleration is $a=dv/dt=4s_0/t_0^2$.

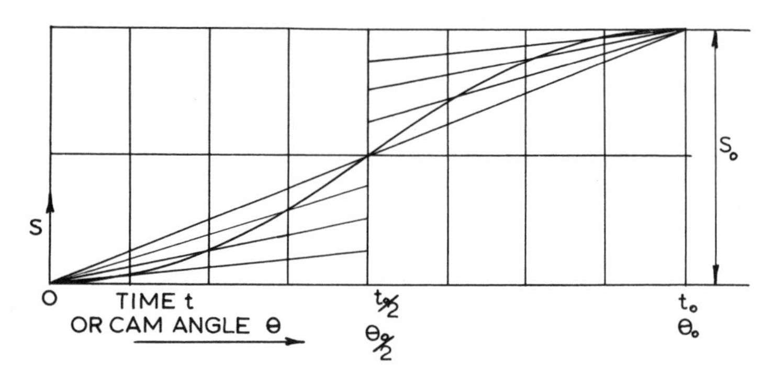

Fig. 11.24

Example 11.11. Determine graphically the contour of a disc cam which is to give a reciprocating motion to a point or knife-edge follower. The cam is rotating clockwise at constant angular velocity ω, and the minimum radius is to be 5 cm. The follower is to be offset by 2 cm.

The motion of the follower is to have constant and equal acceleration and retardation on both the upstroke and the downstroke. The maximum displacement is to be 2·50 cm after 120° of cam rotation, the follower is then to dwell during 60° of rotation, return during the next 90° of rotation and dwell during the remaining 90° of cam rotation.

Solution. The lift curve is determined as explained under parabolic motion. The curve is shown in Fig. 11.25(a).

The cam contour is shown in Fig. 11.25(b). The offset circle with centre O and radius 2 cm is drawn and a vertical tangent O_1A introduced as shown. The minimum cam radius 5 cm is measured from O to A, and this determines the starting point for the follower. The offset circle is divided into arcs of 30° to give the points

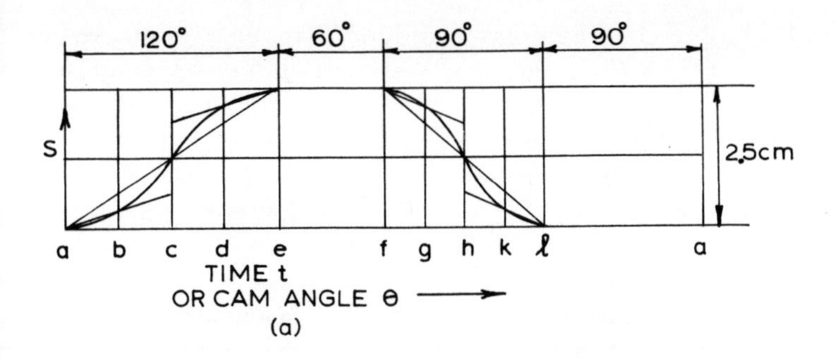

120°
60°
90°
90°
S
2.5cm
a b c d e f g h k l a
TIME t
OR CAM ANGLE θ
(a)

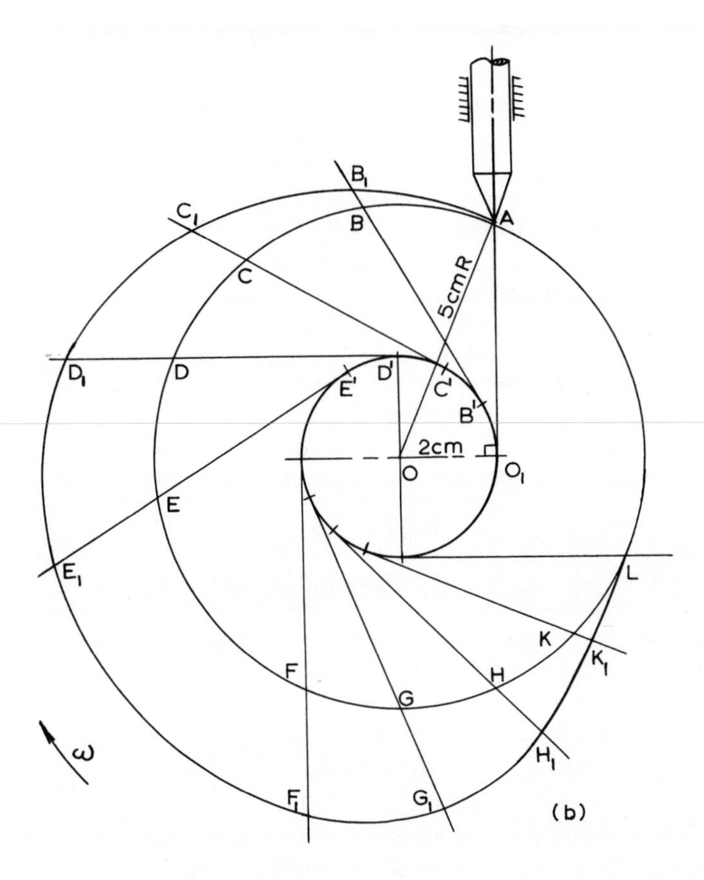

B1
B
A
C1
C
5cm R
D1
D
D'
E'
C'
B'
2cm
O
O1
E
E1
L
K
K1
F
H
G
H1
ω
F1
G1
(b)

Fig. 11.25

B', C', D' and E'. By drawing tangents to the circle at these points, the points B, C, D and E are determined. The lift at positions b, c, d and e are now taken from the lift curve, and measured out along the tangential lines to give the points B_1, C_1, D_1 and E_1. The contour of the cam is drawn as a smooth curve through the points A, B_1, C_1, D_1 to E_1. The rest of the contour is determined in a similar way.

It will be appreciated from the figure that as the cam rotates clockwise through 30°, the point B will coincide with point A, and the follower will have been lifted through a distance BB_1 as it should be. It would be necessary to determine more points of the cam contour in the actual production case.

For a roller follower the minimum cam radius must first be increased by the radius of the follower to find the starting position A of the roller centre. The position of the centre of the roller is determined in the same way as points B_1, C_1 etc. in Fig. 11.25(b). Finally the roller circle is drawn with centres B_1, C_1 etc. and the cam contour is fitted as a tangential curve to the roller circles.

For a flat face follower the construction is similar to Fig. 11.25(b). Lines are drawn perpendicular to the tangential lines $B'B$, $C'C$ etc. through the points B_1, C_1 etc. The cam contour is fitted as a tangential curve to those lines representing the flat face of the follower. The construction gives the contact point between the flat face and the cam, and the minimum width of the flat face.

11.7 In-line engine balancing

11.7.1 Balancing of a single-cylinder engine

A close approximation to the piston acceleration was determined in Chapter 1, Example 1.3, eq. (1). With reference to Figure 11.26(a), the piston acceleration is

$$\ddot{x} = r\omega^2\left(\cos\theta + \frac{\cos 2\theta}{n}\right)$$

where ω is the *constant* angular velocity of the crank, $\theta = \omega t$ and $n = l/r$. The value of n is generally between 2·4 and 7.

Denoting the total reciprocating masses by m_0, this includes the mass of the piston (including cooling oil, if any, piston rings and piston pin), the mass of the small end of the connecting rod and the top half of the connecting-rod shaft. For cross-head engines, the mass of the cross head and piston rod must also be included.

The rotating masses m include the mass of the crank pin, the big end of the connecting rod, the lower half of the connecting rod shaft and the mass of the two crank arms after the total mass of the two arms has been multiplied by the fraction r_1/r, where r_1 is the radius to the centre of mass of the arms.

A better mass distribution of the connecting rod may be obtained by placing the rod horizontally on two scales supporting each end. This process leaves the total mass and the position of the centre of mass unchanged, so that this equivalent rod is dynamically equivalent to the actual rod as far as forces are concerned. The moment of inertia about an axis perpendicular to the rod through the centre of mass, will however, be too large.

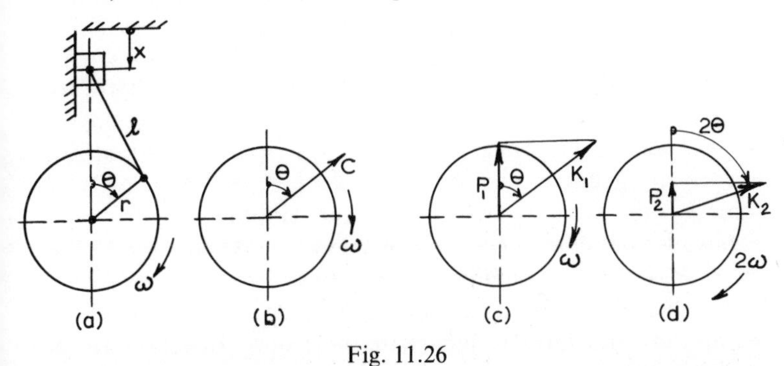

Fig. 11.26

We have now the following inertia forces on a piston and crank: (a) reciprocating force

$$P = m_0 r\omega^2\left(\cos\theta + \frac{\cos 2\theta}{n}\right)$$

in the direction of the cylinder axis and upwards if the expression is positive, and (b) a rotating force, the centrifugal force, $C = mr\omega^2$, always directed radially outward in the direction of the crank arms.

The two terms of the reciprocating force are now $P_1 = m_0 r\omega^2 \cos\theta$ and $P_2 = m_0(r\omega^2/n)\cos 2\theta$. For one crank we have thus the three inertial forces C, P_1 and P_2.

$C = mr\omega^2$ is a constant-magnitude rotating force in the direction of the crank as shown in Fig. 11.26(b). The force $P_1 = m_0 r\omega^2$

$\cos\theta = K_1 \cos\theta$ is called the primary reciprocating force. The magnitude of P_1 may be found as the vertical projection of a constant force $K_1 = m_0 r\omega^2$ acting in the direction of the crank, as shown in Fig. 11.26(c). The force $P_2 = m_0(r\omega^2/n)\cos 2\theta = K_2 \cos 2\theta$ is called the secondary reciprocating force. It may be determined as the vertical projection of a constant force $K_2 = m_0 r\omega^2/n$ acting in a radial direction and always making an angle with the vertical of twice that of the crank. P_2 is shown in Fig. 11.26(d).

The inertia forces may be partially balanced by counterweights fixed to the crank arms. Using counter weights of total mass m_1 and with a centre of mass at a radius R_1 and opposite to the crank arms, the counter weights introduce a centrifugal force $C_1 = m_1 R_1 \omega^2$ which acts in the opposite direction to C.

If we take $C_1 = C$, or $m_1 R_1 = mr$, the rotating force will be completely balanced. The reciprocating forces P_1 and P_2 are, however, still operating.

We may also choose instead to take $C_1 = C + K_1$, or $m_1 R_1 = mr + m_0 r$. In this case both the rotating and primary reciprocating forces are balanced, but a new horizontal force equal and opposite to the horizontal projection of K_1 has been introduced. This reciprocating horizontal force may be just as damaging as the force P_1. The secondary force P_2 is unbalanced.

It is also possible to make a compromise and take $C_1 = C + \frac{1}{2}K_1$, or $m_1 R_1 = mr + \frac{1}{2}m_0 r$. Here the rotating force and half of the primary force have been balanced. In the vertical direction $\frac{1}{2}P_1$ and P_2 are left unbalanced, and in the horizontal direction $\frac{1}{2}P_1$ has been introduced.

The secondary inertia force P_2 cannot be balanced by counter weights rotating with the crank.

Since all the forces are concurrent, no moment investigation is called for.

11.7.2 Balancing of multi-cylinder engines

As an example of an engine with more than one cylinder, consider the three-cylinder engine in Fig. 11.27(a) and (b).

The rotating forces are $C = mr\omega^2$, and these forces are shown in Fig. 11.28(a), projected onto a plane perpendicular to the crank-shaft centre line. The force polygon for the forces C is shown in Fig. 11.28(b). The resultant $R_C = 0$.

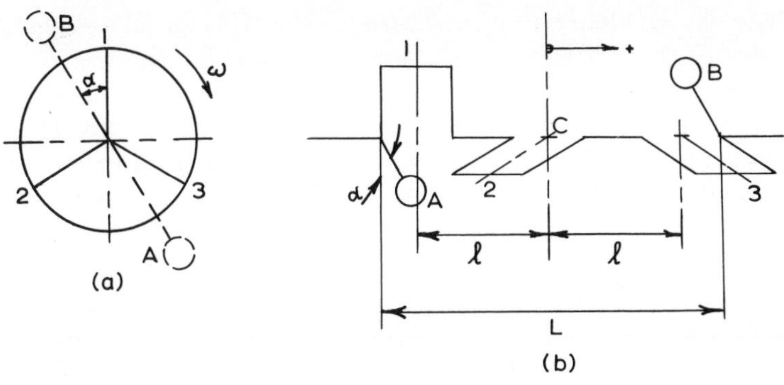

Fig. 11.27

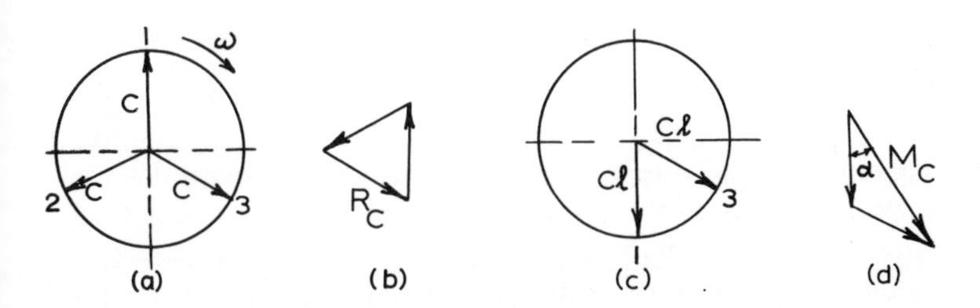

Fig. 11.28

The moments of the forces are taken about point C in Fig. 11.27(b). The moment arms are taken as positive to the right of C and negative to the left. Since all forces are perpendicular to the crank centre line, we may use a simplified convention, called Dalby's convention, for the moment vectors, which is to represent them as vectors in the direction of the force for positive moment arms and opposite to the force for negative moment arms.

For crank 1, the moment arm is $-l$ and the moment $-\mathbf{Cl}$ is taken in the direction opposite to crank 1 as shown in Fig. 11.28(c). For crank 2, the moment is zero, and for crank 3 the moment is $+\mathbf{Cl}$. The resultant moment **Mc** is shown in Fig. 11.28(d). **Mc** is rotating with the crank shaft and forms an angle α with the direction of crank 1 as shown.

A constant magnitude force $K_1 = m_0 r\omega^2$ in the direction of each crank is used for the investigation of the primary forces. The result is shown in Fig. 11.29(a) and (b). The summation of the forces is

$\mathbf{R}_{K_1}=\mathbf{0}$, and the vertical projection of this force is the resultant primary force which, of course, is also zero in this case. The resultant moment of the primary forces is determined in the same way as for the rotating forces. This is shown in Fig. 11.29(c) and (d). The vertical projection of the moment $\mathbf{M}_{K_1}$ is the moment of the primary forces.

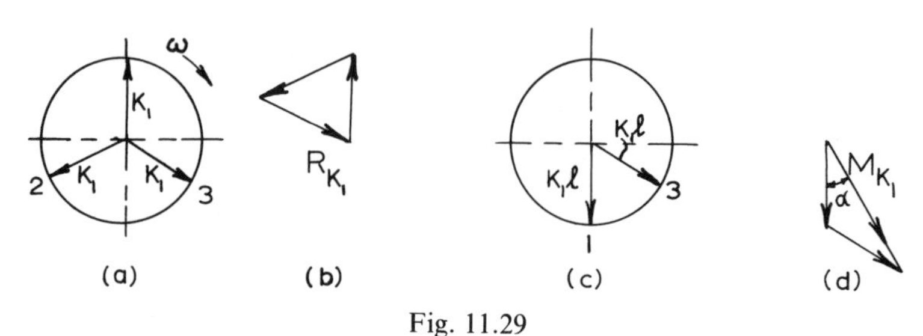

Fig. 11.29

For the investigation of the secondary forces, we use a constant force $K_2=m_0r\omega^2/n$ in a radial direction forming twice the angle with the vertical of the actual crank. This is done by imagining a crank shaft as shown in Fig. 11.30(a), in which each crank has been rotated twice the angle of the actual crank. This imaginary crank is assumed to rotate at an angular velocity 2ω. The force and moment investigation is done in the same way as for the primary and rotating forces. The result is shown in Fig. 11.30(b), (c), (d) and (e).

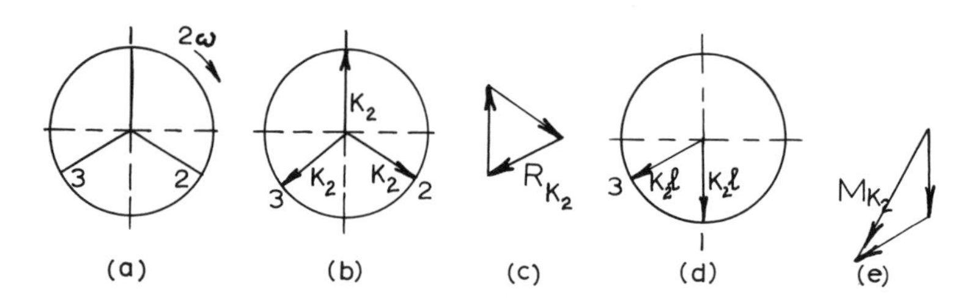

Fig. 11.30

The resultant force, and its vertical projection, is $\mathbf{R}_{K_2}=\mathbf{0}$. The resultant moment is $\mathbf{M}_{K_2}$, which rotates at an angular velocity 2ω, and the vertical projection of $\mathbf{M}_{K_2}$ is the actual moment of the secondary forces.

To balance the engine, two counter weights A and B may be used on the two outer crank arms.

Since $\mathbf{R}_C = \mathbf{0}$, the counter weights must be equal and opposite, so that no resultant rotating force is introduced. The plane of the counter weights is determined by the direction of the vectors $\mathbf{M}_C$ and $\mathbf{M}_{K_1}$ in Figs. 11.28(d) and 11.29(d).

Taking the mass of each counter weight as m_1 and the radius to the centre of mass as R_1, the centrifugal force on each weight is $m_1 R_1 \omega^2$.

Taking moments of the forces on the counter weights in the same manner as for the **C** forces, we find the resultant moment as shown in Fig. 11.31 equal to $M_R = 2m_1 R_1 \omega^2 (l+b) = m_1 R_1 \omega^2 L$. This is in a direction opposite to $\mathbf{M}_C$ and $\mathbf{M}_{K_1}$.

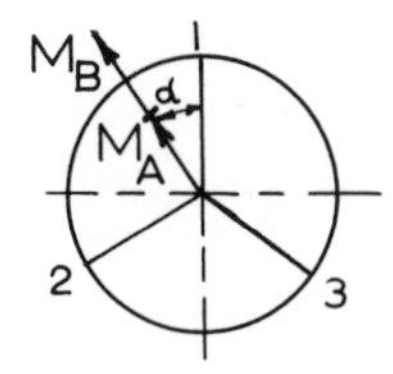

Fig. 11.31

If we want to balance the rotating forces only, the product $m_1 R_1$ for the counter weights is determined from $m_1 R_1 \omega^2 L = M_C$. If both the rotating and primary reciprocating forces are to be balanced, we have $m_1 R_1 \omega^2 L = M_C + M_{K_1}$. In this case a horizontal reciprocating moment, equal to M_{K_1} in magnitude, is introduced.

In some cases horizontal and vertical moments are equally objectionable, and it is better to balance the rotating moment and half of the primary reciprocating moment. In this case we must take

$$m_1 R_1 \omega^2 L = M_C + \tfrac{1}{2} M_{K_1}$$

This results in an unbalanced primary moment in both the horizontal and vertical direction of magnitude $\frac{1}{2} M_{K_1}$.

The secondary forces and moments cannot be balanced by counter weights fixed to the counter shaft. A similar vector method may be used for the balance investigation of motors with any number of cylinders and various crank configurations.

Problems

11.1 Find the part of the total beam mass m that must be added to a concentrated mass M at the centre of the beam for use in the simple formula:

$$f=(1/2\pi)\sqrt{(K/M)},$$

for the case of a *fixed–fixed beam.* Use Rayleigh's method and assume a full cosine wave $y=(y_0/2)[1-\cos(2\pi x/l)]$ for the deflection curve. $K=192EI/l^3$.

11.2 A solid cylindrical steel shaft is simply supported in bearings A and B (Fig. 11.32) and carries two discs of weights $W_1=267$ N and $W_2=445$ N.

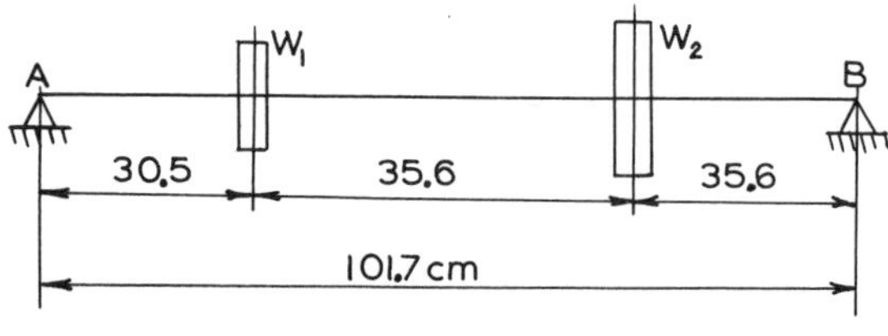

Fig. 11.32

The second moment of area of the cross-sectional area with respect to a diameter is $I=5$ cm^4 and the modulus of elasticity $E=21\times10^7$ kN/m^2.

Using Rayleigh's method, calculate the critical speed of the shaft assuming that the deflection curve is a parabola. Repeat the calculation on the assumption that the deflection curve is a sine curve.

Comment on the accuracy of the two values and explain why one is a better result than the other.

The effect of the mass of the shaft may be neglected in this calculation.

11.3 A thin ring of weight $W_1=13{\cdot}35$ N and mean radius $r_1=12{\cdot}7$ cm contains a solid wheel of weight $W_2=22{\cdot}25$ N and radius $r_2=10{\cdot}17$ cm (Fig. 11.33). Across a diameter of the ring is a steel rod of $d=0{\cdot}318$ cm to which both ring and wheel are rigidly attached. The whole system floats freely in bearings as shown.

If the wheel is initially twisted through a small angle relative to the plane of the ring and released, calculate the natural torsional frequency of the system. $G=8{\cdot}27\times10^{7}$ kN/m^2.

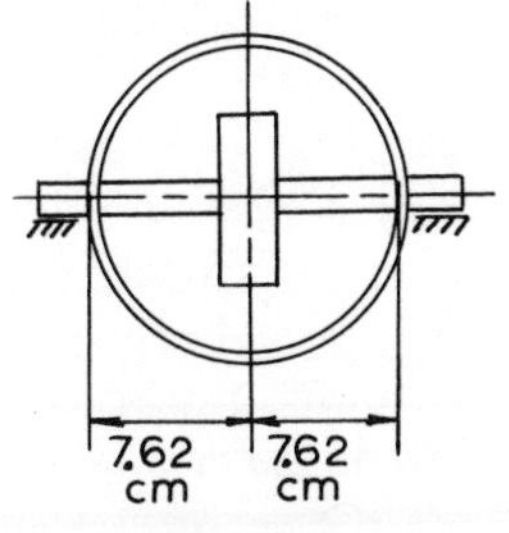

Fig. 11.33

11.4 The propeller shaft of a ship (Fig. 11.34) has a length $l=3{\cdot}66$ m and diameter $d=10{\cdot}17$ cm. The shaft is of steel with $G=7{\cdot}93\times10^{7}$ kN/m^2. The steam turbine rotor is of weight $W_1=8{\cdot}90$ kN with radius of gyration 76·2 cm. The propeller is $W_2=4{\cdot}45$ kN and radius of gyration 50·8 cm. Find the frequency of torsional vibrations. Find the percentage increase in the frequency if the diameter of the shaft is increased by 10% along half its length. Neglect the shaft inertia.

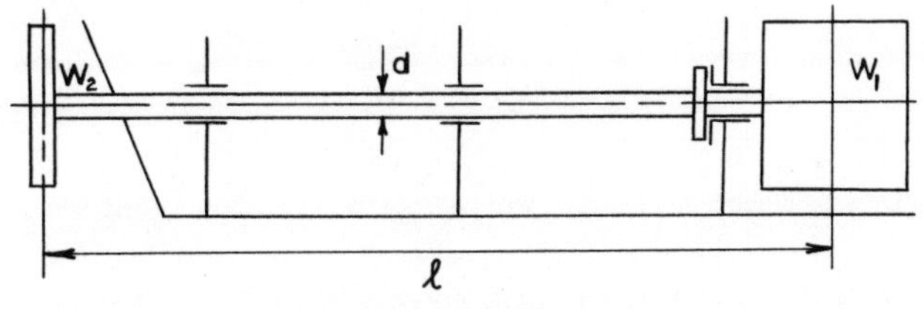

Fig. 11.34

11.5 The driving shaft X of the reduction gear shown in Fig. 11.35 rotates at 1000 rev/min and its arm carries a shaft onto which are

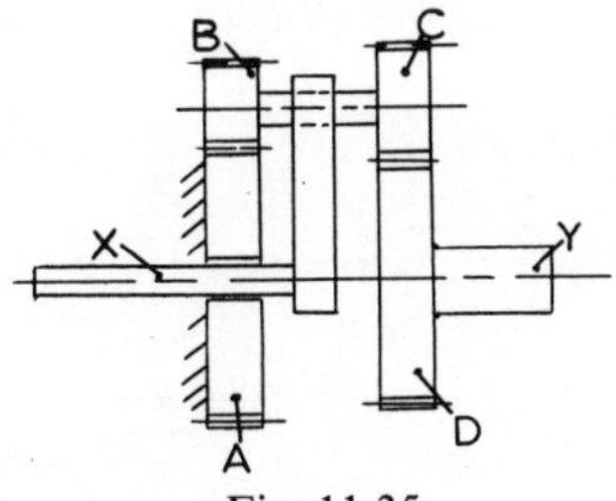

Fig. 11.35

keyed wheels B (13 teeth) and C (14 teeth). B and C engage respectively with the wheel A (73 teeth) which is fixed with wheel D (72 teeth) which is on the driven shaft. Find the speed of Y and its direction of rotation compared with that of X.

11.6 In an epicyclic gear, Fig. 11.36, wheel A with 100 internal teeth is keyed to a horizontal shaft F. A fixed wheel B with 102 internal teeth is concentric with A. Two wheels C (30 teeth) and D are integral with one another and revolve freely on a horizontal pin projecting from a rotating arm keyed to a shaft G which is coaxial with F. The length of the arm is such that C engages with A and D with B.

Given that G is driven at 800 rev/min and all teeth have the same pitch, find the speed and direction of rotation of F.

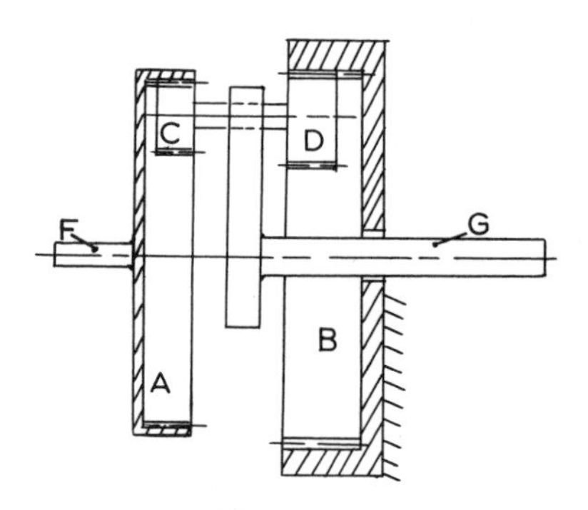

Fig. 11.36

11.7 In the epicyclic gear train shown in Fig. 11.37, A is a fixed circular gear with 60 internal teeth. BC is a double intermediate wheel mounted on an eccentric E which is keyed to the shaft F.

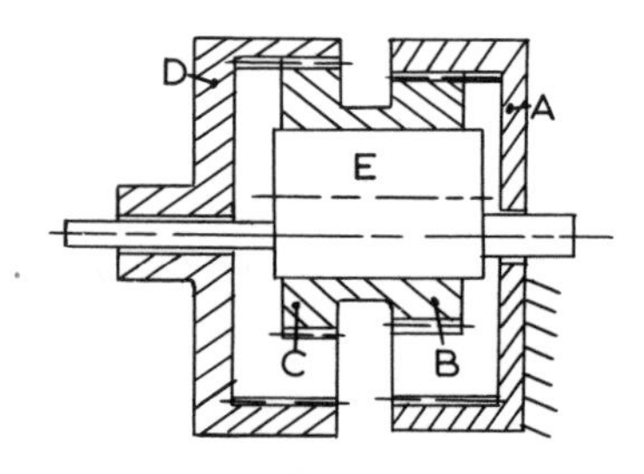

Fig. 11.37

Gear B has 55 and C has 59 teeth. B meshes with A, and C with the annular wheel D of 64 internal teeth. The gear D is loose on the shaft F. Find the number of revolutions of F and its direction of rotation compared with that of D, for one revolution of D.

11.8 In the planetary gear train shown in Fig. 11.38, D is a fixed circular gear with 80 internal teeth, B and C are integral with one another and revolve freely on a horizontal pin projecting from a rotating arm integral with the shaft H. Wheel C has 15 teeth and engages with D. The driving shaft carries a gear A with 18 teeth which engages with wheel B. All the gears have the same pitch.

Given that the driving shaft rotates at 240 rev/min, find the speed of the driven shaft H and its direction of rotation compared to that of A.

If the input power is 37·3 kW and the overall efficiency is 97%, what is the output torque on H?

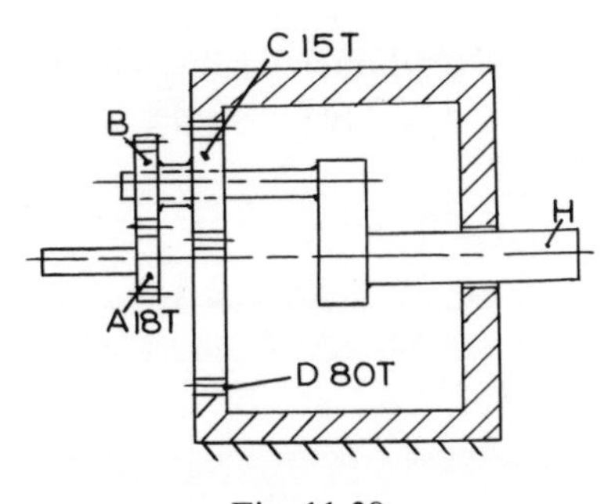

Fig. 11.38

11.9 A plate cutter is driven by a constant-torque motor through a flywheel. The required torque varies linearly with the angle of rotation from zero to 1356 N m through $\frac{1}{2}$ revolution and is zero through the remaining part of one revolution. The mean speed is 400 rev/min.

Determine the degree of irregularity, if the speed variation is $\pm 2\%$, and the moment of inertia of the flywheel.

11.10 The axis of a reciprocating follower is 1·91 cm to the right of the cam centre, and is vertical. The roller is 2·54 cm in diameter and its centre is 5·08 cm from the cam centre when on a minimum radius. Design the cam for S.H.M. along a 5·08 cm stroke, both up and down, with no rest periods. The cam rotates through 180° during the upward motion.

11.11 A cam follower, with a plane horizontal face at its lower end, is to reciprocate along a vertical line passing through the horizontal axis of its steadily rotating cam which is arranged beneath it and rotating in the anticlockwise direction.

The follower is to have uniform acceleration during the first half of its lift and uniform retardation during the second half, the total lift being 3.81 cm and corresponding to the first 150° of cam rotation. A rest period is then to occur for 30° of cam rotation, followed by a fall of a similar nature to that of the rise, the final 30° of cam rotation being occupied by a rest period at the lowest follower position.

Using a minimum cam circle of radius 2·54 cm, set out, full size, a suitable cam profile.

11.12 Show that a six cylinder engine with cranks at 0°, 120°, 240°, 240°, 120° and 0° and equal cylinder distances has all forces and moments balanced.

Answers to problems

Chapter 1

1.1 $x = gt/n + g/2n^2$.

1.2 $t = \dfrac{s}{V} + \dfrac{V}{2}\left(\dfrac{1}{a_1} + \dfrac{1}{a_2}\right)$, $t_{\min} = \dfrac{s}{V} + \dfrac{V}{a}$.

1.3 (a) $\mathbf{a} = 2\mathbf{i} + 6\mathbf{j}$, 6·32 m/s^2; (b) $y^2 = x^3$.

1.4 $\mathbf{a} = \mathbf{a}_n + \mathbf{a}_t = 8\downarrow + 6\leftarrow$; 10 m/s^2.

1.5 $\mathbf{V} = 2\mathbf{r}_1 + 6\boldsymbol{\varphi}_1$; 6·32 m/s; $\mathbf{a} = -18\mathbf{r}_1 + 14\boldsymbol{\varphi}_1$, 22·8 m/s^2.

1.6 $\omega = 1{\cdot}305$ rad/s$\circlearrowright$, $\dot{\omega} = 1{\cdot}430$ rad/s$^2\circlearrowright$.

1.7 3 m/s; 11·46 m, $-\pi/2$ m/s^2.

1.9 $\mathbf{V} = -214\mathbf{i} + 92{\cdot}8\mathbf{k}$; 234 m/s; $\mathbf{A} = -23\,200\ \mathbf{r}_1$ m/s^2.

1.11 $\mathbf{V}_P = r\dot{\varphi}\boldsymbol{\varphi}_1 + r\dot{\theta}(\cos\varphi)\boldsymbol{\theta}_1$. $|\mathbf{V}_P| = r\sqrt{(\dot{\varphi}^2 + \dot{\theta}^2\cos^2\varphi)}$

Chapter 2

2.1 $S = \dfrac{M_1 g(\sin\alpha + \mu\cos\alpha + 2)}{1 + 4M_1/M_2}$; 932 N.

2.2 $\ddot{x}_A = g/17$; $\ddot{x}_B = 5g/17$; $\ddot{x}_C = -7g/17$.

2.3 $f = \dfrac{1}{2\pi}\sqrt{\left(\dfrac{T}{m}\dfrac{a+b}{ab}\right)}$.

2.4 $\dot{x} = (g/A)(1 - e^{-At})$; terminal g/A; $x = (g/A^2)(At + e^{-At} - 1)$.

2.5 2·65 km/s; 2·6R = 15·95 N.

2.7 $V = -\gamma m_1 m_2/r$; $r^2 = x^2 + y^2 + z^2$.

2.8 $\Delta = 3$ cm; 2.9 9 blows.

Chapter 3

3.1 0·5 m/s↓; 2·52 rad/s↺.
3.2 $-5{\cdot}41\mathbf{i}-9{\cdot}39\mathbf{j}$; 10·8 m/s; $4{\cdot}78\mathbf{i}-8{\cdot}28\mathbf{j}$; 9·56 m/s.
3.3 0·238 m/s↖; 2·34 rad/s↺.
3.4 $-1{\cdot}22\mathbf{i}$ m/s; $-15{\cdot}4\mathbf{i}-9{\cdot}7\mathbf{j}$ m/s^2; 4 rad/s↺, 22·7 rad/s^2↺.
3.5 2·05 m/s at 188°, 12·8 m/s^2 at 333·8°; 8·75 rad/s↺, 59 rad/s^2↻.
3.6 3·60 m/s at 9°; 70 m/s^2 at 120°; 11·4 rad/s↺; 632 rad/s^2↺.

Chapter 4

4.1 $\mathbf{V}=3\mathbf{i}+18{\cdot}86\mathbf{j}$; 19·1 m/s; $\mathbf{A}=-18{\cdot}86\mathbf{i}+6\mathbf{j}$; 19·8 m/s^2.
4.2 $13{\cdot}28\mathbf{j}$ m/s; $1064\mathbf{i}$ m/s^2.
4.3 $3000\mathbf{i}+9{\cdot}43\mathbf{j}$ m/s; $-5{\cdot}92\mathbf{i}+3770\mathbf{j}$ m/s^2.
4.4 4·67 m/s.
4.5 0·28 m/s at 342°; 4·7 m/s^2 at 309°; 0·40 m/s at 297·5°; 3.43 m/s^2 at 302·5°; 4·65 rad/s↻; 6·90 rad/s^2↺.
4.6 0·94 $r\omega$; 248 N; 4·7 $\dfrac{1}{2\pi}\sqrt{\left(\dfrac{2K}{m}-\omega^2\right)}$.

Chapter 6

6.1 11·83 mm from flange outer face; (a) $8{\cdot}83\times10^4$ mm^4. (b) $2{\cdot}27\times10^4$ mm^4.
6.2 $-\frac{1}{72}a^2b^2$; $\frac{1}{24}a^2b^2$.
6.3 $\varphi=18{\cdot}5°$; $I_{x_1}=1211\times10^4$ mm^4; $I_{y_1}=85\times10^4$ mm^4.
6.4 $\frac{5}{3}b^2M$; $\frac{3}{2}r^2M$.
6.5 $\frac{1}{4}h$ from base.
6.6 $I_x=\frac{1}{3}M(b^2+d^2)$.

Chapter 7

7.1 $\ddot{\theta}+(g/l)\sin\alpha\sin\theta=0$; $\dfrac{1}{2\pi}\sqrt{\left(\dfrac{g}{l}\sin\alpha\right)}$.
7.2 $\dfrac{r_1}{r_2^2}\dfrac{I_2\omega_1}{\mu P}$; $\dfrac{I_1I_2r_1\omega_1}{\mu P(I_1r_2^2+I_2r_1^2)}$; $\dfrac{I_1\omega_1r_2^2}{I_1r_2^2+I_2r_1^2}$.
7.3 (a) 157 N; (b) 528 N; 274 N.

7.4 $M_L = 154$ g, $M_R = 52{\cdot}5$ g, both in 4th quadrant, angles to horizontal 83·1° and 50·5°.

7.5 (a) $A_X = -B_X = (Ml^2/24L)\dot{\omega} \sin 2\beta$;

$$\left.\begin{matrix} A_Y \\ B_Y \end{matrix}\right\} = \tfrac{1}{2}M(g \pm (l^2\omega^2/12L) \sin 2\beta);\ A_Z = B_Z = 0,$$

(b) $T = \frac{1}{12}Ml^2\dot{\omega} \sin^2 \beta$.

7.6 73·2 rad/s²; $A_X = -B_X = -123{\cdot}8$ N; $A_Y = -33{\cdot}1$ N, $B_Y = -78$ N, $A_Z = B_Z = 0$.

7.7 (a) $A_Z + B_Z = 0$, $A_X + B_X = 0$, $A_Y + B_Y - Mg = 0$.
(b) $I_{XZ} = 0$, $I_{YZ} = \frac{1}{2}Ma^2$, $I_Z = \frac{1}{3}Ma^2$; (c) $\dot{\omega} = 3T/Ma^2$;

$$A_X = -B_X = -\frac{T}{2a};\ \left.\begin{matrix} A_Y \\ B_Y \end{matrix}\right\} = \frac{M}{2}\left(g \mp \frac{a\omega^2}{3}\right);\ A_Z = B_Z = 0.$$

Chapter 8

8.1 $8M_1g/(8M_1 + 3M_2)$; $3M_1M_2/(8M_1 + 3M_2)g$.

8.2 (a) $\dfrac{P(\cos \alpha - r/R)}{M[1 + (rc/R)^2]}$; (b) α limited to 60°, motion to right; (c) 1·612 m/s²; (d) 0·355.

8.3 29·6°.

8.4 1·37 s; 1.46 s.

8.5 $\frac{1}{4}V(1 + \cos \alpha)^2$.

8.6 16 rad/s, 14·2 m/s at 338·7°.

8.7 Left bearing 1215 N, right bearing 1455 N, thrust bearing 332 N.

8.8 28 600 N m in a horizontal plane, anticlockwise seen from above for bow moving down.

8.9 5410 N m twisting AC clockwise seen from above.

8.10 $A_X = B_X = 0$; $A_Y = -448$ N, $B_Y = 0$; $A_Z = -245{\cdot}7$ N, $B_Z = 314{\cdot}4$ N.

Chapter 9

9.1 $m\ddot{x} - mx\dot{\theta}^2 + mg(1 - \cos \theta) + K(x - x_0) = 0$;
$mx^2\ddot{\theta} + 2mx\dot{\theta}\dot{x} + mgx \sin \theta = 0$.

9.2 $\sin \varphi\ddot{\theta} + 2\dot{\theta}\dot{\varphi} \cos \varphi = 0$; $\ddot{\varphi} - \dot{\theta}^2 \sin \varphi \cos \varphi + (g/l) \sin \varphi = 0$
$\ddot{\varphi} + (g/l) \sin \varphi = 0$, simple pendulum; $\omega^2 = g/(l \cos \alpha)$ for $\varphi = \alpha$, conical pendulum.

9.3 $m\ddot{x}+(AE/l^3)x^3=0$; $\ddot{x}+(2S/ml)x=0$.

9.4 $\ddot{\theta}+\frac{1}{2}\ddot{\varphi}\cos(\varphi-\theta)-\frac{1}{2}\dot{\varphi}^2\sin(\varphi-\theta)+(g/l)\sin\theta=0$;
$\ddot{\varphi}+\ddot{\theta}\cos(\varphi-\theta)+\dot{\theta}^2\sin(\varphi-\theta)+(g/l)\sin\varphi=0$.

9.5 (a) $T=\frac{1}{2}(M_1l^2+4M_2r^2)\dot{\theta}^2$; $V=M_1gl(1-\cos\theta)+2M_2rg\sin^2\theta$;
(b) $(M_1l^2+4M_2r^2)\ddot{\theta}+M_1gl\sin\theta+2M_2rg\sin 2\theta=0$;
(d) 1·41 cycles/s.

9.6 $\overleftarrow{\ddot{x}}_C=-\ddot{x}=\dfrac{g\sin\alpha\cos\alpha}{3-\cos^2\alpha}$; $\ddot{y}_C\!\downarrow=2g/(3+2\cot^2\alpha)$.

9.7 (a) $(M+m)\ddot{x}+m(R-r)\dot{\theta}^2\cos\theta+m(R-r)\ddot{\theta}\sin\theta+Kx=0$,
$\frac{3}{2}(R-r)\ddot{\theta}+\ddot{x}\sin\theta+g\sin\theta=0$;
(c) $\sqrt{[\frac{2}{3}g/(R-r)]}$; (d) $\sqrt{[K/(M+m)]}$.

9.8 $(M+m)\ddot{x}-ml\dot{\theta}^2\sin\theta+ml\cos\theta\,\ddot{\theta}+Kx=P(t)$;
$ml^2\ddot{\theta}+ml\cos\theta\,\ddot{x}+mgl\sin\theta=0$; S.H.M.

9.9 (a) $T=\frac{1}{2}m(l_0+x)^2.\dot{\theta}^2+\frac{1}{2}m\dot{x}^2$;
$V=(\frac{1}{2}Kx^2+mgx)-mg[(l_0+x)\cos\theta-l_0]$;
(b) $\ddot{x}+(c/m)\dot{x}+(K/m)x-(l_0+x)\dot{\theta}^2+g(1-\cos\theta)=0$;
$(l_0+x)\ddot{\theta}+2\dot{x}\dot{\theta}+g\sin\theta=0$.

9.10 $c\dot{x}_1+(K_1+K_2)x_1-K_2x_2=0$
$m\ddot{x}_2+K_2x_2-K_2x_1=F_0\cos\omega t$.

Chapter 10

10.1 1·28 s;

10·2 $2\pi\sqrt{\left(\dfrac{I+Mr^2}{Ka^2}\right)}$.

10.3 $\dfrac{1}{2\pi l}\sqrt{\left(\dfrac{2Ka^2}{m}-gl\right)}$;

10.4 $\dfrac{1}{2\pi}\sqrt{\left[\dfrac{K}{M}\left(\dfrac{l}{b}\right)^2-\left(\dfrac{c}{2M}\right)^2\right]}$; $2\dfrac{l}{b}\sqrt{(KM)}$.

10.5 0·428.

10.6 0·505 N s/cm; 3·82 cycles/s, 4.41 cycles/s.

10.7 512 N; 1030 N/cm.

10.8 39 800 N/cm.

Chapter 11

11.1 $\frac{3}{8}$ m.

11.2 1006 rev/min,
901 rev/min better.

11.3 9·86 cycles/s.

11.4 7·75 cycles/s; 9%.

11.5 −91·8 rev/min

11.6 +35 rev/min

11.7 −176 revolutions

11.8 +16·1 rev/min;
21 450 N m.

11.9 $\frac{1}{25}$; 17·1 kg m^2.

Appendix SI Units in Mechanics

Definitions

Basic SI Units

Metre (m); unit of length. The metre is the length equal to 1 650 763·73 wavelengths in vacuum of the radiation corresponding to the transition between the energy levels $2p_{10}$ and $5d_5$ of the krypton-86 atom (Eleventh General Conference of Weights and Measures, 1960; XI CGPM, 1960).
Kilogram (kg); unit of mass. The kilogram is represented by the mass of the international prototype kilogram (III CGPM, 1901). The international prototype is in the custody of the Bureau International des Poids et Mésures (BIPM), Sèvres, near Paris.
Second (s); unit of time interval. The second is the interval occupied by 9 192 631 770 cycles of the radiation corresponding to the $(F=4, M_F=0)-(F=3, M_F=0)$ transition of the caesium-133 atom when unperturbed by exterior fields. (Nominated for present use by XII CGPM, 1964).

Derived SI units having special names

Hertz (Hz); unit of frequency. The number of repetitions of a regular occurrence in one second.
Newton (N); unit of force. That force which, applied to a mass of 1 kilogram, gives it an acceleration of 1 metre per second per second.

Pascal (*Pa*);* *unit of pressure*. The pressure produced by a force of 1 newton applied, uniformly distributed, over an area of 1 square metre.
Joule (*J*); *unit of energy, including work and quantity of heat*. The work done when the point of application of a force of 1 newton is displaced through a distance of 1 metre in the direction of the force.
Watt (*W*); *unit of power*. The watt is equal to 1 joule per second.
Standard gravity or standard acceleration. Denoted by $g = 9{\cdot}80665$ m/s^2; 9·81 for slide-rule work.

List of SI units

Plane angle	rad (radian)
solid angle	sr (steradian)
length	m (metre)
area	m^2
volume	m^3
time	s (second)
angular velocity	rad/s
velocity	m/s
frequency	Hz (hertz)
rotational frequency	1/s
mass	kg (kilogram)
density (mass density)	kg/m^3
momentum	kg m/s
moment of momentum, angular momentum	$kg\ m^2/s$
moment of inertia	$kg\ m^2$
force	N (newton)
moment of force	N m
pressure and stress	N/m^2
viscosity (dynamic)	$N\ s/m^2$
surface tension	N/m
energy, work	J (joule)
power	W (watt)

* The name pascal has not been accepted internationally, but is often used, particularly in France.

Conversion of common British units to equivalent values in SI units

Length

1 mile = 1·609 34 km
1 yd = 0·914 4 m
1 ft = 0·304 8 m
1 in = 2·54 cm = 25·4 mm

Area

1 sq mile = 2·589 99 km^2 (258·999 ha)
1 yd^2 (square yard) = 0·836 127 m^2
1 ft^2 (square foot) = 0·092 903 0 m^2 = 9·290 30 dm^2
1 in^2 (square inch) = 6·451 6 cm^2

Volume

I yd^3 (cubic yard) = 0·764 555 m^3
1 ft^3 (cubic foot) = 28·316 8 dm^3
1 in^3 (cubic inch) = 16·387 1 cm^3

Capacity

1 gal = 4·546 09 dm^3 = 4·546 1 (litre)
1 US gal = 3·785 41 dm^3 = 3·785 1
1 qt (quart) = 1·136 52 dm^3
1 pt (pint) = 0·568 261 dm^3

Moment of section (*second moment of area*)

1 ft^4 = 86·309 7 dm^4
1 in^4 = 41·623 1 cm^4

Velocity

1 mile/h (m.p.h.) = 1·609 34 km/h
1 ft/s = 0·304 8 m/s
1 ft/min = 0·304 8 m/min
1 in/s = 2·54 cm/s
1 in/min = 2·54 cm/min
1 UK knot = 1·853 18 km/h

Acceleration

1 ft/s^2 (foot per second per second) = 0·304 8 m/s^2

Mass

1 ton = 1016·05 kg (1·016 05 t)*
1 cwt = 50·802 3 kg
1 lb = 0·453 592 37 kg
1 oz = 28·349 5 g
1 slug = 14·593 9 kg

Density

1 ton/yd^3 = 1328·94 kg/m^3 (1·328 94 t/m^3)
1 lb/ft^3 = 16·018 5 kg/m^3
1 lb/in^3 = 27·679 9 g/cm^3
1 lb/gal = 0·099 776 3 kg/dm^3 = 0·099 78 kg/1
1 slug/ft^3 = 515·379 kg/m^3

* t = tonne (metric ton) equal to 1000 kg.

Moment of inertia

1 lb ft^2 = 0·042 140 1 kg m^2 1 oz in^2 = 0·182 900 kg cm^2
1 lb in^2 = 2·926 40 kg cm^2 1 slug ft^2 = 1·355 82 kg m^2

Momentum

1 lb ft/s = 0·138 255 kg m/s

Angular momentum

1 lb ft^2/s = 0·042 140 1 kg m^2/s

Force

1 tonf* = 9964·02 N 1 pdl (poundal) = 0·138 255 N
1 lbf* = 4·448 22 N

Force (weight)/unit length

1 tonf/ft* = 32·690 3 kN/m 1 lbf/in* = 1·751 27 N/cm
1 lbf/ft* = 14·593 9 N/m

Moment of force (torque)

1 tonf ft* = 3037·03 N m 1 pdl ft = 0·042 140 1 N m
1 lbf ft* = 1·355 82 N m 1 lbf in* = 0·112 985 N m

Pressure, stress

1 tonf/ft^2* = 107·252 kN/m^2 1 lbf/in^2* = 6894·76 N/m^2 (68·947 6 mb)
1 tonf/in^2* = 15·444 3 MN/m^2
1 lbf/ft^2* = 47·880 3 N/m^2 1 pdl/ft^2 = 1·488 16 N/m^2

Viscosity (dynamic viscosity)

1 lbf h/ft^2* = 0·172 369 MN s/m^2 1 lb/ft s = 1·488 16 kg/m s
1 lbf s/ft^2* = 47·880 3 N s/m^2 1 slug/ft s = 47·880 3 kg/m s
1 pdl s/ft^2 = 1·488 16 N s/m^2

Viscosity, kinematic

1 ft^2/h = 0·092 903 0 m^2/h 1 in^2/h = 6·451 6 cm^2/h
1 ft^2/s = 0·092 903 0 m^2/s 1 in^2/s = 6·451 6 cm^2/s

Energy (work, heat)

1 therm = 105·506 MJ or MN m 1 ft lbf* = 1·355 82 J
1 hp h = 2·684 52 MJ (horsepower × hour) 1 ft pdl = 0·042 140 1 J
1 Btu (British thermal unit) = 1·055 06 kJ

Power

1 hp (horse power) = 745·700 W (J/s) 1 ft lbf/s = 1·355 82 W

* 1 tonf (ton-force), 1 lbf (pound-force), etc., are often loosely referred to as 1 ton, 1 lb (pound).